AF510830

MEDICALLY ASSISTED CONCEPTION

An Agenda for Research

Report of a Study by a Committee of the

INSTITUTE OF MEDICINE
Division of Health Sciences Policy

NATIONAL RESEARCH COUNCIL
Board on Agriculture

National Academy Press
Washington D.C. 1989

NOTICE: The project that is the subject of this report was approved by the Governing Board of the National Research Council, whose members are drawn from the councils of the National Academy of Sciences, the National Academy of Engineering, and the Institute of Medicine. The members of the committee responsible for the report were chosen for their special competencies and with regard for appropriate balance. This report has been reviewed by a group other than the authors according to procedures approved by a Report Review Committee consisting of members of the National Academy of Sciences, the National Academy of Engineering, and the Institute of Medicine.

The Institute of Medicine was chartered in 1970 by the National Academy of Sciences to enlist distinguished members of the appropriate professions in the examination of policy matters pertaining to the health of the public. In this, the Institute acts under both the Academy's 1863 congressional charter responsibility to be an advisor to the federal government and its own initiative in identifying issues of medical care, research, and education.

The National Research Council was organized by the National Academy of Sciences in 1916 to associate the broad community of science and technology with the Academy's purposes of furthering knowledge and advising the federal government. Functioning in accordance with general policies determined by the Academy, the Council has become the principal operating agency of both the National Academy of Sciences and the National Academy of Engineering in providing services to the government, the public, and the scientific and engineering communities. The Council is administered jointly by both Academies and the Institute of Medicine.

This project received support from the National Academy of Sciences Basic Science Fund and the U.S. Department of Agriculture (Contract 59-32U4-8-63/R).

2101 Constitution Avenue, N.W.
Washington, D.C. 20418
(202) 334-3300

Library of Congress Catalog Card No. 89-63252
International Standard Book Number 0-309-04128-7

Additional copies of this report are available from:
National Academy Press
2101 Constitution Avenue, NW
Washington, DC 20418

S041

Printed in the United States of America

IOM 89-08

INSTITUTE OF MEDICINE

NATIONAL RESEARCH COUNCIL
Board on Agriculture

COMMITTEE ON THE BASIC SCIENCE FOUNDATIONS OF
MEDICALLY ASSISTED CONCEPTION

KENNETH J. RYAN,* (Chairman), Professor and Chairman, Department of Obstetrics and Gynecology, Harvard Medical School, Boston, Massachusetts

JOHN D. BIGGERS, Professor of Physiology, Harvard Medical School, Boston, Massachusetts

MARIAN D. DAMEWOOD, Assistant Professor, Department of Gynecology and Obstetrics, The Johns Hopkins Medical Institutions, Baltimore, Maryland

CLAUDE DESJARDINS, Professor of Physiology, Department of Physiology, University of Virginia School of Medicine, Charlottesville, Virginia

NEAL L. FIRST,** Professor of Reproductive Physiology, Meat and Animal Science Department, University of Wisconsin, Madison, Wisconsin

GARY D. HODGEN, Professor and Scientific Director, Eastern Virginia Medical School, Norfolk, Virginia

ROBERT B. JAFFE,* Professor and Chairman, Department of Obstetrics, Gynecology and Reproductive Science, School of Medicine, University of California, San Francisco

HOWARD W. JONES, Jr., Professor of Obstetrics and Gynecology, Eastern Virginia Medical School, Norfolk, Virginia

STANLEY P. LEIBO,*** Research Associate Professor, Departments of Obstetrics, Gynecology and of Urology, Baylor College of Medicine, Houston, Texas

---

*Member, Institute of Medicine
**Member, National Academy of Sciences
***During the study, Dr. Leibo was in the Research and Development Division of Rio Vista International, Inc.

JONATHAN VAN BLERKOM, Professor of Molecular, Cellular, and Developmental Biology, Department of Molecular, Cellular and Developmental Biology, University of Colorado at Boulder, Boulder, Colorado

EDWARD E. WALLACH, Professor and Chairman, Department of Gynecology, The Johns Hopkins Medical Institutions, Baltimore, Maryland

STUDY STAFF

Institute of Medicine

RUTH ELLEN BULGER, Director, Division of Health Sciences Policy

JESSICA TOWNSEND, Study Director

CONSTANCE M. PECHURA, Meeting Rapporteur

WALLACE K. WATERFALL, Editor, Institute of Medicine

Board on Agriculture
JAMES TAVARES, Associate Executive Director

TABLE OF CONTENTS

ACKNOWLEDGEMENTS

The committee appreciates the generous contribution of time and knowledge of those who presented papers at the committee's workshop at the Arnold and Mabel Beckman Center, Irvine, California, August 21-23, 1988. The thought-provoking papers contributed to the workshop, and the presence of exceptional scientists and clinicians stimulated outstanding discussions. The committee gained invaluable knowledge and insight into issues in research in reproductive and developmental biology that provided the basis for its deliberations. I want to extend the gratitude of the committee to all who attended the workshop. By offering those resources that we all value -- time and knowledge -- the committee's work was greatly facilitated.

Kenneth J. Ryan
Chairman
Committee on the Basic Science Foundations
of Medically Assisted Conception

PREFACE

This report is the result of a study by a committee of the Institute of Medicine (IOM) and the National Research Council's Board on Agriculture. The study results from a 1984 meeting of an ad hoc group convened by the National Academy of Sciences to discuss aspects of genetic engineering of the human germline. It was noted that because of a policy that, in effect, prohibited the use of federal funds for research involving human embryos, the clinical practice of in vitro fertilization and embryo transfer (IVFET) was in danger of outstripping its scientific foundations. Moreover, the United States had not systematically reviewed the current state of knowledge and practice of IVFET. In 1986, the IOM's Board on Health Sciences Policy convened a meeting of experts in the fields of human and animal research, clinical practice, law, ethics, and public policy to consider some of the issues raised at the earlier meeting. This group was asked to advise the IOM on whether it should pursue a study in any of the areas in which difficult issues had been raised by the practice of medically assisted conception--namely, professional, social, legal, and ethical issues, in addition to questions of science policy. The group identified several topics to which the IOM could make a substantial contribution. An important result of the meeting of this group was to highlight the scientific contribution of research relating to animal uses of medically assisted conception, and the lack of communication between those working to further human clinical IVFET and those working in the animal sciences. This study of the basic science foundations of medically assisted conception results from the recommendations made by this group of experts.

Scope of the Study

A committee was appointed whose membership included individuals with expertise in the clinical practice of in vitro fertilization, research in animal and human reproductive and developmental biology, and physiology. The committee was asked to examine the basic science foundations of medically assisted conception, and develop an agenda for basic research in reproductive and developmental biology that would contribute to advances in the clinical and agricultural practice of IVFET. This research would also be applicable to other concerns in human and animal reproduction such as male and female infertility and contraception. Because of limitations on research using human embryos, the committee was also asked to identify animal systems that provide useful models for specific aspects of medically assisted conception. In addition, the committee was asked to address ways of diminishing barriers to progress in reproductive research and consider ways of fostering communication between investigators oriented to human clinical questions and those interested in studies of animals.

Although the charge to the committee includes consideration of ways of lowering barriers to progress in research, some of which are caused by legal and ethical concerns, the committee was not constituted to resolve or make recommendations on such issues. Rather, it was to confine its deliberations to consideration of ways that scientists might contribute to the furtherance of the debate on legal and ethical issues related to medically assisted conception.

The committee held two meetings. At the first, a plan was developed for a workshop that would bring together clinicians and investigators to review the status of assisted conception in animals and humans, and suggest productive areas for future research. The committee carefully reviewed current research directions in reproductive science, and selected research areas and investigators to present them at the workshop. The committee was aware that it was neither feasible nor desirable to provide a comprehensive view of reproductive research. Rather, the committee selected areas of research that show particular promise or describe particularly problematical areas in relation to IVFET. This workshop was also conceived as a major effort in bringing together representatives from the human and animal research worlds to enable them to exchange ideas, enhance their understanding of ways in which they can contribute to each other's work, and together participate in a joint activity that might establish continuing ties.

The committee's second meeting, after the workshop, was spent in developing this report and its recommendations.

SUMMARY

This study was prompted by a concern that, because of a policy that in effect prohibits the use of federal funds for research involving human embryos, the clinical practice of in vitro fertilization and embryo transfer (IVFET) was pushing to the limit of its scientific foundations. In addition, there was a perception that animal science had made substantial progress in the development of IVFET but, because of a lack of communication among those working to further human clinical IVFET and those working in the animal science area, the knowledge developed in one sector was not being conveyed to the other. As a result progress in each sector was seen as being slowed.

An Institute of Medicine committee was appointed to examine the basic science foundations of medically assisted conception, to develop an agenda for basic science research that would contribute to advances in the clinical and agricultural practice of IVFET, to suggest animal systems that provide useful models for specific research areas, to identify ways of diminishing barriers to progress, and to recommend ways of bringing together the veterinary and human reproductive research workers.

The centerpiece of the study was a workshop at which clinicians and investigators reviewed the status of assisted conception and the related basic research in humans and animals, and suggested productive areas for future research. This workshop brought together representatives from the human and animal research worlds to enable them to exchange ideas, enhance their understanding of ways in which they can contribute to each other's work, and participate in a joint activity that could establish continuing ties.

Social Concerns That Can be Addressed By IVFET

Research directed at pushing forward the frontiers of medically assisted conception has the potential to provide benefits beyond the limited number of couples whose infertility may be solved by IVFET. There are expectations that such research would enable practitioners to identify genetic defects in embryos without damaging them, and to determine the sex of embryos without damaging them, so that those with sex-linked genetic diseases can be identified at a very early stage. Major areas to which advances in basic reproductive biology would make large contributions include:

o Infertility. The magnitude of the problem of human infertility is reflected by the number of women with "impaired fecundity" — 4.4 million or 8.2 percent of women of childbearing age in 1982 (National Center for Health Statistics, 1985). By one estimate, more than half of the 4.2 million women who have been surgically sterilized for non-contraceptive reasons, and half of the 4.4 million subfecund women would like to become pregnant. Furthermore, one million women between the ages of 15 and 44 who were or had been married reported at least one medical visit for infertility in 1982 (Fuchs and Perrault, 1986). Although data cannot describe the emotional toll of infertility, the communications media are beginning to portray some of the distress. The lengths to which couples will go in attempting to conceive and the formation of nationwide support groups for childless people are indicators of the pain of childlessness. Although IVFET is a solution for only limited numbers of infertile couples, research that advances the practice of IVFET also has the potential of advancing other forms of infertility treatment.

Numerous infertility treatments exist, including education to give couples sufficient knowledge of reproductive biology, surgical repair, artificial insemination, and the use of drugs to induce ovulation. Two major new technologies are IVFET and gamete intrafallopian transfer (GIFT). These are complicated technologies. The simple description that follows will facilitate understanding of the research agenda developed by the committee.

For IVFET, eggs are removed from the woman either during a natural cycle or after growth and maturation of oocytes has been stimulated by such drugs as human menopausal gonadotrophin. This latter method has the advantage of allowing more than one oocyte to be harvested. The egg is placed in a petri dish together with washed sperm that have been treated to ensure capacitation. If fertilization is achieved, the process of cleavage starts, and somewhere between the 2- and 16-cell stage, the embryo is transferred to the uterus. Pregnancy is established when the developing embryo implants itself into the wall of the uterus. More than one embryo may be transferred to the uterus.

For GIFT, growth and retrieval of eggs are performed in a manner similar to that used for IVFET. Semen is collected and placed in a catheter with the eggs, and they are then transferred to the fallopian tube. Fertilization takes place in vivo.

Sometimes donated sperm, eggs, or fertilized zygotes are used in assisted conception. For example, excess zygotes collected from a patient undergoing IVFET can be fertilized and implanted in a recipient uterus that has been synchronized with the donor's cycle.

o Contraception. Advances in the basic science that would improve the clinical practice of assisted conception would, at the same time, help in the search for better contraceptive technologies. Despite widespread use of such contraceptive methods as sponges, surgical sterilization, intrauterine devices, and birth control pills, there remain unresolved problems of safety and efficacy. The search for improved forms of contraception is spurred not only by the desire of individuals to gain control over their reproductive lives, but also by the social cost of unwanted pregnancies and the problems caused by fast-growing populations in countries unable to provide an adequate standard of living for the present population.

- Agriculture. The application of assisted conception techniques has made rapid inroads in the domestic cattle industry. Artificial insemination is the norm, with 70 percent of dairy cows conceiving in this manner in 1985. In less than two decades, a multimillion dollar IVFET bovine industry has developed. About 25 percent of embryo transfers in 1984 were of frozen embryos. Artificial insemination has resulted in genetic improvements in dairy cattle that have doubled milk production per cow in thirty years (First, Crister, and Robl, 1985). Embryo transfer technology increases the rate of production of valuable cows. The adoption of new reproductive technologies to enhance the production of food-producing animals has the potential for lowering the cost of food and quickening the process by which animals genetically suited to difficult climates can be created.

- Biodiversity. Advances in reproductive technologies may sustain biodiversity by improving the reproductive efficiency of endangered species.

- Primates for Research. A limited number of primates are in captivity and available for research, and there is a possibility that the capture of more may be halted because of concerns for the future of the species. It will be increasingly important to maximize the reproductive capabilities of the primates available to science.

Barriers to Progress in In Vitro Fertilization and Embryo Transfer

Since the birth of Louise Brown in England in 1978, in vitro fertilization with embryo replacement has become an established method of treatment for certain types of infertility that do not respond to alternative methods of treatment. However, the chances of success in IVF are relatively low. In 1985, 14.1 percent of stimulation cycles resulted in clinical pregnancies. In 1986 this figure rose to 16.9 percent (Fertility and Sterility, 1988). But, the proportion of women entering treatment who attain a live birth is far lower -- only 8.9 percent of oocyte retrievals ended in live birth (Journal of the American Medical Association, 1988). Why are the odds for successful IVFET so low? The state of clinical practice of IVFET today is limited by lack of knowledge of some of the basic reproductive biology involved. This is caused, in part, by the many ethical questions raised by research in pursuit of the needed information. Difficulties in resolving these issues have caused the research to be deprived of federal funding.

Ethical and Social Issues

Some of the ethical or social issues that arise from the various forms of assisted conception are unrelated to decisions about the progress of research. Examples of these are questions about the protection of the rights of gamete donors, gestational parents, and social parents; the ownership of cryopreserved embryos; and the sale of gametes and embryos.

Some ethical questions have a direct bearing on research, and have had important consequences for the funding of research. The major questions focus on the status of the embryo at each stage of its development. How the embryo is regarded dictates what is morally acceptable to do to it.

At one end of this spectrum of thought is the position taken by the Roman Catholic Church. The Vatican's Instruction on Respect for Human Life states that "from the first moment of its existence until birth . . . no moral distinction is considered between zygotes, pre-embryos, embryos or fetuses" (cited in Fertility and Sterility, 1988b). Therefore, the absolute sanctity that is accorded to post-natal human life begins with the zygote. This concept makes it impossible to discard spare embryos or use them for research purposes. At the other end of the spectrum is the position that an embryo is merely biological material like any other group of living cells. The special value that might be attached to that material results from the expectations or aspirations of others (Office of Technology Assessment, 1988).

Midway between these two positions is one that holds that "the human embryo is entitled to profound respect; but this respect does not necessarily encompass the full legal and moral rights attributed to a person" (Department of Health, Education, and Welfare, 1979). Holding this position, the Ethics Advisory Board (established by the Department of Health, Education and Welfare (DHEW) in 1979) concluded that research was acceptable on embryos up to 14 days after fertilization.

The Federal Government and Embryo Research

Policy concerning research on human subjects has been slowly evolving since the 1960s. A study group was convened at NIH to develop guidelines, and a National Advisory Commission on Health Science and Society was proposed by Senator Walter Mondale in 1968 to examine developments in medical research. Following reports of the infamous Tuskegee syphilis experiments, DHEW recommended that Congress establish a permanent body to regulate federally funded research using human subjects.

In the 1970s the abortion issue became linked to the issue of embryo research. After the Roe v. Wade decision legalized abortion under certain conditions, concern developed that women would be pressured into having abortions and the sale of aborted embryos might occur. In 1974, the federal government created the National Commission for the Protection of

Human Subjects (P.L. 93-348). Until this commission reported to Congress, research on the living fetus was prohibited unless it was used to help that fetus survive. In 1975, DHEW issued regulations based on the findings of the commission. These regulations did not cover embryo research. The commission also recommended establishing an Ethics Advisory Board (EAB) to review requests for research on embryos and in vitro fertilization. However, in 1980, the Secretary of DHHS allowed the EAB charter to expire. Thus, no research could be approved, and federal funding of embryo research was de facto prohibited. As a result, embryo research has relied on private funding from patient care revenues, pharmaceutical companies, and university budgets.

Since 1985, efforts have been made that, if successful, might establish some rules under which embryo research could proceed. However, the chances of such an outcome in the near future appears to be slim. A Congressional Biomedical Ethics Board, composed of six senators and six representatives, has been appointed. This group established a Biomedical Ethics Advisory Committee. In 1988 the Department of Health and Human Services announced its intention to revive the Ethics Advisory Board and publish a proposed charter. A final charter is awaited.

Domestic and Foreign Decisions Concerning Embryo Research

The two professional societies in the United States that represent the physicians most involved in human IVFET have considered ethical questions about the practice of IVFET and embryo research. In 1986 the Committee on Ethics of the American College of Obstetricians and Gynecologists (ACOG) (1986) issued a statement that acknowledged the ethical issues posed by the creation of embryos outside a uterus, the dilemma of surplus embryos, and the acceptability of research using early human embryos. The ACOG committee recommended that human embryos should be used only if nonhuman embryos could not provide the needed knowledge. It also recommended banning research on embryos that had reached the age of 14 days. The American Fertility Society (AFS) also issued a report in 1986, approving experiments on embryos up to 14 days (Fertility and Sterility, 1986). A year later, after consideration of the Vatican's Instruction for Human Life in its Origin and on the Dignity of Procreation, issued by the Congregation for the Doctrine of Faith, the AFS issued another report. This report stated that progressive degrees of respect are due with progressive development of embryos, and that experimentation can be justified and is necessary if the human condition is to be improved (Fertility and Sterility, 1988b).

The government of the United States, since 1979, has not followed the lead of nations that have systematically examined issues related to human IVFET. Since 1979, at least 85 statements have been prepared by committees representing at least 25 countries. Four Australian committees found research on early (preimplantation) embryos to be ethically unacceptable. Eleven committees approved at least some kinds of early embryo research. Six of these accept such research only on embryos left over from clinical activities. Five committee statements (including the 1979 DHEW Ethics Advisory Board) would allow the creation of embryos for research purposes. Although the majority of committees favor limiting research on embryos to up to fourteen days, one committee allowed it only to seven days, and one only through the first cleavage (Walters, 1987).

In sum, numerous groups have wrestled with questions related to the ethical problems of embryo or fetal research. Some have based their conclusions on religious tenets, some on an interpretation of scientific knowledge, some on a mixture of both.

Other Barriers to Scientific Progress

Other factors besides ethical considerations are slowing the progress of research in areas of reproductive biology related to assisted conception.

Deficiencies in the Science Base Papers presented at the committee's workshop and the research agenda developed from that workshop indicate deficiencies in the scientific underpinnings of reproductive biology, and identify many areas in which further research efforts would make major contributions to improvements in medically assisted conception. The deficiencies are on three levels: basic science knowledge; knowledge needed to improve the technologies being used for medically assisted conception, such as cryobiology; and knowledge needed to improve both human and animal clinical practice of IVFET.

Research Funding Approximately $155 million annually is spend on research in reproductive processes. Federal agencies are the principal support for research. In 1986 they provided $109 million for research in reproductive processes (National Institutes of Health, undated). Federal funds for research relating to agricultural animal reproduction are available from the U.S. Department of Agriculture.

Funding for basic research in reproductive biology is undoubtedly constrained by the lack of vocal and focused advocacy groups. Lacking such a voice a major increase in federal support is unlikely.

Lack of Communication Among Researchers Discussion with the scientists and clinicians at the committee's workshop revealed an underuse of available mechanisms for communications among the individuals involved with various aspects of research in reproductive biology -- basic, clinical, animal sciences, etc. Also revealed was a desire for greater communication to allow cross-fertilization of ideas and development of ongoing relationships among investigators pursuing similar approaches to problems.

<u>Sources of Research Material for Experiments with Humans and Other Primates</u> The committee's workshop provided many excellent examples of instances in which information about reproductive physiology derived from animal models has been useful in understanding human physiology. However, animal models cannot suffice for investigating all central questions; progress is some areas requires the use of human tissue. An example of this is investigation of reasons for developmental failure of human embryos.

Although specific primates are good models for some aspects of human reproductive physiology, there are only a limited number of monkeys of desirable species in captivity and many of them are presently being used for AIDS research.

Research Agenda

A workshop was held August 21-23, 1988 at the Arnold and Mabel Beckman Center in Irvine, California. Overviews of the experience gained by the clinical practice of IVFET and of the practice of assisted conception in food-producing animals directed attention to unanswered questions that will require basic science research for their resolution. These questions reflect important gaps in our knowledge of the biology of all the stages of reproduction from the development of male and female gametes to the process of embryo implantation. The topics listed below are areas in which further research was recommended by workshop participants and committee members. Work in these areas is expected to increase understanding of the biology of reproduction with the hope that increased knowledge will eventually lead to improvements in the practice of IVFET in humans and other animals, or to advances in contraception. Research areas are listed here in summary form and apply both to lower animals and human beings unless specifically noted. The complete summary of the workshop is contained in Chapter Two of the full report.

Basic Science

<u>Male Gametogenesis</u>

o Definition of the role of cell adhesion molecules in interactions between Sertoli cells and developing sperm cells.

o Understanding the function of differential protein synthesis in different stages of sperm development.

o Determination of the role of paracrine factors including fibroblast growth factor, somatomedin C, epidermal growth factor, and interleukin-1 on the development and differentiation of male gametes.

o Structural analysis to identify normal and abnormal sperm and the development of markers for abnormal sperm.

o Understanding of the biochemistry of sperm capacitation.

Female Gametogenesis

o Analysis of the effects of superovulation or hormonal stimulation protocols on oocyte development and maturation. This work should also examine differences between species.

o Development of ways to mature oocytes in vitro.

o Investigation of ways to naturally stimulate oocyte and follicular development.

o Investigation into the biochemistry of meiotic arrest and the factors, such as cyclic AMP, purines, calcium, and maturation-promoting factor, that may mediate this process.

o Development of ways to produce or synthesize hormones from non-human primates to be used in ovarian stimulation.

o Definition of the role of ovarian estrogen in oocyte maturation and ovulation and the interactions between estrogen and paracrine factors including fibroblast and epidermal growth factors, insulin-like growth factor, transforming growth factor, and inhibin.

o Definition of the point at which oocytes become sensitive to factors that influence their development.

o Elucidation of the processes that underlie oocyte depletion, to determine why oocytes are lost at a predictable rate throughout life.

o Investigation into ways to augment natural hormone release.

o Investigation into the biochemistry of protein synthesis and modification in ovarian cells.

Fertilization

o Investigation into the biophysics of cell membranes as it relates to sperm and egg interactions at fertilization.

- Continued investigation to identify the genes for zona proteins in various species, especially humans.

- Further delineation of the role of zona proteins, especially ZP2 and ZP3, in sperm binding.

- Understanding of the biochemistry of the modification of zona proteins in preventing polyspermy.

- Elucidation of the molecular determinants of antibody formation to zona proteins and their possible role in contraceptive strategies.

- Definition of the biochemical mechanisms of the cortical reaction in the egg and the effects of this reaction on zona proteins.

- Determination of the physiological significance of germinal vesicle breakdown and the biochemistry of sperm chromatin decondensation.

- Definition of the molecular events associated with formation of the male and female pronuclei.

- Definition of the molecular events during zygote formation and the first cleavage.

Preimplantation Development

- Definition of the metabolic requirements of early embryos at different stages.

- Determination of embryonic gene expression.

- Assessing the potential of individual embryonic cells and defining the point at which embryonic cells are committed to particular fates.

- Identification of substances produced by early embryos that signal changes in the uterus prior to implantation.

- Improvements in embryo multiplication and embryo splitting, especially for food producing animals.

Implantation

- Definition of the biochemical events that make the uterus permissive to implantation.

o Definition of the factors released by embryos that cause endometrial changes at the site of implantation.

o Identification of the role of embryo-released factors in suppressing the immune responses of the mother.

o Isolation and analysis of substances released by endometrial cells and their effects on embryos.

o Continued work with in vitro models of human implantation to study the biochemistry and mechanisms of embryo-endometrial interactions, especially the role of extracellular matrix proteins and the biochemistry of trophoblast invasion of the endometrium.

Technological Advances

o Improved cryopreservation techniques, including freezing and thawing protocols for eggs and embryos.

o Improved resolution of ultrasonography for localization and noninvasive harvest of oocytes, eggs, embryos--would have particular usefulness for non-human primates and food producing animals.

o Development of new culture media and methods for in vitro maturation of oocytes.

o Development of safe methods of biopsy of early embryos for preimplantation diagnosis of genetic diseases.

Clinical Research Opportunities

The following areas are those in which a coordinated data collection effort across IVFET clinical centers would help improve the quality and success rates of IVFET nationally and, possibly, internationally.

o Evaluation of hormonal stimulation protocols in terms of number of oocytes harvested, quality of oocytes, and rate of fertilization success.

o Documentation on the incidence of abnormal implantation rates in IVFET practice and correlation of incidence with particular stimulation protocol used.

o Collection of information regarding the incidence of abnormal zygotes and embryos, failed fertilization, and developmental arrest of embryos.

o Analysis of data pertaining to synchronization of embryonic stage with endometrial stage and development of methods to improve synchronization.

o Collection of information on sharing of spare eggs and arrested embryos for research purposes.

Conclusions and Recommendations

Developing Research Policy

Lack of a mechanism for dealing with ethical disagreement over the use of embryos in research has slowed the rate of progress in research by, in effect, placing a moratorium on the use of federal funds for eight years. This has had undesirable results: the human clinical practice of IVFET is less effective than it might have been had research progressed at a faster pace; other socially desirable goals such as improved contraception, better techniques to preserve endangered species, and more cost-effective methods of producing food have developed at a pace slower than optimal.

The recent appointment of the Biomedical Advisory Committee by the Biomedical Ethics Board, to report to Congress by November 1990 on embryo research issues, could be a step toward a solution. The committee applauds the intention to revive the Ethics Advisory Board of the Department of Health and Human Services to rule on the ethical acceptability of research relating to human embryos, which is required before federal funding of such a research grant can be considered. However, until these groups become fully functional and show evidence of progress, their impact must remain in question.

If these groups can assume leadership roles in resolving the difficult issues of reproductive research, and develop guidelines for research that are based on information provided by science, and on concepts that are ethically acceptable to society, research in reproduction will be able to move forward. But if these groups become paralyzed because of political considerations or an inability to develop a framework for the resolution of differences of opinion, another organization should take over the role. The committee recommends that, <u>if the groups currently being formed fail to come to conclusions concerning embryo and fetal research, a non-governmental organization should be established to develop guidelines for embryo and fetal research that are based on the most advanced knowledge that science can muster, and with serious consideration of the expressed values of society. The group should be composed of individuals with expertise in the relevant scientific disciplines, representatives of the lay public, and experts in the legal, ethical, and social issues. The organization should be housed in an institution that would allow it to conduct its deliberations free from any undue pressures from political and special interest groups.</u> A model for such activities can be found in the Voluntary Licensing Authority of Great Britain.

Basic Science Foundations

The number and range of topics included in the research agenda indicate the exciting potential for productive scientific exploration. The committee believes that fundamental research to enhance the basic science foundations of reproductive biology should be stimulated and supported. This includes studies of human beings, laboratory animal models, and food-producing animals. The knowledge that would be generated is fundamental to an understanding of how to reverse infertility, to new approaches in the area of contraception, and to increasing the world's food supply.

It is important that male as well as female reproductive biology be studied and that investigators make use of some opportunities that are largely ignored today. These opportunities occur as a result of clinical activities as well as research activities.

The committee recommends that a vigorous program of funding for a basic science agenda in reproductive biology be maintained in a coordinated fashion by an appropriate office in the National Institutes of Health.

Applied Research

Research needs to be stimulated concerning technologies used in medically assisted conception in food producing animals and in human beings. Lack of support in these areas is leading to inadequate scientific underpinnings for safe and effective clinical practice. An example of a technique used, but not carefully evaluated for possible detrimental effects, is freezing eggs or zygotes. Further experiments should be conducted to assess the effects on safety and viability of this technology which is standard practice in many IVFET clinics. Other areas of technology that need to be developed include less invasive ways to retrieve oocytes, ways to mature oocytes in vitro, and ways to assess the quality of spermatozoa or eggs to be used for fertilization.

The committee recommends that applied research into technologies used in medically-assisted conception be undertaken to provide a firm foundation for the safe and effective practice of in vitro fertilization and embryo transfer. Such applied research should be coordinated by the appropriate office at the National Institutes of Health.

Clinical Research Opportunities

Perhaps the most obvious missed opportunity is the failure to learn from the diverse experiences of the approximately 160 clinical programs that provide IVFET. In addition to scientific questions, there are questions to do with the organization of clinics and the outcomes of procedures, the answers to which would enable practitioners to work more effectively, and policy makers to make decisions on the basis of the best available information.

Clinical IVFET centers can provide unique opportunities for important studies. For example, human oocytes that fail to fertilize in vitro could be used to investigate the phenomenon of failed fertilization. Research that seeks to understand the basis of reproductive failure, and its relationship to hyperstimulation should be encouraged. Coordinated studies utilizing the mass of material and experience from IVFET centers could begin to answer these and other questions.

The committee applauds the activities of the various professional societies that have issued non-binding statements about the quality of practice of IVF. The American Fertility Society has also provided a voluntary registry for centers.

The committee believes that a mechanism is needed to monitor and evaluate clinical practice so that existing information that is relatively easy and inexpensive to collect can be disseminated. This would enable clinicians to build on the broadly based experience of the community and help ensure that patients have access to information about developments in IVFET and to well-informed physicians. The committee recommends that a mechanism for multi-centered data collection be established to monitor and evaluate human and veterinary practices of medically assisted conception in order to improve the safety, effectiveness, and quality of clinical practice. A cooperative group composed of the relevant professional societies should be established to fund and initiate data collection under the direction of an inter-society council composed of representatives of each participating organization.

Improving Communications

The IOM Workshop on Medically Assisted Conception brought together researchers from basic science, clinical practice, and animal sciences. The resulting interaction was viewed as extremely helpful by investigators from each of these communities. The committee recommends that a mechanism (or multiple mechanisms) be found for fostering continued communication between researchers in diverse areas of reproductive science. The initiative should come both from NIH research administrators who could sponsor additional workshop opportunities, as well as from the professional societies, either individually or through an intersociety council.

REFERENCES

American College of Obstetricians and Gynecologists. 1986. Ethical Issues in Human In Vitro Fertilization and Embryo Placement. Committee on Ethics ACOG Committee Opinion Number 47. Washington, D.C.

Department of Health, Education and Welfare. 1979. HEW Support of Research Involving Human In Vitro Fertilization and Embryo Transfer. Report and Conclusions. May 4. Washington, D.C., U.S. Government Printing Office.

Fertility and Sterility. 1988a. Invitro Fertilization/Embryo Transfer in the United States: 1985 and 1986 Results from the National IVF/ET Registry. 49(2):212-215.

Fertility and Sterility. 1988b. Ethical considerations of the new reproductive technologies. By the Ethics Committee (1986-7) of the American Fertility Society in light of Instruction on the Respect for Human Life in its Origin and on the Dignity of Procreation issued by the Congregation for the Doctrine of the Faith. Feb;49(2 Suppl 1):I-7S

Fertility and Sterility. 1986. Ethical Considerations of the New Reproductive Technologies. The Ethics Committee of The American Fertility Society. Sep;46(3 Suppl 1):IS-94S.

First, N.L., Critser, E.S., and Robl, J.M. 1985. Boving Embryo: Development, Cloning, Sexing and Transfer of Genes for Immunology of Reproduction and Contraception, P. Talwas, ed. Elsevier, 1985.

Fuchs, V.R. and Perreault, L. 1986. Expenditures for Reproduction-Related Health Care. Journal of the American Medical Association. Vol. 225, No.1. January 3:76-81.

Journal of the American Medical Association. 1988. IVF Registry Notes More Centers, More Births, slightly Improved Odds. 259(13):1920-1921.

National Institutes of Health. Undated. Inventory and Analysis of Federal Population Research. Fiscal Year 1986. U.S. Department of Health and Human Services, Public Health Service, National Institutes of Health.

Office of Technology Assessment. 1988. Infertility: Medical and Social Choices. OTA BA BP 48. August. Washington, D.C.

Walters, LeRoy. 1987. Ethics and New Reproductive Technologies: An International Review of Committee Statements. Hastings Center Report. June:3-9.

CHAPTER 1

IN VITRO FERTILIZATION AND EMBRYO TRANSFER AND SOCIAL CONCERNS

The scientific advances that today permit clinicians and veterinarians to use such procedures as drug therapy, laser surgery, artificial insemination, in vitro fertilization, and embryo transfer to combat human infertility or improve the productive capabilities of valuable animals are the results of the work of generations of investigators. Sometimes these achievements came about because of research aimed at resolving specific problems related to human infertility or toward enabling livestock owners to improve the return on their investments. More often, however, scientific advances result from the inquiries of scientists investigating fundamental biological processes in the absence of specific applications. Thus, the work of investigators in areas such as physiology, morphology, endocrinology, molecular biology, developmental biology, and biochemistry have contributed to improvements in the treatment of infertility and to progress in assisted conception in humans and other animals.

Just as the findings of scientists pursuing answers to basic science questions often contribute in unexpected ways, the work of those pushing the frontiers of medically assisted conception has applications beyond the limited number of couples with infertility problems who can benefit directly. There are expectations that, in the future, the work being done to improve the practice of in vitro fertilization and embryo transfer (IVFET) will enable practitioners to identify genetic defects in embryos without damaging them, and to quickly determine the sex of embryos so that those with sex-linked genetic diseases can be identified at a very early stage. The stress of abortion later in a pregnancy could be avoided. This chapter briefly outlines some major areas to which advances in research in basic reproductive biology and IVFET would make large contributions.

Infertility

Infertility is defined in many ways. Most often the word is used to denote the inability of a woman to conceive after some months (12 to 24) of intercourse without contraception, or the inability to carry a pregnancy to term. According to a summary by the National Center for Health Statistics (NCHS), some 8.2 percent of women of childbearing age (4.4 million) suffered from "impaired fecundity" in 1982—the latest data available. This category includes over 800,000 women who said that it was impossible for them to have a baby because of accidents or other unexplained reasons. Also included are 2.9 million women defined as "subfecund" who said that it was physically difficult for them to conceive or deliver a baby, and 650,000 sexually active women who reported that they did not use contraception and did not become pregnant within 36 months. Defining as infertile couples those who were continuously married, had not used contraception, and had not conceived during the preceding 12 months, NCHS reported that 2.3 million women were infertile in 1982. This represents a decline from 11 percent of women in 1965 to 8 percent in 1982. However, problems of infertility are not diminishing for all sections of the population. The overall decline masks an increase in infertility among women under 30 years of age, and an increase from approximately 700,000 to nearly one million infertile couples with no children (National Center for Health Statistics, 1985).

Although these figures give a rough approximation of the fecundity of women in 1982, they do not indicate the number of men and women who want a child and are experiencing difficulty conceiving. More than half of the over 4.2 million women who have been surgically sterilized for non-contraceptive reasons, and half of the 4.4 million who have impaired fecundity, say they would like to become pregnant (Fuchs and Perreault, 1986).

The magnitude of the problem of human infertility is further reflected in data on medical care for infertility. Although not all infertile individuals seek treatment, by one estimate $1 billion was spent in 1980 on medical services for infertility treatment (Office of Technology Assessment, 1988b). In 1982, 1 million women between the ages of 15 and 44 who were or had been married reported at least one infertility visit during the past year (Fuchs and Perreault, 1986). A similar estimate is produced by a survey of primary care and reproductive care specialists in 1987 which indicated that they treated approximately 1.2 million patients for infertility (Office of Technology Assessment 1988a).

Infertility is not confined to problems experienced by women. By one estimate, about half of infertility problems are due in whole or in part to problems of the male (Fuchs and Perreault, 1986). Indeed, over 80 percent of the women seeking artificial insemination did so because of the infertility of their male partner, and approximately 65,000 children were conceived by artificial insemination during a 12-month period in 1986-1987 (Office of Technology Assessment, 1988a).

Although the emotional toll of infertility cannot be appreciated from these data, in recent years the communications media have begun to portray some of the distress of childlessness, and groups such as Resolve, begun in 1973, have responded to the emotional support needed by childless couples. Equally telling are the lengths to which individuals will go in attempting to conceive. The many months spent undergoing diagnostic procedures can be followed by additional efforts of drug therapy, surgery, and finally the physical, financial, and psychological stress of IVFET.

IVFET is a possible solution to infertility for only a small number of couples. By one estimate, as much as 10 to 15 percent of infertile couples who could not be successfully treated by other means could be offered hope through IVFET or by a related technology, gamete intrafallopian transfer (GIFT) (Office of Technology Assessment, 1988b). Indications for use of IVFET are quite specific and include tubal disease that has not responded to other therapies, endometriosis, oligospermia, cervical mucus abnormalities and unexplained infertility. Indications for GIFT are more limited because it can only be used when fallopian tubes are normal. As knowledge expands, assisted conception of all sorts is expected to become applicable to a wider range of indications. IVFET can be applied regardless of whether fallopian tubes are present, therefore it is possible that IVFET will become a major therapy relative to the large number of individuals with infertility problems (Jones, 1989). However, research that advances the practice of IVFET will at the same time promote understanding of human reproduction and has the potential of advancing other forms of infertility treatment as well as providing better methods of contraception. As noted by a recent comprehensive study of infertility, "Even as infertility treatments become more sophisticated and complex, basic knowledge of the male and female reproductive process remains lacking. Further research stands as a prerequisite in order for dramatic improvements in infertility treatment to occur" (Office of Technology Assessment, 1988b).

Treatments for Infertility

Infertility can be treated in a variety of ways, including ensuring that the infertile couple know how to pinpoint the time of ovulation, eliminating causes of infertility such as infectious diseases or endometriosis, evaluating sperm seminal fluid, using fertility drugs to induce ovulation, performing surgical repair procedures in the male or female, and employing artificial insemination. Before assisted reproduction technologies are attempted a standard evaluation is conducted which includes hormonal evaluation, endometrial biopsy, hysterosalpingogram, diagnostic laparoscopy. Alternate therapy such as microsurgical corrections of tubal disease or endometriosis might be attempted. At least two noncoital reproductive technologies have been introduced in the last ten years. The major technologies are in vitro fertilization and embryo transfer (IVFET) and gamete intrafallopian transfer (GIFT). These technologies for establishing a pregnancy are reviewed in detail in a variety of recent publications (Office of Technology Assessment, 1988b; Seibel, 1988). To facilitate the understanding of the research agenda proposed in this report, a brief review of the steps utilized during IVFET, GIFT and some other methods of assisted conception follows.

In Vitro Fertilization and Embryo Transfer

IVFET can be used to overcome infertility caused by numerous conditions including tubal disease, endometriosis and oligospermia. A first step in IVFET is to prepare the woman for removal of eggs (oocytes). Two methods are used to accomplish this. Sometimes oocytes can be obtained during a natural cycle of a woman by determining the time of the marked increase in the luteinizing hormone level in the blood, which precedes ovulation by about 1 1/2 days. Using a natural cycle, however, frequent blood samples must be analyzed to exactly pinpoint the increase in this hormone level. Only one mature egg is usually obtained by this method. Alternatively, follicular growth and maturation, which leads to ovulation, can be induced by the use of various fertility drugs such as human menopausal gonadotrophin. The subsequent development of ovarian follicles can be monitored by ultrasound and by measuring blood estrogen levels. By this method, which is most commonly used today, more than one oocyte is stimulated to develop and can be obtained for fertilization.

Just before the timed ovulation would occur, oocytes are removed from the ovary either laparoscopy or by needle aspiration guided by ultrasonography. The eggs, with their adherent nurse cells, are placed in a petri dish so that their state of maturation can be assessed using the state of dispersion of the attached cells as a marker. Fertilization of the mature egg is accomplished by incubation for approximately 24 hours in the petri dish with washed sperm that have been treated to ensure capacitation. Fertilization is defined by the visible presence of two pronuclei in the newly formed zygote.

The first cleavage of the zygote occurs approximately 1 1/2 days after insemination. A catheter is used to transfer the dividing embryo into the lumen of the uterus at somewhere between the 2- and 16-cell stage. To supplement the natural luteal phase, hormones such as progesterone are sometimes administered after transfer of the embryo, (or embryos if more than one oocyte has been fertilized) to the uterus. Pregnancy is established when the developing embryo implants itself into the wall of the uterus. Implantation can be documented by a measured increase in blood levels of human chorionic gonadotrophin.

Sometimes, a greater number of mature eggs are harvested than can usefully be implanted. Increasingly, these excess eggs are fertilized and preserved by cryopreservation for subsequent use.

Gamete Intrafallopian Transfer

In 1985, Asch et al. (1985) reported on gamete intrafallopian transfer (GIFT) as a new treatment for infertility. In 1987, the Lancet noted that GIFT had been readily accepted in to clinical practice (Lancet, 1987). GIFT involves the transfer of eggs and sperm into patent fallopian tubes so that fertilization may take place in vivo. Follicular growth of oocytes and retrieval are performed in a manner similar to that used for IVFET. Semen is collected and placed in a catheter with the eggs, which are then transferred to the fallopian tubes. In 1987, GIFT was achieving a higher success rate than IVFET. Although this might have been due to the better conditions of in vivo fertilization compared to in vitro, it may also have been due to patient selection. GIFT requires that at least one fallopian tube be patent and that a sufficient number of normal sperm can be obtained (Lancet, 1987). By 1987 there was a report of successful use of GIFT with donated oocytes (Craft et al. 1987). Gift can be used when infertility is caused by such factors as endometriosis, premature ovarian failure, oligospermia, and unexplained infertility (Office of Technology Assessment, 1988b).

Donated Gametes or Concepti

Sometimes the donation of spermatozoa, eggs, or in some cases fertilized zygotes, are necessary. Excess eggs collected from one female donor patient undergoing IVFET can be fertilized and implanted in a recipient uterus which has been synchronized with the donor's cycle. Artificial insemination using donor spermatozoa is a common technique. The results of one survey indicate that each year about 30,000 babies are born from artificial insemination using donor spermatoza (Office of Technology Assessment, 1988a).

Two less frequently used methods of treatment for infertility that also involve the manipulation of eggs or embryos are tubal ovum transfer and embryo lavage and transfer. Egg stimulation and harvesting are undertaken as in IVFET and GIFT. The egg is then reinserted below fallopian tube blockage or other damage and fertilization takes place in vivo (Office of Technology Assessment, 1988b).

In embryo lavage and transfer fertilized eggs are flushed out and removed by a special catheter. They are then transferred to a recipient whose cycle has been synchronized to be ready for the introduced egg. This technique is becoming less frequently used, partly because of fears about transmission of virus and the risk of retained embryos resulting in pregnancy in the donor.

Contraception

Advances in the basic science that would improve the clinical practice of assisted conception, such as improved understanding of the mechanics of egg implantation, would be likely, at the same time, to help in the search for better contraceptive technologies.

Contraceptive methods range from rhythm methods, the contraceptive sponge, birth control pills, and intrauterine devices to surgical sterilization and barrier methods, including condoms, spermicides, cervical caps, and diaphragms. More than half of American women aged 15 to 44 years used some form of contraception, at a cost of $2.4 billion in 1982 (Fuchs and Perreault, 1986). Despite this widespread use of contraception, there remain unresolved problems of safety, efficacy, and acceptability for each type of contraceptive.

The search for improved forms of contraception is spurred not only by the desire of individuals for control over their reproductive lives, but also by the social costs of unwanted pregnancies and the pressures of fast-growing populations in countries whose economies are unable to provide an adequate standard of living for their present population. Countries have several options of how to cut their rate of population growth, according to a study by the Office of Technology Assessment (1988c), but the only solution that is both morally tenable and feasible is to lower fertility rates. Contraceptive use is by far the most important means of attaining that goal.

Agriculture

In 1890, Walter Heape wrote the first paper on transfer of a fertilized egg, stating, "In this preliminary note I wish merely to record an experiment by which it is shown that it is possible to make use of the uterus of one variety of rabbit as a medium for the growth and complete fetal development of fertilized ova of another variety of rabbit." (Heape cited in Adams, 1982). It was not until 1932 that a successful transfer in a larger animal, a goat, was reported. Only after the second World War was the potential for application of the techniques for livestock improvement and production realized (Adams, 1982). In 1981, it was reported in Science that a "multimillion dollar industry centered on recovery, in vitro culture, and transfer of bovine embryos has evolved over the last decade." This fast growing activity had developed in less than a decade into a $20 million a year industry (Seidel, 1981). By 1985, assisted conception was the norm in dairy cows, with 70 percent fertilized by artificial insemination; 100,000 embryo transfers were performed in the United States in 1984, and 200,000 worldwide, of which 25 percent were with frozen embryos. Artificial insemination resulted in genetic improvement in dairy cattle to an extent that milk production per cow doubled in 30 years (First, Critser, and Robl, 1985). The major use of embryo transfer technology in the food-producing industry is to increase the rate of reproduction of valuable cows. The techniques of superovulation, recovery of embryos, storage in vitro, and transfer to a recipient cow enable some cows to be the egg donors for 50 calves in a year. IVFET can also be used to enable infertile but genetically sound cows to reproduce. The new technologies make the export of cattle breeding stock more economical because it is cheaper to transport embryos (frozen or unfrozen) than mature animals. Moreover, the resulting calves have immunity to local pathogens (received via the foster mother's colostrum), which imported animals lack (Seidel 1981). In 1988 the possibility arose of further improvements in the reproductive efficiency of food-producing animals; the Granada Corporation claimed that techniques for cloning animals were nearing commercial application (Schneider, 1988).

The development of procedures for the control of reproduction in domestic animals has come from universities, nonprofit research institutions, and commercial organizations. Growth of commercial interest has been recent and rapid. In 1986, more than 115 commercial companies and 100 veterinary practices offered embryo transfer services. Some of these commercial organizations also contribute to the research effort by establishing research laboratories (Dresser and Leibo, 1986).

The adoption of the new reproductive technologies to enhance the production of food-producing animals has potential for lowering the cost of food and for increasing the speed with which animals genetically suited to difficult climates can be created. The impact of artificial insemination on the productivity of cows has already been seen. There is reason to expect that further advances in reproductive technologies could improve production of other food animals.

Biodiversity

Advances in reproductive technologies are potentially important in sustaining biodiversity by improving the reproductive efficiency of endangered species. The new reproductive technologies are being studied by zoo researchers interested in conservation of species. These researchers see IVFET and artificial insemination as a way of improving the reproductive processes of endangered species. Breeding of animals is a new role for zoos, which have in the past regarded themselves mainly as a place to display animals. However, as zoos become the last repository for some endangered species, and as genetic diversity is lost because of inbreeding, maintaining diversity has become an important goal. Thus, the development of reproductive technologies that can be used for endangered exotic species takes on a new urgency as it is increasingly realized that captive breeding programs can prevent extinction (Dresser, 1988).

Primates for Research

The ongoing battle between wildlife preservationists and scientists who use primates for research purposes is a final illustration of the far reaching implications of developments in reproductive technologies.

The U.S. Fish and Wildlife service has been asked to put chimpanzees on the endangered species list. Some of those making this request believe that the use of chimpanzees in biomedical research is one cause of their endangerment. This assertion is disputed by officials at the National Institutes of Health. If the chimpanzee is declared endangered, new prohibitions on capture, transport, and use of this species will be imposed. The question remains whether, with such restrictions, the 950 chimpanzees in government facilities at the present time are enough to meet the needs of biomedical research, especially in light of their important role in AIDS research (Science, 1988).

It is here that the potential of new reproductive technologies might play a role. With a limited number of available animals it will become increasingly important to maximize the reproductive capabilities of the 350 chimpanzees that have been set aside for breeding in government facilities.

REFERENCES

Adams, Cyril E. 1982. Mammalian Egg Transfer. Boca Raton, Florida: CRC Press, Inc.

Asch, R.H. et al. 1985. Gamete intra-fallopian transfer (GIFT): A new treatment for infertility. International Journal of Fertility. 30: 41-45.

Biggers, John D. 1988. Human Generation: Fact, Foible and Fable. Plenary Lecture to the American Association for the Advancement of Science. Boston, Mass. February 2.

Braude, Peter. 1988. Gene Expression in Early Embryonic Development Paper Presented at the Institute of Medicine, Board on Agriculture, Workshop on the Basic Science Foundations of Medically Assisted Conception, Irvine, California, August.

Craft, Ian. et al. 1987. Successful Births After Ovum Donation. Letter. Lancet. April 18. p. 916-917.

Dresser, Betsy L. 1988. Biodiversity. E.O. Wilson ed. Cryobiology, Embryo Transfer and Artificial Insemination in Ex Situ Animal Conservation Programs. Washington, D.C.: National Academy Press.

Dresser, Betsy L. and Leibo, S.P. 1986. Technologies to Maintain Animal Germplasm in Domestic and Wild Species. In Evaluation of Technologies to Maintain Biological Diversity. Vol.1, Contract Papers, Part B. Animal Technologies. Washington, D.C.: Office of Technology Assessment.

First, N.L., Critser, E.S., and Robl, J.M. 1985. Bovine Embryo: Development, Cloning, Sexing and Transfer of Genes for Immunology of Reproduction and Contraception, P. Talwas, ed. Elsevier, 1985

Fuchs, Victor, R. and Perreault, Leslie. 1986. Expenditures for Reproduction-Related Health Care. Journal of the American Medical Association, Vol. 225, No. 1, Jan 3, pp 76-81.

Jones, Howard E. Jr., 1989. Howard and Georgeanna Jones, Institute for Reproductive Medicine. Personal Communication. February 23.

Lancet. 1987. Clinical Status of IVF, GIFT and Related Techniques. Lancet. October 26, pp. 945-947.

National Center for Health Statistics. 1985. Fecundity and Infertility in the United States, 1965-82. Advancedata, No. 104. Feb. 11, Washington, D.C.: U.S. Department of Health and Human Services, Public Health Service.

Office of Technology Assessment. 1988a. Artificial Insemination Practice in the United States. OTA BA BP 48. August. Washington, D.C.: Office of Technology Assessment.

Office of Technology Assessment. 1988b. Infertility: Medical and Social Choices. OTA BA 358. May. Washington, D.C.: Office of Technology Assessment.

Office of Technology Assessment. 1988c. World Population and Fertility Planning Technologies: The Next 20 Years. Summary. OTA HR 158. February. Washington, D.C.: Office of Technology Assessment.

Schneider, Keith. 1988. Better Farm Animals Duplicated by Cloning. New York Times, Section D, Page 1, Feb 16.

Seibel, Machelle M. 1988. A New Era in Reproductive Technology. In Vitro Fertilization, Gamete Intrafallopian Transfer, and Donated Gametes and Embryos. New England Journal of Medicine. 318(130): 828-834.

Seidel, George E., Jr. 1981. Superovulation and Embryo Transfer in Cattle. Science. 211 (4479): 251-357.

Science. 1988. Chimps and Research: Endangered? News and Comment. Science 241: 777-778.

CHAPTER 2

ADVANCES IN THE PRACTICE AND SCIENCE BASE OF MEDICALLY ASSISTED CONCEPTION

This chapter is based on papers presented at a workshop on Basic Science Foundations of Medically Assisted Conception sponsored by the Institute of Medicine (IOM) and the Board on Agriculture of the National Research Council. It was held August 21-23, 1988 at the Arnold and Mabel Beckman Center in Irvine, California. For the workshop, the organizing committee developed a program that explored the recent advances in reproductive and developmental biology that apply to medically assisted conception. This chapter represents a report of science topics selected by the committee. It is therefore not a comprehensive review of recent advances in reproductive research. The workshop also initiated an interchange of ideas among those involved in patient-related clinical practice, animal IVFET, and those working in basic research as it applies to humans and other animals. Thus, this chapter indicates research areas that promise improvements in the practice of IVFET. First are some developments in human and animal IVFET; subsequently, the processes and recent advances relating to gametogenesis, fertilization, preimplantation development, and implantation are discussed. This chapter summarizes each talk given at the workshop. The full papers contributed by each author are found in Appendix A.

Developments in Human In Vitro Fertilization [1]

Eastern Virginia Medical School, one of the premier centers for the clinical practice of in vitro fertilization, has a pregnancy rate of 18.3 percent, based on the number of attempts used to stimulate the ovaries with either follicle-stimulating hormone (FSH) or human menopausal gonadotropin (hMG), and a pregnancy rate of 37.6 percent based on the number of patients. Both patient age and the cause of female infertility have an effect on the outcome of in vitro fertilization. For women infertile because of tubal ligations or endometriosis, the rate of viable pregnancies obtained by IVFET at Eastern Virginia is 16.4 percent. Although the pregnancy rate did not vary according to whether infertility was due to tubal ligation or endometriosis, age has an effect. After the age of 40, it becomes more difficult to stimulate the ovaries to produce mature eggs.

Because couples often seek in vitro fertilization to overcome male infertility, the evaluation of semen is important. Currently, this evaluation is descriptive and relatively imprecise. Factors such as number of sperm, sperm motility, and general shape of sperm components(e.g. head shape and tail shape) are important variables in achieving pregnancy. For example, if 14 percent of the sperm are "normal," and there are 50,000 sperm per cubic centimeter of ejaculate, in vitro fertilization produces ongoing pregnancies in 30 percent of all attempts. However, even if there are 100,000 sperm per cubic centimeter, but only 4 percent are "normal," the ongoing pregnancy rate drops greatly. Abnormal sperm often fail to fertilize an egg because of their inability to penetrate the egg's protective covering, the zona pellucida, therefore workers have tried a technique known as "zona drilling" in which holes are produced in the zona pellucida to permit direct access of sperm to the oocyte plasma membrane. Thus far, this technique has not been successful. Another technique, known as zona splitting, has been used at Emory University, Atlanta. By this technique, the zona pellucida is split mechanically. Pregnancies have been reported by use of this method. Basic science questions need to be answered in order to develop better markers for normal and abnormal sperm and to improve the performance of sperm in IVFET.

Cryopreservation (freezing) of in vitro fertilized embryos[1] is another new and promising technique that will be discussed in greater detail later in this chapter. If embryos are frozen for later use, the stage of their development at the time they are placed in the uterus can be matched with the stage of the uterine wall (endometrium), increasing the likelihood of a successful pregnancy. Before cryopreservation became an option, embryos had either to be placed in the uterus or discarded. In order to avoid the ethical dilemma of what to do with excess embryos, more than the optimal number of embryos were sometimes transferred into the uterus. By allowing the preservation of embryos for later use, the technique of cryopreservation reduces not only the chance of multiple pregnancies, but also the number of times a woman's ovaries might be subjected to hormonal stimulation to produce oocytes for additional attempts of IVFET. The use of cryopreservation has resulted in increased rate of pregnancy at the Eastern Virginia clinic. Nevertheless, basic science research is needed to assess the necessary parameters for successful cryopreservation and the possible deleterious effects of freezing on the embryo.

In cases of ovarian failure, failure of in vitro fertilization, poor quality of eggs, genetic abnormality, or inaccessible ovaries, the only option available to women wanting to bear a child is to use eggs from a donor to perform in vitro fertilization with the husband's sperm. The donated eggs most often come from IVFET patients who have received hormonal stimulation and produced more eggs than necessary for their own use. Sometimes, however, women who are to undergo a tubal ligation agree to ovarian stimulation before the surgery so that eggs can be harvested simultaneously. These eggs are then donated for in vitro fertilization and transfer. Donors are phenotypically matched with recipients and are screened for psychological problems and infectious diseases. Patients who

---

[1] The term embryo in this chapter often refers to in vitro fertilized eggs that developed to two- to four-cell staged embryos. Although there are more technically precise terms for various early stages of development, the more precise terms are used here only when the distinctions are important to the concepts under discussion.

conceive through egg donation may need supplemental hormone therapy to replace the estrogen and progesterone normally released from the ovaries. In such a patient, however, the placenta produces detectable estrogen by week 6 of the pregnancy and progesterone by week 7. Fertilized eggs from oocyte donation, if transferred to the uterus by day 17-19 of the recipient's cycle, resulted in a 36 percent pregnancy rate and 80 percent of those pregnancies going to term. One benefit of oocyte donation, in addition to overcoming the problems mentioned above, is that a time period of optimum implantation can be achieved such that the stage of the embryo and endometrium are synchronized. This technique has been found to increase the success rate of IVFET.

Good quality embryos are critically important to the success of IVFET. Experience at many clinics shows that a number of factors determine the quality of an embryo. Some important factors include the way ovaries are stimulated with hormone supplements and the synchrony between the age of the embryo and the endometrial stage. More research is needed to understand fully these factors and to establish unambiguous and unbiased criteria with which to distinguish poor quality and good quality embryos.

Because fertilization is more likely with mature oocytes than immature ones, research to learn how to recruit a more synchronous population of follicles is important. Such follicles produce a more synchronous population of oocytes capable of responding to the maturational stimuli. Failing the recruitment of mature oocytes, research is needed to develop reliable methods of maturing oocytes in vitro. All of these areas, if improved by increased knowledge of the cell biology of early reproductive events, should greatly increase the ability to identify couples with high probability of success and may increase the success rates of IVFET in human clinical practice.

Developments in Assisted Conception in Food-Producing Animals [2]

There are critical differences between the goals of assisted conception in animals and human beings. In human beings, the goal is to increase the reproductive ability of those with impaired fertility or, perhaps in the future, to avoid the transmittal of genetic disorders. In contrast, the goal of assisted conception in food-producing animals often is to increase the yield of milk or meat. Because the ethical barriers are lower and the financial stakes higher, a number of advanced technologies are available for use in animals that are not available in the clinical practice of human IVFET. Techniques in commercial or research use include artificial insemination, superovulation, embryo transfer, freezing of embryos, sexing of embryos, multiplication of embryos by bisection and cloning, in vitro fertilization, and the modification of embryos by gene transfer. All western European dairy cows and 60 percent of U.S. dairy cows are impregnated by artificial insemination (AI). AI allows bulls, selected for their transmission of milk producing genes, to achieve up to 500 inseminations from one ejaculate. Thirty years of AI has helped to double the milk production of each cow.

Another technique used widely in humans as well as in other animals is superovulation, which involves ovarian stimulation by the administration of hormone supplements. In both animals and humans, superovulation is unpredictable and often unsuccessful. There has been some success when cattle are given priming doses of follicle stimulating hormone (FSH), and when highly purified forms of FSH are used, but the reasons for these effects are unclear. In addition, studies of cattle and other species have found that the oocytes produced by superovulation sometimes exhibit abnormal characteristics. Increased understanding of hormonal cycles and the effects of hormones on the cell biology of maturing oocytes would likely have direct application to the solution of some of these problems.

The process of embryo transfer in cattle starts with superovulation of a cow, and subsequent mating to a desirable bull. The resultant embryos are collected nonsurgically and transferred to another cow whose estrous cycle has been synchronized with that of the donor cow. Although the success rate with embryo transfer is respectable (about 60 percent of transfers result in pregnancy), the technique has not produced great increases in milk or meat production. The production of many identical copies of embryos from cattle with highly desirable traits may improve this situation. For this reason, embryo freezing and embryo multiplication procedures are receiving a great deal of attention in the commercial breeding of cattle.

Embryo freezing, in concert with embryo transfer, has been generally successful in terms of pregnancy rates. Freezing embryos allows for the storage of rare breeds and preservation of a cattle surplus. However, embryo multiplication has greater potential for the production of large numbers of highly desirable cattle. Two methods of embryo multiplication are used--embryo bisection and nuclear transfer. Embryo bisection is performed at a very early embryonic stage and yields at least two cell masses which are genetically identical. The bisected embryos can then be transferred by normal procedures to a recipient cow. There is a species-specific limit on how many bisections of a given cell mass can be done without compromising viability; with cattle, the maximum yield is four embryos from one. In nuclear transfer, a blastomere (embryonic cell), or its nucleus, from a valuable embryo is placed into an oocyte from which the nucleus has been removed. The transferred embryo nucleus promotes development of a multicellular mass that can be used to make a number of copies or clones. There is a great deal of interest in this method both in academic and commercial research. Since these techniques allow more precise selection of desirable traits, both have the potential to effect rapid changes in the prevalence of animals with those traits.

While the techniques discussed above are frequently and successfully used, other methods of assisted reproduction are also being investigated. One of the newer approaches is embryo sexing. Sexing of embryos is of particular importance to the dairy industry since only female offspring are needed for milk production. Sexing is done by three methods. The first is karyotyping of embryos by bisecting them and using one half for cytogenetic analysis. The chief disadvantages of this method are the damage that often occurs to the embryo and the variable reliability of the karyotypic identification of sex. The second method employs antibodies to male-specific antigens on the embryos to identify the male embryos. The accuracy of this immunological method is greater than that of karyotyping and may become widely used, depending on the availability of antisera. Finally, DNA hybridization techniques have been adapted from molecular biology to sex embryos by labeling specific parts of the male Y chromosome obtained by embryo biopsy.

There are compelling reasons to develop more fully IVFET procedures in food-producing animals. Such techniques provide a large number of embryos for transfer, research, and embryo multiplication. However, as in human IVFET, the success of in vitro fertilization and embryo transfer in the cattle industry suffers from a lack of basic knowledge of the cell biology and biochemistry of early gamete maturation and fertilization. Among the areas in which research is lacking is oocyte maturation. As in humans, IVFET in cattle is more successful if it begins with mature rather than immature oocytes. The biochemistry of sperm capacitation and entry into the eggs also needs to be better understood. A number of chemicals have been identified that are necessary for these processes, but precise identification of the role each chemical plays in fertilization is needed for further progress. Finally, being able to maintain the growth of an embryo in culture for a longer period than is now possible would be helpful. Such an ability would increase the chances of a favorable match between the embryonic and endometrial stages.

Gene transfer, if developed further, also shows potential for enhancing the productivity of cattle. With the development of techniques by which genes can be microinjected into embryos, the ability to alter the phenotypic characteristics of food-producing animals has become a possibility. In an early demonstration of transgenic technology, workers injected mouse eggs with the gene for human growth hormone. The resulting mice grew to almost twice the size of normal mice. Such technology could be used in cattle to alter the genes for skeletal muscle in order to produce higher quality meat products, or to alter genes in such a way that biologically active substances would be secreted in milk.

There are many areas in which the results of basic research would further the practice of assisted conception and embryo transfer in both human beings and other animals. These include oocyte maturation, sperm morphology, the biochemistry of fertilization, cryopreservation, ovarian stimulation, and molecular genetics. In addition, topics such as membrane biochemistry, hormonal control of testicular and ovarian function, gene expression in early development, and the cell biology of implantation are identified as areas of exploration that would make major contributions to the success of human and animal IVFET.

Gametogenesis and Gametes

Induction of Gametogenesis and Superovulation [3]

Comparative research on ovarian stimulation of non-human primates and domestic animals such as cattle, horses, sheep, and pigs raises interesting issues. Some work highlights the differences between the normal estrous cycles of animals such as cattle, and the cycles of human and non-human primates. Yet there are problems that are common among species in the stimulation of follicular development and the induction of mature oocytes. Ovarian stimulation and superovulation are exogenous interventions into a highly regulated physiological process. The normal physiology is not always easily manipulated. Developing strategies to circumvent the differences between normal and induced ovulation is a major goal of research proposed for this area.

Non-Human Primates Non-human primates exhibit a menstrual cycle that closely approximates that of humans. In monkeys, the hormonal events of the cycle include a geometric increase in serum estradiol, midcycle surges of luteinizing hormone (LH), follicle stimulating hormone (FSH), and a later increase in serum progesterone after LH and FSH decrease in concentration. These hormone levels are precisely synchronized in the normal cycle and lead to the development of one follicle, the dominant follicle, from which ovulation occurs. One of the purposes of superovulation is to encourage more than one follicle to develop fully, thereby producing multiple oocytes for in vitro fertilization. In monkeys, one of four chemicals is administered in combination with human chorionic gonadotropin to stimulate ovulation. These are human menopausal gonadotropin, human FSH, pregnant mare's serum gonadotropin, or porcine FSH. However, monkeys sometimes produce antibodies that block the activity of hormones obtained from other species activity, and ovarian stimulation fails. Thus, there is a need to develop stores of monkey hormones to be used for ovarian stimulation in this species. The response to the administration of hormones to stimulate ovulation is also not uniform in non-human primates. In general, estradiol levels are higher than in a natural cycle, the LH surge is lacking, LH concentration remains high over a long period, FSH is low, and progesterone remains high in concentration. However, in some subjects there is a premature, spontaneous LH surge that is not synchronized with follicular maturation; this surge stops the follicles and oocytes from maturing. Even if such an LH surge does not occur, there is a wide range of responses to gonadotropin administration. It is impossible to predict the response of any individual animal. The LH surge and the heterogenous responses, however, can often be prevented by additional administration of gonadotropin releasing hormone. Another difficulty often observed with superovulation protocols is an overstimulation of prolactin secretion. These departures from the normal hormonal levels raise important questions relating to the efficacy and unknown effects of ovarian stimulation.

Despite the problems of ovarian stimulation in non-human primates, reasonable success rates have been achieved. Nevertheless, there are additional barriers to progress in this field. The supply of non-human primates is severely limited. In addition, these animals often cannot be used more than once for ovarian stimulation because of the immunological responses to human, equine, and porcine gonadotropins used in the stimulation procedures. Finally, eggs must be recovered from monkeys by either laparotomy or laparoscopy, surgical techniques that are restricted by ethical considerations and formal legal constraints to a low number of repetitions per animal. Although these problems of availability, reuse, and the lack of non-human gonadotropins are more difficult to overcome than the biological problems of overriding normal physiology, numerous questions about the basic biological processes remain unanswered. Resolution of the problems associated with the use of non-human primates in this type of research could be facilitated by an increased ability to augment natural hormone and gonadotropin release, by improved resolution of ultrasound imaging to identify mature oocytes and guide their collection by non-surgical means, and by the development of cell lines that could produce larger quantities of non-human primate gonadotropins. Finally, it is important to note that many of the same questions remain unanswered regarding human ovarian stimulation and that particular non-human primate species serve as the best model for human reproductive physiology.

Domestic Animals Extensive use of artificial insemination in cattle has increased the genetic contribution of desirable bulls to the overall supply of cattle. However, there is little likelihood that the desirable characteristics of females will be further increased by augmenting the male genetic contributions. Superovulation, on the other hand, combined with embryo freezing and embryo transfer, has the potential to increase directly the gene pool of desirable female traits by increasing the number of offspring from a valuable cow. There are a number of differences between non-human primates and domestic animals in the ways ovarian stimulation is accomplished. Some of these differences arise from differences in the reproductive cycles of the two groups. In dairy and beef cattle, stimulation is accomplished with the administration of either pregnant mare's serum gonadotropin (PMSG) or porcine FSH. The latter is preferred to PMSG because PMSG has a long half-life and often results in asynchronous ovulations. Porcine FSH is not without problems either since it is contaminated by variably high levels of luteinizing hormone. Human chorionic gonadotropin, routinely used in non-human primates, is not used in the commercial practice of ovarian stimulation of cattle. To control the variability and lack of predictability in the estrous cycles of cattle, current protocols often include an injection of a specific type of prostaglandin (PGF-2 alpha). This treatment causes the regression of the corpus luteum, thereby artificially restarting the cycle.

The success rate of superovulation in cattle has been variable. This is caused by a number of factors including seasonal variability, breed differences, dose and timing of gonadotropin administration, and history of previous superovulations. Such differences contribute to difficulty in comparing studies performed on different breeds and, at times, in comparing studies in the same breeds, but done with varied techniques or in animals of varying age. It has been found that, even if all the factors possible are controlled, there remains a great deal of variability in the success rate of superovulation. A number of strategies to improve the success rate are currently being tried. One is an attempt to purify porcine FSH preparations to exclude the contaminating LH and, further, to determine the FSH to LH ratios that are optimum for successful superovulation. Another strategy is to try to neutralize PMSG with antibodies so that the complications that arise from the long half-life of PMSG can be prevented. Because the lack of normal coordination of endocrine events is a common occurrence in superovulation, investigators are also trying to define the sources of asynchrony and develop ways of normalizing a coordinated series of biological events.

Superovulation in domestic animals other than cattle also has had mixed results. In goats, the technique has been tremendously successful. Superovulation of sheep has also been largely successful, and a few innovative strategies have been applied in this species. Sheep are seasonal breeders and attempts have been made, with initial success, to induce ovulation in this species during times when they are normally anestrous. In order to bypass the use of exogenous gonadotropins, workers have employed antibodies to inhibin in sheep. Inhibin is a substance made in an ovarian follicle that suppresses FSH secretion. The blocking of inhibin with antibodies raises FSH levels, thereby increasing the natural stimulus for follicular development. Antibodies against steroids have also been tried in sheep to achieve the same end.

In horses, superovulation has proven to be difficult, and to require nonstandard protocols. The period of estrous or "heat" in mares is unusually long and variable. In addition, there are preprogrammed physiological mechanisms in mares to prevent the development of twins. These two aspects of equine reproductive physiology greatly complicate the successful application of ovarian stimulation techniques. Also of critical importance is the fact that PMSG, even in high doses, does not adequately stimulate the development of follicles in mares. Unlike other animals, where PMSG acts as both FSH and LH receptors, in mares PMSG has only LH activity. Some success has been achieved in mares with administration of porcine FSH, especially if combined with human chorionic gonadotropin. In general, however, progress in successful superovulation in horses lags well behind that in other domestic animals, except for the pig.

The most limited of all commercial applications of superovulation in animals has been in the pig. Although possible, the technique does not confer many of the advantages that it holds for other species. One reason is that pigs deliver litters as opposed to one offspring at a time. Therefore, the need to increase the offspring from one individual, while helpful in certain circumstances, is not as compelling in pigs as in other domestic animals. Nevertheless, superovulation and embryo transfer may be able to lower the frequency of diseases by increasing the number of offspring from disease-free or disease-resistant populations. More frequent reasons for the use of embryo transfer in pigs are to obtain disease-free embryos from infected pigs[2] and to introduce new genetic material into specific pathogen-free herds. Thus, further research into the normal reproductive physiology of pigs will be useful.

To improve IVFET in domestic animals as well as in human and non-human primates, more needs to be known about the normal regulatory events of ovulation, including the physiology and biochemistry of the development of a dominant follicle. Improved ultrasonography has potential for increasing the success of superovulation in all species. As with monkeys, developing ways to bypass the use of exogenous agents like PMSG and porcine FSH would be of enormous use in domestic animals. One way of doing this may be by blocking inhibin as has been done in sheep. In order to improve the success rate of assisted conception in animals, many of these research areas must, nevertheless, be pursued in each species separately with realization of the inter-species differences. It is not always possible to predict when the results obtained from one species can contribute knowledge applicable to other species.

Biology and Maturation [4]

This section examines three topics relating to the biology and maturation of oocytes: 1) The study of membrane biochemistry, which has relevance to the complex membrane interactions that occur at fertilization; 2) the study of molecules that are important in maturation events, is also important to understanding the possible reasons for failure in IVFET; and 3) the biological and physical properties of eggs and embryos that are affected by preservation techniques such as freezing, known as cryopreservation.

Membrane Biochemistry The membranes of all cells of the body share certain standard structural characteristics. The basis of a cell membrane is called a bilayer, which is a two-layered collection of phospholipid molecules (lipids containing phosphorus). Phospholipid molecules are polar in that each has one end that is stable in water (hydrophilic), and one end that repels water (hydrophobic). In a bilayer, then, the lipid molecules line up so that the hydrophobic ends meet each other on the inside of the bilayer. The hydrophilic ends are thus oriented on the outside of the bilayer such that one group faces the intracellular space and the other faces the extracellular space. Such an arrangement makes sense when one considers that both the cytoplasm of a cell and the extracellular spaces are composed largely of water. It is, in fact, a quite natural response of lipids to form bilayers in an aqueous environment; such a process underlies soap bubble and oil droplet formations.

---

2 Embryos can be rinsed free of pathogens or treated with enzymes to destroy pathogens making embryos the safest method of moving germ plasm without pathogens around the world.

Proteins are part of the membrane also. Some proteins span the thickness of the membrane and act as hydrophilic channels for ions or participate in molecular signal transduction mechanisms. Other proteins are partly embedded in the lipid bilayer and partly exposed to the extracellular space. Attached to such proteins are long, often branching, chains of carbohydrate molecules that also extend into the extracellular space. Finally, large extracellular proteins often associate with or stick to the most external parts of the carbohydrate molecules. Thus, the cell membrane, even in its standard form, is a complex structure made up of different types of molecules arranged in regions that each exhibit special properties. The possible interactions of membrane molecules are immense since the various molecules are mobile within the bilayer. These interactions underly the equally immense range of biological responses of cells including cell-cell interactions, receptor-mediated signal processing, and transmembrane transport of nutrients or toxins.

Reproductive events in which membrane biochemistry is of particular importance include sperm capacitation and fusion of sperm and egg membranes for fertilization. Basic aspects of membrane biochemistry are most often studied using membranes from cells such as red blood cells or membranes manufactured in the laboratory, called liposomes. Such membranes are simpler than the membranes from eggs or sperm and, therefore, easier to study. However, data obtained from these model membranes must be extrapolated to eggs and sperm with great caution.

<u>Mammalian Oocyte Maturation</u> In a normal follicle there is communication between the oocytes and the surrounding nurse cells, called granulosa cells, through specialized junctions (gap junctions) through which small molecules pass from one cell to another. Developing oocytes undergo a type of cell division in which only half the normal number of chromosomes are retained. This process is known as meiosis and has a number of stages. Until just before maturation, however, the growing oocytes remain in a state of suspended or arrested meiosis. It is thought that the granulosa cells help to maintain this arrest of meiosis by producing substances that enter the oocyte through the gap junctions. It is also possible that substances that induce final maturation of the oocytes are produced in the surrounding cells and are transmitted to the oocytes in the same manner. Since the ability to mature oocytes in vitro would be of enormous value in IVFET, the understanding of the substances that either maintain or abolish meiotic arrest is particularly desirable. It is important to note that many of the studies summarized in this section were done with mouse oocytes and these data were compared to those obtained from other mammals including rabbits and rats. Investigators have also studied frog oocytes, but the mouse is the most common animal model. Eventually, of course, certain critical experiments will have to be replicated using oocytes from human or non-human primates.

Several substances are thought to maintain meiotic arrest. One, cyclic adenosine monophosphate (cyclic AMP), is a molecule present ubiquitously in the body. Another, hypoxanthine, is common in the body and belongs to a group of molecules known as purines. Guanosine, another purine, is also a potent inhibitor of meiosis and the purine, adenosine, has been shown to augment the activity of hypoxanthine. By examining the effects of some of these substances on each other, a hypothesis has arisen which suggests that hypoxanthine and other purines may increase the level of cyclic AMP, which then inhibits meiosis. Future research, focused upon the molecules of the cyclic AMP pathway inhibiting meiosis, may elucidate ways in which oocytes can be maintained in meiotic arrest in vitro.

While knowledge regarding the inhibition of oocyte maturation is important, knowing how to mature oocytes in culture would have many applications that are described later. In vivo, oocyte maturation is produced by the surge of LH immediately prior to ovulation. The mechanism of action of LH is unclear, however. One might assume that, if cyclic AMP maintains meiotic arrest, then LH somehow causes a decrease in cyclic AMP in the oocyte or the surrounding nurse cells. It has been shown in vivo that administration of LH in combination with human menopausal gonadotropin reduces cyclic AMP in the oocyte, but not in the surrounding nurse cells. This effect is puzzling since the gap junctions between the oocytes and the other cells still appear to be functional, so cyclic AMP should pass through these junctions easily. An alternative hypothesis is that LH acts indirectly by causing some maturation-inducing substance to be produced in the surrounding cells that decreases the cyclic AMP in the oocyte. Since an increase in intracellular calcium has been shown to be essential in somatic cell division, it is possible that a maturation-inducing substance operates in the oocytes by causing an increase in intracellular calcium. The search for these substances and their mechanism of action is a major area for future research.

The study of oocyte maturation has led to extremely successful methods for maturing mouse oocytes in vitro. Such work has shown that the in vitro culture of oocytes and their surrounding cells can be accomplished in ways that allow subsequent fertilization, implantation, and delivery of normal young to occur. Successful in vitro fertilization of oocytes from other species, however, depends upon continued basic research on oocyte nutrition and metabolism and on the factors or substances that control differentiation and maturation. The potential applications of an increased ability to mature oocytes in vitro are numerous. As discussed earlier, such an ability would be of enormous value in the production of agriculturally important animals. These techniques also have important applications in attempts to preserve endangered species. For example, it has been demonstrated that oocytes removed from mice up to six hours after death can be matured in vitro (Lazarus effect). Yet another application can be found in basic science research into genetic diseases through the use of transgenic mice. Such mice are produced by introducing known gene fragments into the genome of a mouse. These gene fragments have been inserted by microinjection into eggs. Alternatively, the fragments, incorporated into a harmless virus, have been injected into neonatal female mice. In this way, such fragments are incorporated into the DNA of developing oocytes in these females. It would be extremely helpful to be able to transfer gene fragments by viral infection of oocytes, which could then be matured in vitro for subsequent fertilization. Finally, the ability to mature human oocytes may have some application for human IVFET, since matured oocytes from donors would obviate the need for hormonal stimulation therapy and permit synchronization of the developmental state of the embryo with the state of the uterine lining. The degree to which data obtained from experiments with mouse oocytes can be applied to humans is not known. Therefore, for such applications to be realized in human IVFET, it is probably necessary that basic research be done using oocytes from human ovaries removed for justified clinical reasons and/or oocytes from non-human primates.

Cryopreservation of Oocytes and Embryos Cryobiology is the study of how living tissues can be frozen and stored for later revival. There are a number of uses for cryopreservation of oocytes and embryos in connection with IVFET and other reproductive technologies. Cryopreservation of human and cattle sperm has been a routine and successful procedure for many years. It is thought that, properly stored, frozen embryos can be preserved for many decades before thawing and transfer to the uterine cavity. To date, embryo cryopreservation has been successful in many mammalian species, including mouse, sheep, cow, human, and baboon. Rare mutant animals have also been preserved for later study.

The Mouse Embryo Banks at the Jackson Laboratory in Maine contain many examples of cryopreserved mutant mouse embryos. Such mutants are valuable for research into genetic and other diseases. Cryopreservation allows easy transportation of frozen embryos by researchers who move frozen mice embryos from one lab to another, and for the national and international transportation of cattle embryos.

If proved safe for use in clinical practice, routine cryopreservation of eggs or embryos could reduce substantially the number of times a woman would have to undergo hormone stimulation, because the eggs harvested could be frozen for later fertilization or fertilized in vitro and the extra embryos stored for future transfer. These stored embryos could be used if the first pregnancy failed or used at a later time so that an infertile couple could have more than one child from one stimulation/IVFET procedure. Thus, such preservation techniques could also minimize the waste of extra eggs and extra embryos produced by superovulation and other IVFET procedures.

Although the specific methods of cryopreservation vary, the basic steps involved are the same. The first is chemical protection of the tissue against damage from freezing. The tissue is then sometimes cooled in sequential steps to a subzero temperature before being rapidly cooled to -196° C for long-term storage. The procedures used for warming and cooling are variable. After rewarming, the chemical protective solutions are removed by rinsing. Protection from damage by freezing is critical to the survival of an embryo because, without protection, the formation of ice crystals causes tissue damage that can not be repaired.

Cryoprotective solutions, which are made from chemicals such as glycerol or dimethyl sulfoxide, function in a manner similar to antifreeze in a car. They replace the water in the embryo.

Because of the importance of this step in cryopreservation, much basic research is being conducted to find appropriate cryoprotective solutions. Certain specific properties of the solutions are important determinants of the survival rate of eggs and embryos. For example, some chemicals, such as sucrose, cannot enter cells and, therefore, do not protect from freezing damage in the same way, or with the same efficiancy, as other molecules that can enter cells. Glycerol, on the other hand, can permeate a cell, partially replace the water, and impart the cryoprotection necessary. The osmotic characteristics of particular solutions are important also. In the case of glycerol, if one exposes a cell to a solution of water and glycerol, glycerol moves into the cell by osmosis, a process by which molecules in a high concentration will move across a semi-permeable barrier, such as a cell membrane, into a region of low concentration of that molecule. Likewise, water moves out of the cell by osmosis into the surrounding fluid. Eventually, the concentration of glycerol will equilibrate and be the same on both the inside and outside of the cell. Since glycerol moves into the cell more slowly than water moves out of the cell, the cryoprotection procedure must be carefully controlled so that the cell does not lose water so quickly as to shrink and die.

It is especially important to control the movement of molecules across the cell membrane when thawing ova and embryos. After thawing, it is important to remove the glycerol from the cells and to return the intracellular contents to normal. However, a cell protected with glycerol, if suddenly exposed to a normal saline solution, will take up a great deal of water very quickly. This influx of water causes the cell to swell and the membrane may burst. Although this problem can be avoided by exposing the cells to solutions of decreasing concentration of glycerol in a stepwise fashion, this is not always easy to accomplish. Research has shown that adding the impermeable solute, sucrose, to the saline solution results in a more even exchange of glycerol and water across the membrane and obviates the rapid influx of water into the cell. This one-step method has proven to be of particular use in the application of these techniques under field conditions such as in work with endangered species.

The method used to freeze and thaw the ova and embryos is also important. Until recently, the tissue was first cooled to a few degrees below zero so that ice formation could be controlled. The tissue was then cooled to some intermediate subzero temperature before being rapidly cooled to the storage temperature of -196° C. Research with mouse ova showed that the cooling procedure used could affect the survival rate in important ways. For example, it was demonstrated that, if ova lose a considerable amount of their original volume during cooling, the survival rate is greater. Cells that retain most of their original volume usually do not survive. This is probably because a cell is primarily composed of water from which the damaging ice crystals form. The method used to freeze the tissue also determines the subsequent thawing procedure. Thawing is performed stepwise in reverse order to the freezing steps. New freezing protocols have been developed that simplify the multiple step methods. For example, mouse embryos have been successfully frozen in one step after incubation in glycerol and sucrose. This method uses the glycerol to cryoprotect the tissue and the sucrose to dehydrate it before freezing, eliminating the need to slowly cool the tissue before freezing. Variations of this method have also been successful. One novel method is called vitrification, which means freezing without crystal formation. Sometimes, this can be accomplished by extremely rapid freezing. However,in a variation of this approach, embryos are exposed to a solution of chemicals that form a glassy solid substance when cooled. No ice crystals form in the embryos exposed to this solution so freezing can be rapid and done in one step. These new freezing procedures are still being developed, but they have the potential to increase the success rate of ova and embryo freezing.

Improvements in cryobiology would have significant impact. Improved preservation and storage of embryos from valuable cattle and other agriculturally important animals would greatly increase the yield from superovulation and IVFET procedures. The preservation of endangered species and rare genetic breeds of laboratory animals would be enhanced. Finally, successful storage of human embryos that result from IVFET techniques can increase the chances of an infertile couple to have a baby.

Intratesticular and Intraovarian Paracrine Control [5]

Development of male and female gametes occurs over varied periods of time in special body compartments, the ovaries and the testes. In this section, the biochemical control mechanisms that affect gamete development will be discussed. In contrast to hormonal control mechanisms in which bioactive substances are released into the bloodstream and act at distant sites, paracrine control mechanisms involve chemical interactions between neighboring cells. For example, one cell can release an active substance into the extracellular space, the substance then diffuses around other cells in the region and affects the activity of these other cells in some way. These effects most often are mediated by cell surface receptors for the substance released.

<u>Intragonadal Control of Testis Function</u> The testes are composed of a number of different cell types, including germ cells in their various developmental stages (spermatogonia, spermatocytes, spermatids, and mature sperm) each of which displays unique characteristics. Other cell types include Sertoli cells, which are nurse cells for the developing spermatogonia and spermatocytes, and Leydig cells, which release the hormone testosterone, required for sperm maturation. Many bioactive hormones and peptides have also been found in the testis and research has begun to determine the cell sources of these chemicals and their function in spermatogenesis. Some of the molecules found in the testis have been shown to affect cell cycles and cell differentiation. For example, somatomedin C (also called insulin-like growth factor, IGF) and epidermal

growth factor stimulate cells to begin DNA synthesis for mitotic cell division. Fibroblast growth factor stimulates quiescent cells to reenter the cell cycle in order to replicate and divide. Various hormones and growth factors act as stimulators to cell differentiation, thus causing the cells to exit permanently from the cycle of chromosome replication and mitotic division. Cells are inhibited from exiting the cell cycle by interleukin-1, which is a molecule made by cells of the immune system. The functions of most of the hormones and growth factors in the testis are not known. For a few, however, certain functions are beginning to be described.

In the rat, a common animal model for investigations of sperm development, Sertoli cells start to divide and proliferate around the 22nd day of embryonic life, and complete this proliferation by 20 days after birth. Leydig cells, in contrast, do not start to divide until 2 or 3 days after birth and continue to proliferate until 35 days after birth. By the end of these processes, almost equivalent numbers of Sertoli and Leydig cells are found in the testis, despite very different developmental timetables. A hypothesis has been proposed that these cells may regulate each other by complicated interactions during development. Fibroblast growth factor stimulates replication of Sertoli cells by causing an increase in the number of receptors for follicle stimulating hormone (FSH). As this occurs, the numbers of Sertoli cells increase and release somatomedin C in greater and greater amounts. Somatomedin C increases the number of receptors for luteinizing hormone (LH), which causes the Leydig cells to divide and increase in number. The increasing numbers of Leydig cells leads to an increase in beta-endorphin released by the Leydig cells. A negative feedback occurs such that beta-endorphin decreases the proliferation of Sertoli cells.

Such paracrine interactions are probably not unusual and may occur between other types of cells in the testis. In addition, non-paracrine mechanisms of communication probably occur, including cell to cell adhesion and communication through gap junctions between cells. Very little is known regarding the mechanisms of interaction between developing sperm cells and their nurse cells, the Sertoli cells. Cell adhesion molecules have been found on spermatocytes that, some suggest, promote continued meiosis of the spermatocytes bound to the Sertoli cells. Experiments done in culture have shown that sperm cells synthesize more DNA and RNA when they are cultured with Sertoli cells (especially if FSH is in the culture medium), than if the sperm cells are cultured alone, or alone with FSH. Such data suggest that the Sertoli cells are important to sperm cell development.

Investigators have looked for proteins that can be related to specific stages of sperm cell development. The stages of sperm cell development are quite complex, but can be generally grouped into four main stages. The primordial germ cells are called spermatogonia, and undergo six mitotic divisions before becoming spermatocytes. Meiosis begins in spermatocytes. Further development forms spermatids, which are immature germ cells in which meiosis is arrested. Spermatogonia, spermatocytes, and early spermatids are in close apposition to Sertoli cells in the long tubular structures of the testis where sperm are formed, called the seminiferous tubules. The seminiferous tubules open into other tubes through which sperm are transported. As spermatids mature and differentiate into early sperm cells, they are released from the Sertoli cells into the seminferous tubule. An amazing feature of the seminiferous tubule in some species is that, in any given segment of the tubule, the germ cells form a similar pattern of stages. While this arrangement argues for some kind of cellular control of developmental stage, it also provides a definable tissue state with which to experiment.

Scientists have isolated specific segments of seminiferous tubules and extracted the proteins contained in them. Maps of these proteins have shown that the different stages are associated with different proteins. By isolating some of these proteins and producing monoclonal antibodies to them, it has also been possible to determine which cells contain the protein of interest. One such protein has been found in Sertoli cells, and its synthesis and secretion have been shown to be stage-specific. These experiments have also found that such proteins diffuse into the extracellular spaces of the seminiferous tubules, but because of the slow flow rate within the tubules, remain in spaces near developing sperm cells for quite a long time before being degraded by proteolytic enzymes. Therefore, there seems to be strong evidence that Sertoli cell products greatly influence the development of sperm cells.

Continued research into the molecular events of male gamete development would fill some of the large gaps in knowledge about the normal development of sperm cells and would further the search for causes and cures of male infertility.

Paracrine Control of the Ovary Although it is generally agreed that paracrine control is important to the development of follicles in the ovary, much of the evidence is circumstantial. This state of affairs derives from the nature of autocrine/paracrine regulation. It has proved difficult to manipulate experimentally systems in which the cell of origin of an observed signal is the same as or adjacent to the target cell.

Early experiments to examine the importance of intraovarian chemical control involved removing the pituitaries of rats. Without the pituitary hormones (especially LH and FSH), which normally promote follicular development, investigators could then look at the effects of a number of substances on ovarian follicles. Such experiments found that estrogen, which is synthesized and released from cells in the ovary, was needed for follicular development and that other steroids produced inhibitory effects. Further, it was demonstrated that estrogen could act in concert with or synergistically with LH and FSH to differentiate follicular cells. Later experiments used granulosa cells from pig follicles in culture to show various effects of estrogen. Following a brief inhibitory effect, estrogens stimulated progesterone synthesis in the granulosa cells in a manner that was time-dependent and blockable by pharmacological inhibitors to estrogen. Continued research demonstrated that estrogen action was exerted after the generation of cyclic AMP (see earlier section on oocyte maturation for other actions of cyclic AMP).

There are several chemical forms of estrogen that are made by enzyme modification of a basic structure. One type, catecholestrogen, is found in high concentrations in the walls of large ovarian follicles in the pig and probably enhances, by paracrine mechanisms, the effects of FSH in the follicle. This concept arose from culture experiments which found that catecholestrogen with FSH stimulates progesterone synthesis and production of cyclic AMP. However, antiestrogens do not block this effect. Thus, while experiments have shown that estrogen plays an important role in the control of follicular development, its true role in the natural state remains to be determined.

Estrogen is only one of the substances under investigation in terms of paracrine regulation in the ovary. Tissue culture experiments have shown a number of peptides and growth factors to cause proliferation of granulosa cells. For some of these factors, it is known that they have effects on follicular cells and that they are present in the ovary in vivo. However, whether or not the factors are necessary for follicular development and normal ovarian physiology is not known. The most research has been done on insulin-like growth factors (IGFs). IGFs are synthesized by many tissues and have been found to be a powerful local mediator of growth and differentiation. They seem to stimulate all kinds of effects in granulosa cells including increasing enzyme levels, cell numbers, protein synthesis, glucose utilization, and various secretions. Insulin itself and the IGFs seem essential for FSH induction of receptors for LH, for generation of cyclic AMP, and for FSH stimulation of steroids. IGFs also interact with other growth factors in the ovary to increase cell number. Since most of these findings have been from in vitro studies, it is important to demonstrate the presence of IGFs in follicles in vivo. In fact, IGFs have been shown to be present in follicular fluid in in vivo concentrations similar to those that produce effects in culture. Furthermore, the concentration of IGFs increases during follicular growth and under the influence of FSH, LH, and growth hormone. This is important because any substance active in follicular development and local regulation in the ovary would be expected to change in concentration in response to pituitary hormones.

Another growth factor found in the ovary is a type of transforming growth factor called TGF-alpha, which is a molecule related to epidermal growth factor. TGF-alpha increases the rate of cell division of ovarian cells in vitro and has a negative effect on the secretion of estrogen from granulosa cells. It also seems to decrease the effect of FSH on LH receptors, although this effect varies. TGF-alpha has been found in significant concentrations in follicular fluid and is in highest concentration in small follicles. Based on these findings, it has been proposed that TGF-alpha promotes cell replication in the follicles and inhibits differentiation. The other type of TGF, TGF-beta, is a molecule very similar to inhibin and Mullerian-inhibiting substance, both of which have effects in the ovary. In general, studies have suggested that TGF-beta augments the actions of FSH to increase estrogen secretion, LH receptor activity, and progesterone synthesis. These effects seem to be dependent on both the concentration of TGF-beta and FSH and seem to vary across species.

Fibroblast growth factor is another factor being investigated for its possible role in regulation of ovarian function. This growth factor is also present in many tissues and has been shown to be particularly potent in stimulating the formation of new blood vessels. In vitro, fibroblast growth factor has effects on cells of the corpus luteum and it stimulates granulosa cells to divide. In addition, the factor seems to support the growth of blood vessels in the corpus luteum. Its effects on synthesis of estrogen and progesterone are probably inhibitory. There are many other peptides that have been found in the ovary, but research on these substances is sparse. Many of them have not even been purified enough to allow analysis of their structure and examination of their effects in vitro or in vivo.

This complex area of research promises to provide useful knowledge about regulation in the ovary and could have far-reaching implications for both IVFET and natural reproductive biology. In order to demonstrate that a particular substance has effects in the ovary, the substance must be shown to be secreted by ovarian cells into the local extracellular space and to produce changes in the activity of neighboring cells. Such minimum criteria have been met for some of the factors discussed above; however, elucidation of the mechanisms involved is critical. Especially important for future research is to determine the biochemical regulation of cell replication and differentiation in the ovary. Also, it is necessary to determine how some of these factors are produced by the cells. For example, are there large precursor molecules produced that are modified inside the cells for specific needs? The interaction of these factors with known hormones is also important and may be of special application in evaluating the effects of superovulation. Strategies to block the activity of these factors could help to determine if any of them are absolutely necessary to normal ovarian function as this is entirely unknown. Finally, it is of utmost importance to relate the findings from in vitro studies to what is actually occurring in the natural state.

Assessment of Gametes [6]

There are many reasons why IVFET techniques in human clinical practice do not result in ongoing pregnancies one hundred percent of the time. Certainly, some of the failure rate is attributable to such things as the varied skill levels of the practitioners, the physiological problems of the infertile male or female, or the difficulty in synchronizing the stage of the embryo with the stage of the uterine wall. Nevertheless, a significant contribution to the failure of IVFET derives from poorly understood physiological responses of oocytes and early embryos. Between 20 and 40 percent of apparently mature eggs fail to be fertilized by IVFET methods. Of the 60 to 80 percent that are fertilized, most of the zygotes produced fail to establish a pregnancy after transfer to the uterus. Such early reproductive failure raises critical questions for scientists interested in both improving IVFET and understanding the normal physiology of early reproduction.

Experimental strategies have been developed that are beginning to shed light on some of these problems. One such strategy is to examine the degree of chromosomal abnormality in preovulatory oocytes in an attempt to determine whether these abnormalities contribute to either failure to fertilize or failure to develop after fertilization. There has been a variety of studies assessing the number of eggs obtained from women who had undergone ovarian stimulation, that exhibit a particular type of chromosome abnormality called aneuploidy. A normal oocyte contains 23 chromosomes after meiosis, half of the full human complement of 46 chromosomes. Aneuploid oocytes contain less or more than 23 chromosomes. Comparison between the studies is difficult because the results range from 11 to 65 percent of the oocytes being aneuploid. However, the most exhaustive studies have indicated that the best estimates are likely to be between 20 and 25 percent. The interpretation of such studies is also difficult since the oocytes examined were those that failed to be fertilized. Thus, the possibility exists that the chromosome abnormality is simply an indicator of some other developmental problem. In addition, there are no data on the percentages of normal, unstimulated oocytes which are aneuploid, so how much of the observed abnormality is caused by hormonal stimulation is not known. When human oocytes were matured in vitro, one study found only 1.5 percent to exhibit the chromosome abnormality. This suggests that hormonal stimulation may increase the number of aneuploid oocytes. Finally, in a study of 163 women, a very high percentage of abnormal chromosomes was found in oocytes obtained from particular individual women. The possibility exists, therefore, that some women are especially likely to produce oocytes with chromosome abnormalities.

The extent to which chromosomal abnormalities affect fertilization and pre- or early post-implantation failures is important to know. This is especially true when one considers that sometimes aneuploid eggs are fertilized and the resulting embryos develop to term. Also, findings to date suggest that a significant number of embryos produced by IVFET exhibit some type of chromosmal abnormality. One way to approach these important questions is to develop a focused, multi-center research program to analyze eggs that either failed to fertilize in vitro, or that were produced in excess of the number needed for IVFET methods.

The use of fluorescent stains for chromosomal DNA in such eggs could reveal a variety of abnormalities and answer a variety of questions. Such an approach could determine more precisely the frequency of abnormalities in stimulated cycles. Various stimulation methods could be compared to find ones with the lowest rate of chromosome perturbation. In addition, associations of chromosomal abnormalities with particular patients, or with patients of certain ages, could be made.

While the analysis of oocytes and eggs is important, it is equally necessary to analyze the processes that cause fertilized eggs to develop to varying stages before finally failing to develop. It has been estimated that only 10 to 15 percent of all fertilized eggs develop past a certain stage, and these failures occur at similar frequencies despite differences in ovarian stimulation or in vitro culture conditions. Thus, failures are probably due to processes within the eggs or embryos themselves.

In contrast to the analysis of oocytes, analysis of embryos is an extremely difficult endeavor. Between fertilization and implantation, a myriad of cellular and molecular changes take place in an embryo that have to be coordinated precisely and synchronized both spatially and temporally. Research, then, must focus upon individual stages beginning at the earliest stage of fertilization.

It has been found that 5 to 10 percent of human eggs that appear to be unfertilized actually do contain one or more sperm in the region just outside the egg membrane when viewed microscopically. At higher magnifications with electron microscopy, such oocytes exhibit an absence of binding between the sperm and the egg. Usually the membranes of the sperm form protrusions, called microvilli, which seem to be important for sperm binding to an egg. Sperm associated with these seemingly unfertilized eggs lack this structural feature. Although the reasons for this failure in sperm/egg binding are unknown, a number of possible explanations are likely. For example, cell surface molecules, which function as cell recognition markers, could be absent. Another possible cause is an abnormal organization of specific proteins, called microfilaments, that form the structural framework necessary for microvilli. This possibility is attractive because eggs have been found that lack microvilli except in small restricted regions of their membranes. Since the abnormality has been observed in mature, preovulatory oocytes, it is not likely to be the result of culture conditions. Further, the lack of microvilli is also typical of oocytes from women with a history of failed fertilization by in vitro methods. It may well be necessary to analyze immature but fully grown oocytes from such women to determine if stimulation protocols affect the frequency of this problem. Other research to determine the molecular changes in the egg membrane necessary for fertilization would also be extremely helpful.

There are two other conditions associated with early failure at fertilization. In one condition, the DNA of the sperm fails to decondense. Decondensation is a necessary prior condition to the combining of the male and female chromosomes. While it is possible that there is abnormal packaging of the sperm DNA, it is more likely that there is a problem in the egg. Animal studies have suggested that chemicals in the egg cytoplasm are responsible for causing the sperm DNA to decondense. Clearly, more needs to be known about the process.

Another critical stage in fertilization that can go awry is the migration, juxtaposition, and fusion of the male and female pronuclei, each of which contains the respective chromosomal contribution to the resulting embryo. Electron microscopic analysis has shown that failure here is associated with failure of the membranes between the two pronuclei to dissolve. Therefore, the two pronuclei cannot fuse. Very preliminary data using fluorescent DNA stains have shown that, in these cases, DNA replication is incomplete or absent. It is possible that DNA replication is required for membrane dissolution. Research into this problem could proceed in two ways. First, analysis of fertilized eggs that have arrested development at this stage could be done to further knowledge about the mechanisms of such failure. Second, the mouse could provide an animal model with which to study this problem. It has already been demonstrated in the mouse that chemicals, which specifically inhibit DNA replication, do not affect the formation of pronuclei, their migration, or their juxtaposition. Such chemicals do, however, prevent the dissolution of membranes between the pronuclei. These results also occur when DNA replication is prevented by other means, so probably are not simply a side effect of the inhibiting chemicals.

While this discussion has centered upon very early reproductive failures, there remain many other conditions that cause the demise of early embryos. For example, human development is sometimes arrested when the embryo reaches the four-cell stage. This may be due to a failure in the expression of embryonic genes that results in a lack of proteins required for further development. At the 12-16 cell stage, some normally fertilized eggs develop multiple nuclei. Finally, during the first mitotic division of the zygote to form two cells, a chromosome pair sometimes does not separate or disjoin. This results in different numbers of chromosomes in different cell lines of a developing embryo. These are called "mosaics". Research into the causes of these problems can be of benefit to the clinical practice of IVFET as well as to the understanding of human reproduction in general. While animal models, like the mouse, have great usefulness in this research, the use of developmentally arrested zygotes and embryos produced by IVFET is also necessary. Such zygotes and embryos no longer have the potential to develop into a fetus but may, nevertheless, contain information critical to a better understanding of reproductive success and failure.

Fertilization [7]

Fertilization encompasses sperm and egg maturation as well as the complex cascade of events that occurs once the sperm reaches the egg. Although some aspects of oocyte maturation have been previously discussed, those that directly influence fertilization require further emphasis. The first hurdle for an oocyte is to acquire the ability to continue meiosis. Oocytes from young mice (less than 15 days old) have been shown to be incapable of resuming meiosis if put into tissue culture. Therefore, it seems that there is a growth-related biological process that allows for meiotic competence. Following the breakdown of the germinal vesicle, the chromosome pairs in the oocyte separate with emission of the first polar body and arrest at metaphase II; the polar body frequently degenerates. Resumption of meiosis is initiated by fertilization of the egg.

Before ovulation the oocytes are contained within follicles and, as was previously discussed, it is thought that the follicular cells inhibit the oocyte from maturing. Many experiments with frog and mouse oocytes have been done to elucidate the molecular mechanisms that regulate maturation. Cyclic AMP has been strongly implicated in maintaining meiotic arrest, presumably by activating an enzyme called protein kinase A. Protein kinase A promotes the structural modification of other proteins by adding a phosphorus group; this process is known as protein phosphorylation. Often, proteins are active or inactive biologically, depending upon whether they are phosphorylated or not. In a sense then, modification of proteins by enzymes is one way in which cells turn molecular signals on and off. The proteins that participate in this cascade of reactions associated with oocyte maturation are not known, but are a subject of great research interest. In addition, it must be emphasized that, in addition to cyclic AMP, other types of molecules present in the follicle are likely to participate in maintenance of meiotic arrest.

A molecule that is likely to be involved in oocyte maturation has been isolated from frog oocytes and is called maturation promoting factor (MPF). MPF, when injected into immature oocytes of many species, causes chromosome condensation and breakdown of the nuclear membrane. Its biochemical properties are interesting since it behaves like a protein kinase and is only activated when it has been phosphorylated. Oocytes from mice have a similar substance that has not been fully isolated. There have been innovative experiments performed to measure this MPF-like activity. In one experiment, mature mouse oocytes were fused with immature ones. The mixture of the two cytoplasms resulted in the breakdown of the nuclear membrane of the immature cell. Further experiments using similar strategies have shown that the MPF-like factor's biological activity changes according to the developmental stage of the oocyte. It has also been suggested that oocytes that do not develop meiotic competence may contain a substance which blocks the activity of MPF. Greater understanding of oocyte maturation and the molecules that are directly involved in this process might result in improved ability to mature oocytes from humans and other animals in vitro. Research that tests this hypothesis might also further in vitro oocyte maturation.

Sperm recovered directly from the testis cannot fertilize eggs; like eggs, these haploid cells must also be matured after formation if fertilization of an ovum is to occur. This maturation takes place during sperm transit through the epididymis, a process that usually takes about 10-14 days. During this time many molecular changes in the sperm, particularly on the surface of the sperm cell, occur. The mechanisms which bring about these surface alterations have not yet been elucidated. Similarly, it is not known which of these modifications are related to the acquisition of fertilizing function by sperm. Identification of the molecules used by sperm for gamete interaction is necessary before the relevant mechanisms can be ascertained. One way that has been used to examine sperm at different stages of maturation in the epididymis is to stain the cells with antibodies that recognize particular sperm antigens.

In one case, an antibody recognizes a sperm protein in the anterior region of the sperm head. The protein recognized by the antibody, termed M42, is altered subtly during sperm maturation; prior to the protein's alteration, sperm cannot fertilize eggs, whereas after its modification, fertilization can be achieved readily. It is likely that this change in the M42 protein represents only one example of many molecular changes in the sperm during epididymal transit. Continued examination of individual molecules, like M42, may reveal the mechanisms by which the epididymis causes sperm maturation. This information, in turn, could be used productively in two different ways: to promote fertility by learning how to mature sperm in vitro, or to promote contraceptive development by preventing sperm maturation specifically.

After ejaculation, sperm must undergo a second, and final, maturation phase. This process, termed capacitation, occurs normally in the female reproductive tract, but can be provoked experimentally for many species by incubation in vitro. Like maturation in the epididymis, capacitation is associated with a large number of sperm alterations, the functions of which are largely unknown. But at least two changes in the sperm's surface membrane are now recognized for essentially all mammalian sperm as a function of capacitation: changes in lipid composition and the loss of many surface-associated components. Capacitation is a prerequisite for fertilization, and these alterations apparently permit sperm to penetrate through the physical barriers that surround the egg at the time of fertilization. Clearly, the ability to control this process would assist fertility regulation considerably.

In the ideal circumstance, sperm and egg meet in the fallopian tube and fertilization takes place. However, fertilization is an exceedingly complex process in which the sequence of events is of the utmost importance. At the time of fertilization, the egg is surrounded by two kinds of barriers. Enveloping the egg directly is a jelly-like covering, called the zona pellucida, which is surrounded itself by a cellular layer, called the cumulus. The zona plays a major role during fertilization and many of its molecular features have been defined.

There are three important proteins in the mouse zona--ZP1, ZP2, and ZP3. Each protein has sugar sidechains attached to it and is, thus, a glycoprotein. The structure of the extra-cellular zona pellucida is composed of polymers (compounds produced by the combination of two like molecules) of ZP2 and ZP3 formed into long filaments which are cross-linked by ZP1. ZP2 and ZP3 both function as receptors for sperm. ZP3 also induces the acrosome membrane of the sperm to break down. Both ZP2 and ZP3 are chemically modified following fertilization and play important roles in keeping extra sperm from penetrating the egg.

The genes coding for mouse ZP2 and ZP3 have been isolated. Each is a single-copy gene in the mouse genome and each is highly conserved among mammalian species. The zona genes are expressed only in oocytes where their expression is developmentally restricted to the growth phase of oogenesis that occurs just prior to ovulation. The determination of the structure and the nucleic acid sequence of the ZP2 and ZP3 messenger RNAs has led to the prediction of the amino acid sequence of the ZP2 and ZP3 proteins. Using the cloned genes, the zona proteins can now be expressed by tissue culture cells. By modifying the cloned zona genes, scientists will be able to produce mutant zona proteins for further analysis of how the zona proteins interact with one another as well as the biological function of individual zona proteins.

Antibodies have been produced which bind either to ZP2 or to ZP3. After injection into female mice, these antibodies localize to the zonae pellucidae surrounding growing oocytes. The presence of these antibodies causes highly effective, long-term, but ultimately reversible contraception. Recent studies have demonstrated that immunization of female mice with a synthetic ZP3 peptide (based on the ZP3 DNA sequence) induces the formation of circulating anti-ZP3 antibodies that bind to eggs and effectively preclude fertilization. Because the zona genes are highly conserved, the development of this contraceptive strategy using homologous (same species) peptide as a vaccine may have wide spread application. This method of contraception is desirable because it would prevent fertilization, not implantation, and would avoid the dangers and side effects of the hormones administered in birth control pills. Nevertheless, the reversibility of the strategy remains to be rigorously documented.

The surface changes that sperm undergo during capacitation permit them to navigate through the cumulus cells. A secretory organelle that is essentially an enzyme bomb located at the apex of the sperm head must not be disrupted, however. The sperm will use this bomb, called the acrosome, to digest a path through the zona. If the acrosome breaks down prematurely, sperm stick to the cumulus cells and cannot reach the zona surface. The sperm's recognition site for the zona is located on its surface membrane and binds to the ZP3 protein of the zona. The ZP3 molecule then triggers sperm to release its acrosomal enzymes. Particular components within the acrosome bind to ZP2 and the acrosomal enzymes, together with the motility of the sperm cell, allow the sperm to penetrate the zona and reach the egg membrane. This process is remarkably similar across mammalian species. It is likely that factors from the egg itself are not involved in fertilization up to this point since these events occur even if the zona contains no egg, a so-called zona ghost.

Like the antibodies against zona proteins discussed above, antibodies directed against specific sperm proteins can also inhibit fertilization. The sperm protein M42, described earlier, that is altered during epididymal maturation is one of several candidate sperm proteins for this purpose. Sperm incubated with an anti-M42 antibody cannot fertilize eggs in vitro since the sperm are blocked from losing their acrosomes. In a more natural experimental situation, injection of female mice with the same anti-M42 antibody decreases pregnancy levels markedly. Strong evidence suggests that some forms of infertility in women may be caused by the presence of antisperm antibodies. Perhaps the continued examination of this experimental system--the induction of infertility with specific anti-sperm antibodies--will assist in explaining the mechanisms underlying such naturally occurring types of infertility.

Once the sperm has penetrated through the zona, it associates with the egg surface rapidly. Membrane fusion between the two gametes is initiated at a specific location, near the middle of the sperm head, and proceeds around the cell so that the entire sperm is incorporated into the egg. Once the sperm has fused with the egg, small packages called cortical granules, which contain enzymes that modify the zona pellucida, move to the egg membrane, fuse with it, and release their contents into the space between the egg and the zona pellucida. The released enzymes modify the ZP2 protein so that acrosome-reacted sperm can no longer bind to ZP2. Modification of ZP3 also occurs that affects sperm binding and the induction of an acrosome reaction. In brief, it seems as though nature has devised backup systems to ensure that only one sperm enters the egg.

Very little is known regarding the precise biochemical events that take place during the cortical granule reaction. In vitro experiments have shown that activated protein kinase C inhibits fertilization by modifying ZP2 in the same way as occurs in the natural state. Protein kinase C also modifies ZP3 but it only prevents ZP3 from inducing a complete acrosome reaction. Sperm binding to ZP3 is unaffected. Other experiments have implicated other enzymes and co-factors (inositol phosphate, G proteins, and phospholipase C), all of which participate in cascades of biochemical reactions. Much research needs to be done before any of the actual biochemical events can be known. For example, it is not known what chemicals are released by the granules. The precise changes in the zona proteins are not clear. Since some of these chemical cascades produce changes in calcium storage and release, the role of calcium is not clear. Knowledge of these events could contribute greatly to the ability to fertilize eggs in vitro.

Finally, while the block to polyspermy is taking place, changes occur in the egg that allow the male and female pronuclei to combine. When the sperm first penetrates the egg, its chromosomes are condensed, held together by chromatin-associated proteins called protamines. For the chromosomes of the male to combine with the female, the chromosomes must decondense. It is thought that a chemical change in certain molecules of the protamines is necessary for the chromatin to decondense. Such a chemical change can be induced by glutathione, which has been found in high concentration in mature eggs. Furthermore, depending upon the developmental state of the oocyte, glutathione can cause decondensation of oocyte chromatin when applied in vitro. It is possible that the substance that causes sperm chromatin decondensation is contained in the germinal vesicle, since sperm injected into oocytes before germinal vesicle breakdown retain their condensed chromatin. The substance could be a co-factor required for another reaction, could be an activator of some other molecule, could be solely responsible for the effect, or could actually act as an inhibitor of a molecule whose action is to prevent chromatin decondensation. Further information about these processes may suggest strategies with which IVFET could be enhanced.

To summarize, the sequence of events for gamete interaction in most mammals leading to fertilization is:

1. The capacitated sperm penetrates between the cumulus cells.

2. The acrosome-intact sperm binds to ZP3 in the zona pellucida.

3. ZP3 triggers the loss of the acrosome (the acrosome reaction), causing release of digestive enzymes from the sperm head.

4. The acrosome-reacted sperm binds to ZP2 and penetrates through the zona matrix.

5. The sperm and egg fuse with the eventual formation of two pronuclei and restoration of the diploid state.

6. At the time of sperm fusion with the egg, a reaction is initiated that prevents any other sperm from entering (the block to polyspermy).

Preimplantation Development [8]

In this section, a variety of events that occur between fertilization and implantation are discussed. These include the special metabolic requirements of early embryos and the first expression of embryonic genes. Finally, the results of experiments involving micromanipulation of animal embryos and embryo splitting will be described. The short description that follows reviews the main events of preimplantation development in a general way. It is also offered as an explanation of terms that will be used throughout this section.

The major steps in preimplantation development occur in the fallopian tubes. The fertilized egg, or zygote, undergoes mitotic cell divisions called cleavage. The first cleavage results in the formation of two cells, each of which divide in the second cleavage to form four cells, and so on through 8-cell and 16-cell stages until there are about 50 or 60 cells, at which time blastulation occurs. During the cleavage divisions, the cells are called blastomeres and all the blastomeres are encapsulated by the zona pellucida. Blastulation begins about the fourth day after fertilization, just as the developing embryo reaches the uterus. At blastulation, cavities form in the cell mass and the cells separate into two regions, an inner cell mass, from which the fetus and some extra embryonic membranes will form, and an outer cell layer, called the trophectoderm. Coincident with these changes, the zona pellucida degenerates. The blastocyst, as the embryo is called at this time, can be considered to be a polarized structure in that the inner cell mass is attached to the trophectoderm layer at one side of the blastocyst. If cut through the middle, the blastocyst would look like a ring of cells with a clump of other cells stuck to one part of the inside of that ring. This polarization is important for implantation later since the pole of the blastocyst containing the inner cell mass will first touch the lining of the uterus in preparation for implantation in some species (in others, the opposite is true).

Metabolic Substrates and Pathways

Metabolism encompasses the processes by which cells extract energy from their environment and synthesize the necessary building blocks to form large molecules. Thus, metabolism is an exceedingly complex network of biochemical reactions that can be viewed as intersecting sets of interactions. The sets overlap at key points that regulate which set of reactions will be more or less active, depending on the needs of the cell at the time. The regulators at these key points are enzyme proteins. In most cascades of biochemical reactions, one enzyme's activity will be more important than any other's. Such an enzyme is called a rate-limiting enzyme and is often positioned at the key points of overlap between different pathways of metabolism.

In organisms dependent upon oxygen, glucose is used as a basic fuel for metabolism. There are three linked systems that subserve glucose metabolism--glycolysis, the citric acid cycle, and the oxidative respiratory chain. In glycolysis, glucose is chemically changed in a series of enzyme-catalyzed reactions to pyruvate. In the process, molecules of adenosine triphosphate (ATP) are produced. ATP is the energy source for vast numbers of biochemical reactions. The citric acid cycle uses pyruvate, formed by glycolysis, as a substrate for another series of reactions to produce carbon dioxide and molecules (NADH and FADH) that are required for a particular type of chemical reaction also important in biological systems. Other products of the citric acid cycle are precursors for the synthesis of other molecules. For example, certain amino acids, which are the building blocks for proteins, are formed from the citric acid cycle. Finally, the NADH and FADH from the citric acid cycle are used in the oxidative respiratory chain to transfer electrons to molecular oxygen and form additional molecules of ATP.

Glucose is also necessary in two other systems that synthesize important biological chemicals. Both of these systems intersect with the glycolytic pathway. The first reaction of glycolysis is the conversion of glucose into glucose-6-phosphate (G-6-P) by the enzyme, hexokinase. The G-6-P produced can be diverted when necessary to enter the pentose-phosphate path or to be used to synthesize glycogen. In the pentose-phosphate path, G-6-P is converted to another kind of sugar by a series of reactions and, in the process, molecules of NADPH are formed. NADPH functions as a hydrogen and electron donor in biosynthethic reactions. In glycogen synthesis, the G-6-P is converted to glycogen, which is the way cells store glucose for later use. When cells need ATP, for example, glycogen can be broken down to form G-6-P, which can then continue in the glycolytic pathway. Eventually, through the citric acid cycle and oxidative respiratory chain, every G-6-P molecule leads to the production of 37 ATP molecules. For many years, glucose metabolism was not studied in eggs and early embryos, because it was generally assumed that results of studies from other cells would also apply to eggs. With the advent of tissue culture, it became apparent that eggs and early embryos required unexpected nutrients in their culture media, which seemed to violate this assumption. Specifically, it was shown that, for 2-cell mouse embryos, glucose alone would not promote survival in culture. Rather, lactate was critical for these embryos. Lactate is formed from pyruvate when there is little oxygen in the tissue. Exercised muscles, for example, produce a great deal of lactate from pyruvate. Lactate can also be converted to pyruvate, which may then be used in the citric acid cycle. Thus, the finding of a lactate requirement of these embryos suggested that certain pathways of metabolism were restricted, especially the glycolytic pathway. It was later found that maturing oocytes and embryos up to the 8-cell stage needed pyruvate and other citric acid cycle precursors to survive. While the exact requirements have been shown to differ from one animal species to another, it is interesting that pyruvate is often used in culture media for human and monkey eggs and early embryos despite the lack of experimental data to justify its use.

The nutrient studies described above are not the only evidence that metabolism is restricted in oocytes and early embryos. Morphological studies have demonstrated that the intracellular organelles in which the oxidative respiratory chain takes place are not the same in early cells and embryos of 8 or more cells. These organelles are called mitochondria and are small membrane bound packages of enzymes and other molecules. Inside each mitochondrion is a complex labyrinth of membranes called cristae. It has been known for a long time that the arrangement of the cristae differ among cell types depending on the function of the cell and the metabolic state of the cell. Studies of eggs and early embryos showed a pattern of cristae that included a concentric and transverse arrangement. However, the concentric cristae disappeared by the 8-cell stage in most species. This change in structure correlated well with an observed increase in oxygen consumption by these 8-cell embryos.

Other studies have explored which substrates were taken up by cells. These studies also showed that eggs and early embryos took up pyruvate rather than glucose until they reached the 8-cell stage. To assess whether the cells actually used the pyruvate, studies were done to measure the carbon dioxide produced by metabolism of pyruvate compared to glucose. This is done by labeling the pyruvate or glucose with radioactive tags which, if the pyruvate is metabolized to carbon dioxide, results in the formation of radioactively tagged carbon dioxide. As might be expected, most carbon dioxide was produced from breakdown of pyruvate in early (1 to 2 cell) embryos. At 8 cells, the carbon dioxide was mostly from glucose metabolism.

While the results of these studies seem clear, it is important to note some possible confounding factors in these experiments. The most important of these is that, since much of the work has been done with cells in culture, the differences between the amount of oxygen available in culture and the amount available in the natural state may have affected the results. The amount of oxygen available can, by itself, cause metabolism to be switched from one pathway to another. It is also important to determine if the enzymes responsible for glycolysis are present, and if glucose is being used in a pathway other than glycolysis. Various studies have compared the metabolic processes at play by using the strategy of starvation and refeeding of cells with the substances of interest. It was found that, if one starves embryos of glucose at any stage, the level of G-6-P decreases dramatically. Upon refeeding with glucose, the G-6-P levels increase in all stages also. This suggests that transport of glucose into cells and the first enzymatic reaction by hexokinase to form G-6-P are both normal across different stages of embryonic development. In contrast, the levels of fructose 1,6, biphosphate (an intermediate molecule formed in glycolysis) do not change with starvation of glucose at any embryonic stage. However, the levels of this intermediate increase substantially with refeeding of glucose in late embryos, but not early ones. Such data provide strong evidence that, like all cells, the rate-limiting enzyme for glycolysis (phosphofructokinase) is the one that catalyzes the formation of fructose 1,6, biphosphate. Moreover, the results suggest that this enzyme operates at a very low level in early embryos and does not reach its normal activity level until after the 8-cell stage. A similar starvation/refeeding experiment has suggested that the presence of pyruvate inhibits the formation of fructose 1,6, biphosphate from glucose and that the addition of glucose resulted in no increase in the metabolites of the citric acid cycle, whereas pyruvate did result in an increase in these metabolites at all stages of embryonic development.

There is much that should be known regarding the control of glycolysis in oocytes and early embryos. It is important to continue to investigate the rate-limiting enzymes at the intersection points between different metabolic pathways. Since the energy charge of a cell determines to a great extent the pathway chosen, it is also necessary to consider changes in the total energy charge of oocytes as they proceed through maturation and embryonic development. The energy charge can be expressed as the amount of ATP molecules available. Often, it is measured by determining the ratio of ATP to adenosine diphosphate, ADP, which is a lower energy state form. It is well known that if the energy charge of a cell is high, the pentose-phosphate and glycogen synthesis pathways are favored. This is because more energy in the form of ATP molecules is not needed. If there are not enough ATP molecules, the energy level is low and glycolysis, citric acid cycle, and oxidative respiratory chains are the favored routes. It is known that, in the mouse, zygotes and early embryos have a high energy charge that begins to fall as development proceeds. Such a finding is consistent with the evidence that these early embryos exhibit a restricted metabolism in the glycolytic and citric acid pathways. It is important to know the metabolic requirements of oocytes and early embryos because these requirements are critical to successful maintenance of these cells in culture.

Gene Expression in Early Embryonic Development

The determination of when embryonic genes begin to be expressed for the synthesis of new proteins is important in the detection of genetic diseases in embryos. Of the general population, two percent are at risk to produce children with genetically transmitted diseases. For some the risk is as high as one chance in two. With the development of IVFET techniques, it may become possible to fertilize eggs from high risk couples in vitro to allow for early diagnoses of genetic diseases. In theory, one cell from a multicelled embryo could be removed, analyzed for disease and, if found to be affected, a decision could be made not to implant that embryo. As an alternative, naturally fertilized eggs could be removed from the uterus by flushing, biopsied, and then returned via embryo transfer. For such diagnostic procedures to be realized, however, certain prerequisites are necessary.

First, there must be a realistic chance of maintaining a pregnancy after embryo transfer. Embryo biopsy may adversely affect the chance of a successful transfer. Second, the embryo should not be damaged by biopsy. In cattle, embryo biopsy is a relatively common technique and has been done with embryos at various stages. The optimum stage for biopsy is, nevertheless, important to determine. For example, techniques using 8-cell embryos involve removing the zona pellucida and removing one cell. An advantage of biopsy at this stage is that the embryo can be transferred to the mother when it is well synchronized with the state of the uterine wall. There are some disadvantages at this stage, however. These include the fact that there is only one cell with which to work, and the fact that embryos lacking zonae can stick together during culture or transfer, and can develop into a mosaic or chimera that, at best, is undesirable and, at worst, is fatal to the embryo. In addition to these problems, research has suggested that zona-free embryos do not exhibit normal cleavage and often exhibit cells that vary greatly in size. In later embryonic stages, it is possible to nick the zona and slice off the 5 to 10 cells that herniate out the nick. This approach will likely result in a lower implantation rate, because the optimum time for implantation would be lost. However, the embryos could be frozen for later implantation. Clearly, much more needs to be known about these changes and their effects.

The third prerequisite for preimplantation diagnoses is to have reliable probes with which to identify the presence of genetic diseases. Ideally these probes would make the identification at the level of the genome or DNA, as is possible with new technologies, such as Polymerase Chain Reaction techniques. In situ hybridization is a technique whereby a strand of DNA is made that matches with the strand of DNA one is looking for. To use in situ hybridization, however, the sequence of the gene in question must be known. Some of the sequences for human genetic diseases are known, especially the sex-linked diseases. The problem, however, is that many in situ probes are linked to the Y sex chromosome. This causes an unacceptable rate of false negative readings. A procedure with greater potential is to look at the defective gene products, like hemoglobin, as an indication of genetic blood diseases, or like hypoxanthine phosphoribosyl transferase (HPRT) as an indication of Lesch-Nyhan disease. The last prerequisite is absolutely critical, and that is that the probes must assess the activity of embryonic genes not maternal genes.

Studies in the mouse have suggested that the embryonic genome turns on between the 2- and 4-cell stage. Before this stage, intracellular proteins are determined by RNA which was stored in the oocyte. Investigators at the Medical Research Council Unit of Mammalian Development in London have developed a microassay for the presence of HPRT in embryos that, applied in mouse embryos, has begun to answer some of the questions about the time of appearance of embryonic gene activity. HPRT is an enzyme that is lacking in children born with the sex-linked Lesch-Nyhan disease. The hallmark of the disease is self-mutilation and such children also exhibit strange motor behavior, are often mentally retarded, and usually die by the age of 10. Using the microassay for HPRT, workers have found that the enzyme increases in concentration in 8-cell mouse embryos suggesting that the embryonic genes have become active. So, it seems fairly clear that, in the mouse, gene activity begins at this point. However, before preimplantation diagnoses can be achieved in humans, one must know the timing of human embryonic gene activity.

In Great Britain, research on human embryos is allowed under the control of the Voluntary Licensing Authority. In the Cambridge University clinic, all patients are asked to donate their excess eggs and embryos for research purposes. Of 300 patients, only two refused permission for the research, and these two were still afforded all the clinical services available. Investigators at this clinic have thus conducted a number of studies on human embryos. In a series of experiments, radioactive methionine was added to the culture media of cultured embryos at various stages of development. Methionine is taken up by cells and incorporated into newly synthesized proteins and so is a qualitative measure of protein synthesis. Researchers found new proteins that contained the radioactively tagged methionine were synthesized for the first time between at the 4- and 8-cell stage. In another experiment, when chemicals, which blocked transcription of messenger RNA, were added to the culture medium, they had no effect on unfertilized oocytes or early embryos and development proceeded normally. However, the blockade of RNA, or protein synthesis, from the 4-cell stage caused the embryos to stop developing. While the results suggested that embryonic gene activity, which directs synthesis of new proteins, is important at the 4- to 8-cell transition period, the data are complicated by the fact that embryos have a tendency in vitro to stop development spontaneously at this stage. There is, however, further support for the idea that the embryonic genes are active at this stage. Since earlier experiments had looked at the pattern of proteins present in early and late embryos and found that the pattern changed quite clearly, the protein patterns were measured with and without the blockade of messenger RNA. These results demonstrated that, when RNA synthesis was blocked, the changes in protein pattern characteristic of the 8-cell stage and beyond did not occur, but the pattern remained in the early state.

Because the objective of such research was to develop preimplantation diagnostic techniques, the presence of HPRT was also assayed from human embryos. Unlike the findings in the mouse, there was a large variation in HPRT levels in human embryos, and the massive rise in HPRT at the 8-cell stage characteristic of the onset of gene activity in the mouse was not seen in the human. The reasons for these differences and for discrepancies between these data and that of others are not clear. It could be that HPRT is synthesized by the human embryo later than in the mouse embryo, or that the HPRT from the maternal genes is being broken down as rapidly as new HPRT is produced so that no net increase can bee seen. It is also possible that there are simply unexplained differences in storage, culture conditions, or other variables. Whatever the reasons, the application of HPRT assays to preimplantation diagnosis of Lesch-Nyhan Syndrome in the human has been disappointing. The situation does point out, however, the dangers of extrapolating from animal studies and underscores the need to conduct some research with human embryos that, otherwise, would be wasted.

Regulative Potential of Micromanipulated Embryos

Many of the classic studies on embryonic development involved manipulation of certain parts of animal embryos and subsequent analysis of the effects of that manipulation on final development. Such manipulations included removal of certain parts (limb buds, for example, to study regenerative capacity), transplantation or exchange of parts (as in exchanging upper spinal cord regions with lower spinal cord to see if nerve outgrowth would follow the limb buds in the normal way), and grafting of parts of one embryo to another. Many new technologies allow micro-manipulation of embryos at the level of cells and, sometimes, even individual molecules.

It is clear that the developmental potential of cells gradually narrows as development proceeds and that embryos have remarkable capabilities to readjust to disturbance. Yet little is known about what governs the point at which readjustment can no longer occur because the potential of a given embryonic region has been irreversibly determined. Further, it is expected that that point of determination will vary according to the specific region in question and according to the species involved.

It has long been thought that prior to blastulation an embryo is totipotent, meaning that no cell of the developing embryo is committed to any particular developmental fate. In other words, the cells are undifferentiated and uncommitted. However, it is becoming evident that embryos of different species, even though exhibiting the same morphological state and the same number of cells, may differ in the degree of determination and potential of each cell. Such findings call into question the long-held assumptions of totipotentiality. Since an embryonic cell is totipotent only if it can develop into an entire organism, experiments on embryo splitting can answer questions regarding the real potential of embryonic cells at different stages and in different species.

If a 2-cell stage embryo of a laboratory mouse is bisected, development proceeds normally. Up to the 8-cell stage, individual blastomeres can be dissected from an embryo and aggregated with another embryo successfully. However, such isolated blastomeres cannot develop or organize into viable fetuses on their own. The findings in the rat are similar. In contrast, rabbit blastomeres isolated from 4- and 8-cell stage embryos have been shown to be totipotent. Other experiments have examined the effects of bisecting mouse embryos exactly in half. While 65 percent of half embryos survive from the 2-cell stage, only about 45 percent survive from later stages. The success at later stages may not mean that all the cells are totipotent; rather, it may mean that each half embryo contained the required number and type of cell to complete development.

Information on domestic species is derived mostly from experiments with sheep and cattle embryos. In sheep, isolated blastomeres have been shown to develop normally from embryos up to the 8-cell stage. However, for unknown reasons, only some blastomeres are capable of such development. One case was reported in which an 8-cell embryo was divided into four equal parts. Transfer of these split embryos into ewes resulted in the birth of four lambs. Sheep embryos, which have been halved, seem to survive well from the 2-cell to blastocyst stage. In cattle, similar success has been achieved with quartered and halved embryos from the 8-cell stage to the blastocyst stage. Compared to the mouse, sheep and cattle exhibit blastulation at a later cleavage stage. Therefore, the number of cells is higher at blastulation for sheep and cattle. It is possible that a higher total cell number may allow for greater flexibility in embryonic adjustment to manipulations such as splitting.

Beyond questions of immediate survival of micromanipulated embryos, there are other questions that have been addressed by investigators. Some studies have assessed the size of embryos after splitting and the birth weight of resulting offspring. While one study in mice found no differences in these measurements between control and halved embryos, others have reported differences at various times of development. An examination was made of blastocyst formation following 2-cell stage embryos that had been bisected. One half of the embryo was discarded and the other was cultured, returned to a synchronized female, or placed in an immature oviduct. Those placed into culture conditions exhibited delayed development. A higher percentage of the half embryos developed normally when returned to a synchronized female or an immature oviduct. Such findings indicate that placing the split embryos in vitro somehow compromises their development. However, it is possible that the number of degenerated half embryos were underestimated in the in vivo conditions, since it is more difficult to recover and identify degenerated embryos from the uterus or oviducts.

These studies further showed that viability was correlated with cell number such that the lower the cell number achieved in the first few days after bisection, the lower the chances of survival. Many of the half embryos failed after implantation, indicating that the bisection did not result in a failure to implant. One of the most clear suggestions from these studies is that the cell number at blastulation (which occurs at a particular time, independent of the cell number) is critical and sometimes blastulation results in an embryo with a low number of inner cell mass cells. A reduction in the inner cell mass may then cause the demise of some half embryos. This idea was tested directly by another group of studies that used differential stains to compare the number of cells in the inner cell mass to the total number of cells. Embryos with a low cell number at blastulation had a lower inner cell mass ratio and exhibited a lower viability than embryos with a higher cell number at blastulation. These experiments further showed that maintaining the half embryos in vivo, even in a nonpregnant uterus, increased the formation of the inner cell mass, resulting in a higher inner cell mass ratio and higher viability than that achieved in vitro. In a study examining the long-term effects of embryo splitting, genetically identical embryos were either bisected and immediately placed into foster mothers or were left intact and placed into foster mothers. The half embryos were generally less successful and those born had a higher rate of neonatal mortality than did the controls. However, some of the neonatal mortality might have been because the half embryos were born from smaller litters than the controls. Small litters usually result in longer pregnancies, which are also associated with higher neonatal mortality rates. Comparison of other phenotypic characteristics of the neonates born revealed no significant differences between half and whole embryos in measures such as the age at eye opening or growth curves. Thus, by birth, some type of regulation has taken place. The studies taken together argue strongly for the existence of crisis points for half embryos including blastulation and implantation. Further study of the mechanisms of regulation in micromanipulated embryos not only can add to our basic knowledge of reproductive biology, but can also elucidate important requirements for successful splitting of embryos in different species.

Implantation [9]

Implantation, one of the most poorly understood processes in reproductive biology, involves a complex interaction between the embryo and the cells of the uterus. There are also significant species differences in the process of implantation that can be generally divided into three types. The type of implantation found in rabbits is called fusion; whereas the type found in rodents is called displacement. Implantation in humans and other primates is called intrusive and is the most important to understand for the purposes of this report.

The general model for implantation in the human begins with the attachment of the blastocyst to the endometrial wall. At this point, the outer layer of trophectoderm of the embryo proliferate and differentiate into two types of trophoblast cells. The cells close to the embryo become cytotrophoblasts and the cells in direct contact with the endometrium fuse with each other to become the syncytiotrophoblast. A syncytium is a large, multinucleated, mass of cytoplasm formed by the fusion of many separate cytotrophoblastic cells into one. The syncytiotrophoblast acts to erode a path into the endometrial tissue that, in turn, allows the embryo to burrow into the wall of the uterus. Some cells of the endometrium are induced to build up large stores of nutrient molecules, which are then released into the extracellular spaces (close to the embryo) as these cells degenerate. The changes in endometrial cells can be visualized with a number of experimental techniques and are called the decidual reaction. Eventually, the trophoblastic cells penetrate deeply enough into the uterine wall to contact maternal blood vessels. The interactions between the trophoblast and these blood vessels results in the formation of the vascular supply of the placenta.

The study of implantation involves the search for and identification of factors controlling uterine receptivity and maternal recognition of pregnancy. Some of these factors seem to be produced and released by the cells of the blastocyst prior to implantation. The understanding of the biology of implantation has far-reaching implications for human and non-human reproduction. Both in the natural reproductive process and in the practice of medically assisted conception, there is a huge gap between the number of successfully fertilized eggs and the number of offspring born. A large part of this gap can be accounted for by the loss of embryos at or around the point of implantation. For example, in human IVFET practice, 60 to 70 percent of the eggs are fertilized, but only twenty to thirty percent of the embryos placed in the uterus result in an ongoing pregnancy. Moreover, it has been estimated that, in couples without fertility problems and not practicing birth control, a pregnancy occurs in one out of three menstrual cycles in which a fertilized egg is present. In sheep and cattle, twenty to forty percent of the fertilized eggs do not survive and, in the pig, the loss rate is thirty to forty percent.

Early embryonic loss is common across species. While some early embryonic loss can be explained by factors such as heat stress, nutritional deficiencies, and genetic abnormalities, it has been proposed that much of this loss results from three other possible causes. First, the uterine environment is probably only narrowly permissive to implantation by an embryo. Second, embryos may fail to signal their presence and, consequently, fail to induce the necessary hormonal and uterine changes necessary to maintain pregnancy. Third, embryos may be rejected as a foreign body by the mother's immune system.

The idea that the uterus allows implantation only under precise conditions almost certainly underlies the problems discussed elsewhere in this chapter relating to synchrony between the embryonic stage and the uterine stage. Lack of synchrony has been demonstrated to be a problem in all species examined. It has been shown that if precise synchrony cannot be achieved, it is generally better to transfer embryos that are more advanced than the uterine environment. This makes sense if one considers that, following ovulation, the corpus luteum in the ovary begins to produce progesterone, which acts to prepare the uterus for pregnancy.If the corpus luteum degenerates, the drop in progesterone causes the uterine lining to be shed resulting in menstruation. A hormone secreted by the trophectoderm, chorionic gonadotropin, prevents the degeneration of the corpus luteum and, thus, maintains the progesterone secretion. It is reasonable then that an embryo that is transferred to a uterus more advanced than the embryo may not produce enough chorionic gonadotropin to be able to rescue the corpus luteum from degeneration.

There have been attempts to control the state of the uterine lining through the administration of exogenous hormones. Attempts to advance the uterine environment of pigs by administration of progesterone have been unsuccessful. However, because pig embryos secrete estrogen which also affects the uterine lining, exogenous estrogen was tried. This approach worked, but only during a narrow time window and, if given too early, was actually toxic to the embryos. Other research suggests that it is too simplistic to assume administration of one type of hormone would be sufficient to control the uterine environment. Analysis of proteins synthesized and secreted by embryos indicates that there are a number of different chemicals made by embryos that can affect the uterus or the corpus luteum. It is nevertheless probable that some of the embryo-produced proteins cause changes in the secretions of uterine cells that are necessary and/or supportive of an implanting embryo.

A number of proteins are also secreted by endometrial cells, and much research has focused upon isolation of these proteins, analysis of their functions, and mapping the changing patterns of protein synthesis associated with implantation. One experimental strategy has been to label implantation sites in mouse uterus with a dye, pontamine blue. This dye, when injected into the veins of a pregnant mouse, causes implantation sites to be colored blue without staining the rest of the endometrium. After this labeling, investigators can then remove the uterus and maintain explants of it in culture. Using precursors for protein synthesis that have been radioactively labeled, investigators can label the new proteins synthesized and compare the pattern of proteins from implantation and non-implantation sites. This strategy has shown that the rate of protein accumulation in implantation sites is up to forty percent greater than in other sites. Much of the increase is accounted for by increases in proteins destined to be secreted from endometrial cells. Further work compared the pattern of protein synthesis from natural implantation sites to those that were mechanically induced to look like implantation sites. Although there was an increase in the synthesis of some proteins by the mechanical method, some proteins were not increased. Such a finding argues for there being embryo-dependent protein synthesis in the endometrium at implantation sites. As will be discussed in a later part of this chapter, it is known that embryos release certain factors and proteins that directly affect the cells of the endometrium. At this point, however, it is useful to consider some of the proteins found to be released by endometrial cells.

Endometrial protein 15 has been isolated in humans. This protein is only present in the secretory phase and first trimester of pregnancy. The presence of this protein has been shown to change depending upon the hormonal state, but so far no function has been ascribed to this protein. Endometrial protein 14 has been found in human and mouse endometria. This protein is identical in structure to an IGF, which is present in numerous sites of the body. In rabbits, a protein called uteroglobin has been shown to be induced by progesterone and present in early pregnancy or in pseudopregnancy. This protein has a variety of functions including an anti-inflammatory action. Factors isolated from the mouse include epidermal growth factor and a type of colony stimulating factor, both of which increase in response to progesterone. It is surmised that these, and probably other, growth factors may function to control the proliferation of the placenta.

Studies of endometrial proteins in the pig are of special interest since this species does not exhibit the intrusive type of implantation. In fact, the maternal and embryonic blood supplies never come close to each other in the pig as they do in humans and other primates. Such a situation sets up potential problems in bringing nutrients to the developing embryo. Nutrients must be released from the endometrium and diffuse to the embryo. Proteins have been found in the pig that seem to help in this respect. For example, uteroferrin is a protein that carries iron to the embryo. Another type of protein, which increases in response to progesterone, transports water insoluble molecules to the embryo. These are called retinol-binding proteins. Two other proteins--plasmin and trypsin inhibitors--protect the uterine cells from destruction by embryonic enzymes, and lysozyme protects against infection.

In summary, many classes of proteins are produced and secreted by endometrial cells in response to estrogen, progesterone, or embryo-produced factors. While some of these proteins serve nutritive functions in species in which the embryo either invades the endometrium late or not at all, other functions of these proteins are not clear. It is probable that early embryonic loss is, in some cases, due to abnormal expression of uterine proteins. A poor quality environment for the embryo could result from either excessively low levels of necessary proteins or excessively high levels of proteins, which could be toxic to the embryo.

Less research has been devoted to examining the incidence or mechanisms of immunological rejection of embryos. The uterus is not isolated from the immune system, especially in species in which the maternal and fetal blood are hardly separated. An embryo is like a transplanted organ, which must be transplanted to a site that is protected from immune cells or an immunosuppressive drug must be given. In some species, an implanting embryo causes what looks very much like an allergic inflammatory reaction in the endometrium, that is, dilatation of blood vessels, proliferation of blood capillaries, and fluid accumulation. In addition, it is certain that cell-surface antigens capable of eliciting an immune reaction on the part of the mother are eventually present on embryos. Thus, the reasons why all embryos are not rejected immunologically are interesting and important areas of research. It has been found that embryos release substances, especially interferon, that, in other tissues, act to suppress or modulate immune cell activity. Therefore, it is possible that the embryo acts to control locally the immune response of the mother by secretion of immunomodulating substances. However, more research is needed before such a mechanism can be established.

The study of implantation in human beings is particularly difficult because there is no in vitro model available. Such a barrier has special importance because of the wide species differences that exist in regard to implantation. Findings from other species cannot be assumed to be true for human beings. However, it is possible to examine some aspects of human implantation by using in vitro methods recently developed. One of these methods involves the isolation of cytotrophoblast cells. It has been found that cytotrophoblast cells placed into culture under certain conditions will proliferate and differentiate into syncytiotrophoblast. Such studies have shown conclusively that syncytiotrophoblast is derived from cytotrophoblasts. In a culture dish, this differentiation takes place in two steps. First, the trophoblast cells aggregate. Then they fuse to become a syncytium. It is likely that this process is mediated by cell-adhesion molecules or CAMs, which have important roles in cell-cell interactions during development of the nervous system and other tissues. The CAMs are produced by the trophoblast cells because blockade of proteinsynthesis prevents aggregation of the cells. It is also known that the process depends on calcium, since it does not occur in culture medium lacking calcium. In contrast to the aggregation step, the mechanisms responsible for fusion of the trophoblast are entirely unknown.

The interactions of cultured trophoblastic cells with various components of extracellular matrix have also been studied. Extracellular matrix is simply the intercellular space and its component molecules, which are usually synthesized by the surrounding cells and secreted into the extracellular space. The exact composition of the matrix varies from tissue to tissue, but it is important in the uterus since interactions between it and the blastocyst occur during implantation. For trophoblast aggregation and membrane fusion to occur, there must be serum proteins in the culture media and/or the dishes must be coated with extracellular matrix proteins. It is thought that the serum is required because it contains matrix proteins, (e.g., fibronectin). If these proteins are added, the serum is no longer required. Collagen, fibronectin, and laminin are structural proteins that are plentiful in many extracellular spaces.

The importance of extracellular matrix was underscored by additional experiments that used endometrial explants to co-culture the trophoblast cells. This method is more like the natural situation, since the trophoblast differentiates into syncytiotrophoblast, as before, but now interacts with endometrial tissue. These experiments showed that trophoblast from first trimester and term placentas bind to epithelial cells of secretory endometrium. The cells also bind to cut surfaces of the explants, areas where extracellular matrix was exposed. After 24 to 48 hours, a zone of tissue necrosis developed in the endometrial explants where the syncytiotrophoblast associated with the tissue. Moreover, trophoblastic cells bind to and invade nests of cultured endometrial gland cells. The trophoblastic cells dislodge the endometrial cells and penetrate beneath them in a process resembling intrusive implantation. These experiments suggest that the extracellular matrix always permits attachment and differentiation of the syncytiotrophoblast. That fact has important implications in disease states where the epithelium is eroded and, especially, when the lining of the fallopian tubes is eroded. Such states may cause implantation to occur in either less desirable sites or in totally undesirable sites such as the fallopian tubes.

The mechanisms for trophoblast invasion of the endometrium are not known. There are a variety of proteases (enzymes which break down proteins) that have been implicated, such as plasminogen activator. In some strains of mice, blastocysts are less invasive of the endometrium apparently because they produce less plasminogen activator. Cultured human trophoblasts produce plasminogen activator (urokinase). Urokinase may degrade fibronectin and activate other enzymes (e.g., collagenase). It is possible that the action of proteases like urokinase could be controlled by the presence of cell surface receptors for the enzyme, which localize its actions, and by specific plasminogen activator inhibitors,two of which are known to be generated by trophoblastic cells. This control would be important since the invasion of the endometrium must have an endpoint since embryos do not burrow all the way through the endometrial wall.

It is expected that many of the biochemical interactions between the trophoblast and the endometrium would occur through paracrine mechanisms. Substances released by the syncytiotrophoblast and cytotrophoblast could reach a high local concentration in areas of the endometrium. Substances known to have powerful paracrine effects in other tissues may play a role in implantation including protein and steroid hormones. Substances released from both the trophoblast and the endometrial cells, could account for the inductive and interdependent changes in both these tissues. Further research, however, is clearly necessary to answer these questions.

On the basis of the above studies, investigators have proposed a working model for implantation in the human. The first step of this model holds that the trophoblast binds to specific endometrial cell-adhesion molecules. After binding, the trophoblast penetrates the endometrium and attaches to the extracellular matrix by mechanisms involving proteases and controlled by specific inhibitors. This model, then, forms an adequate starting point for further research. Knowledge of the biochemistry is particularly deficient and could be furthered by the application of technologies such as monoclonal antibodies and DNA probes for particular cell-adhesion molecules. Identification of the molecular mechanisms of placenta formation may determine the factors that regulate the proteases. Such findings could be of great use in understanding disease states like toxemia in which abnormal invasion of trophoblastic cells is suspected to be a cause. Finally, investigators see a real need for the institution of a national or international registry of IVFET programs to track the incidence of abnormal implantation and correlate those data with the types of ovarian stimulation protocols used and hormone replacement therapy given. In this way, the disturbances of implantation caused by various aspects of the procedures of IVFET could be separated from those which occur naturally.

ENDNOTES

1. Based on talk by Dr. Zev Rosenwaks.

2. Summary of talk given by Dr. Neal First; also see paper by this author in Appendix A.

3. Summary of talks given by Dr. Robert F. Williams and Dr. Joanne Fortune; also see papers by these authors in Appendix A.

4. Summary of talks given by Dr. Ray Hammerstedt, Dr. John Eppig, and Dr. Stanley Leibo. Drs. Hammerstedt's and Eppig's papers are included in Appendix A. For additional information on cryobiology, see S.P. Leibo, Cryobiology: Preservation of Mammalian Embryos. In Genetic Engineering of Animals, J.W. Evans and A. Hollaender (Eds.), Plenum Press, New York, 1985.

5. Summary of talks given by Dr. William W. Wright and Dr. James M. Hammond; also see papers by these authors in Appendix A.

6. Summary of talk by Dr. Jonathan Van Blerkom; see paper by this author in Appendix A.

7. Combined summary of talks by Dr. Patricia N. Saling, Dr. Richard Schultz, and Dr. Jurrien Dean; see Appendix A for papers contributed by these authors.

8. Summaries of talks by Dr. John D. Biggers, Dr. Peter Braude, and Dr. Virginia Papaioannou; summary of talk by Dr. Harry M. Weitlauf, which was given in the fertilization section of the workshop, is combined in this chapter with talks included under implantation. See papers by Drs. Biggers and Papaioannou in Appendix A. For additional information, see Braude, P.R., Bolton, V.N., and Moore, S., Human gene expression first occurs between the four- and eight-cell stages of pre-implantation development. Nature 332:459-461, 1988; and Braude, P.R., Bolton, V.N., and Johnson, M.H., The use of human pre-embryos in infertility research. In Embryo Research: Yes or No, Ciba Foundation Study Group, G. Bock and M. O'Connor (Eds.), Tavistock Press, United Kingdom, pp. 63-82, 1986.

9. Combined summaries of talks given by Dr. Harry M. Weitlauf, Dr. Jerome Strauss, and Dr. R. Michael Roberts. See papers by Dr. Roberts and Drs. Coutifaris, Strauss, and Kliman in Appendix A. For additional detail see Weitlauf, H.M. and Suda-Hartman, M., Changes in secreted uterine proteins associated with embryo implantation in the mouse. Journal of Reproduction and Fertility. 84: 539-549, 1988.

CHAPTER 3
BARRIERS TO PROGRESS IN IN VITRO FERTILIZATION AND EMBRYO TRANSFER

Since the birth of Louise Brown in England in 1978, in vitro fertilization with embryo replacement has become an established method of treatment for certain types of infertility that do not respond to alternative methods of treatment. In 1988, 169 centers that offered IVFET were identified in the United States. Appreciable demand for IVFET exists despite some disadvantages, such as a higher rate of ectopic pregnancy than occurs with normal conception, high costs ($6,700 at one of the nation's oldest programs (Jones, 1989) that are often not reimbursed by third-party payers, prolonged treatment time, and uncertain results. Data from 41 clinics that reported to the American Fertility Society indicate the success rate is improving. In 1985, 14.1 percent of stimulation cycles resulted in clinical pregnancies. In 1986 this figure rose to 16.9 percent (Fertility and Sterility, 1988a). However, the proportion of women entering treatment who attain a live birth is far lower -- only 8.9 percent of oocyte retrievals ended in live birth (Journal of the American Medical Association, 1988a). Moreover, some women start treatment, but for a variety of reasons fail to reach the stage of egg retrieval. Improvements in the success rate of IVFET are probably not the result of significant changes in the methods of ovarian stimulation or laboratory procedures. Rather, improvement may be attributed to a decline in the average age of women undergoing IVFET, increasing experience of clinician and laboratory personnel, and more rigorous criteria for selection of patients (Van Blerkom, 1989).

Why are the odds for successful human IVFET so low? The answer lies in part in the state of knowledge of reproductive and developmental processes. There are a large number of unanswered questions whose resolution would have major impacts on the success of IVFET. Some of the most basic questions include: How does one identify a viable embryo? What are the effects of cryopreservation of gametes and embryos? How many embryos should be transferred? Why do some embroys fail to continue to develop after apparently normal development? What are the physiological effects of hormonal treatments? What are the factors that control oocyte maturation? What regulates dominance of ovarian follicles and what are the mechanisms of implantation? One of the most important factors that limits the success of medically assisted conception procedures is the natural frequency of egg and embryo wastage. The most reliable and reproducible estimates of inherent developmental failure in gametes indicate that about 20 to 25 percent of meiotically mature human eggs obtained after hyperstimulation have genetic abnormalities, about 7 percent of spermatozoa are genetically abnormal, and about 10 to 15 percent of unfertilized oocytes contain cytoplasmic aberrations or pathologies. In addition, the vast majority of human embryos derived from fertilization in vivo or in vitro will not develop to a blastocyst. Developmental failure during cleavage is fairly typical. Even embryos that do develop to the blastocyst stage can contain abnormalities in cell structure or number that are not consistent with postimplantation development (blighted ovum, for example). Collectively, developmental failure in the early stages of embryonic development appears to be the normal situation for the human species.

The state of clinical practice of IVFET today is limited by lack of knowledge of some of the basic reproductive biology involved. The reason for this is, at least in part, due to the many ethical questions raised by research in pursuit of the needed information. Difficulties in resolving these issues have caused the research to be deprived of federal funding.

This chapter first reviews the social and ethical barriers to progress in research, and then focuses on the scientific barriers. Brief note is made some of the ethical issues that are raised by the practice of in vitro fertilization, and the history of federal involvement in considerations of human fetal research is described. Some nations have examined the ethical questions and have issued guidelines or regulations. These will be described. Finally, the major science barriers that have slowed progress in human and animal practice of IVFET will be noted.

Ethical and Social Issues

Some of the ethical or social issues that arise from the various forms of assisted conception are unrelated to decisions about the progress of research. Examples of these are questions about the protection of the rights of gamete donors, gestational parents, and social parents; decisions on the fate of concepti, embryos, and fetuses; the moral status of human potential; the ownership of cryopreserved embryos whose parents have divorced or died; the confidentiality of sperm donors; and questions of the right of a child born of medically assisted conception to know his or her complete parentage. Other fundamental questions revolve around concepts of the right of an individual to reproduce, the sale of gametes and embryos, and whether infertility should be defined as a disease.

Some ethical questions have a direct bearing on research, and have had important consequences for the funding of research. The major questions focus on the status of the embryo at each stage of its development. How the embryo is regarded dictates what is morally acceptable to do to it. The implications for what may be done with an embryo differ according to when human life is thought to begin, whether any biological material containing the potential for human life is considered to be sacrosanct, and other such questions.

The determination of the moral status of the embryo drives such decisions as what level of risk to the embryo is acceptable in the practice of IVFET and research; whether it is possible to "discard" embryos, even when placement in a uterus diminishes the probability of other embryos surviving; whether it is permissible to create embryos for research purposes, and up to what stage of development of the embryo should research be allowed, and with what risk to its normal development.

At one end of this spectrum of thought is the position taken by the Roman Catholic Church, which holds that life begins with the union of two haploid gametes to form a diploid zygote. The Vatican's Instruction on Respect for Human Life states that "from the first moment of its existence until birth . . . no moral distinction is considered between zygotes, pre-embryos, embryos or fetuses" (cited in Fertility and Sterility, 1988b). Therefore, the absolute sanctity accorded to post-natal human life begins with the zygote. This concept makes it impossible to discard spare embryos or use them for research purposes. At the other end of the spectrum is the position that an embryo is merely biological material like any other group of living cells. The special value that might be attached to that material results from the expectations or aspirations of others. Those who hold this view also note that a large number of naturally conceived embryos fail to develop after implantation, thus discarding excess embryos created by IVFET can be viewed as a parallel process (Office of Technology Assessment, 1988).

Midway between these two positions is one that holds that "the human embryo is entitled to profound respect; but this respect does not necessarily encompass the full legal and moral rights attributed to a person" (Department of Health, Education, and Welfare, 1979). This is the position taken in 1979 by the Ethics Advisory Board established in the Department of Health Education and Welfare (DHEW) to advise the Secretary on matters concerning embryo research, and to review specific research proposals. Holding this position, the board concluded that research involving IVFET was acceptable on embryos up to 14 days after fertilization. Other constraints on such research included that it should be designed to improve the safety and efficacy of IVFET, and that the information could not be obtained by other means.

The Federal Government and Embryo Research

(Unless otherwise noted, this discussion is derived from a summary of fetal research issues by the Association of American Medical Colleges (1988)).

As mentioned earlier, the Ethics Advisory Board was chartered in 1977 to review applications for federal support of research. This resulted from growing concern about fetal or embryo research, which in turn stemmed from a concern about federal funding of research involving human subjects. Policy concerning research on human subjects had been slowly evolving since the 1960s. A study group was convened at NIH to develop guidelines, and a National Advisory Commission on Health Science and Society was proposed by Senator Walter Mondale in 1968 to examine developments in medical research. This commission was not established, but the impetus from the effort helped put in place the later National Commission for the Protection of Human Subjects. Following reports of the infamous Tuskegee syphilis experiments, DHEW recommended that Congress establish a permanent body to regulate federally funded research using human subjects.

In the 1970s the abortion issue became linked to the issue of embryo research. After the Roe v. Wade decision made legal abortion under certain specific conditions, concern developed that women would be pressured into having abortions and the sale of aborted embryos and fetuses might occur. Many states that had constrained embryo research through abortion statutes proposed separate legislation to ban such research. In 1974, the federal government created the National Commission for the Protection of Human Subjects (P.L. 93-348). Until this commission reported to Congress, research on the living fetus was prohibited unless it was used to help that fetus survive. This Commission did not consider the topic of research on the embryo or IVFET. The Commission however recommended establishing an Ethics Advisory Board (EAB) to review requests for conduct of embryo and IVFET research. Without such review, requests could not be processed. In 1975, DHEW issued regulations based on the findings of the commission. These regulations prohibited federally funded research unless the risk to the embryo was no greater than "those ordinarily encountered in daily life or during the performance of routine physical or psychological examinations or tests," or for therapeutic purposes (cited in Association of American Medical Colleges, 1988).

The EAB met for the first time in 1978. In 1980, the Secretary of Health and Human Services allowed the EAB charter to expire and, without explanation, failed to renew it. Thus, federal funding of embryo and IVFET research was, in effect, prohibited. As a result, embryo research has relied on private funding from patient care revenues, pharmaceutical companies, and university budgets. It is estimated that, were the EAB active today, it would receive more than 100 grant applications (Office of Technology Assessment, 1988). In addition, the federal government has lacked a means for controlling the direction of, or practices used in, such research. As was noted in 1979, "Departmental involvement might help to resolve questions of risk and avoid abuse by encouraging well-designed research by qualified scientists. Such involvement might also help to shape the use of the procedures through regulation and by example." (Department of Health Education and Welfare, 1979). This opportunity has been lost for a number of years.

In 1985, two events occurred in relation to the future of embryo research. Congress created a Biomedical Ethics Board. This board, composed of members of Congress, six senators and six representatives, is to examine the protection of human subjects in biomedical research.

Included in its brief are studies of the ethical implications of embryo research. By 1987, the 12 congressional members of the board had been appointed. The Board established a 14 member Biomedical Ethics Advisory Committee composed of scientists, physicians, clergy, and others. After months of deadlock over disagreements that mainly focused on appointees views on abortion and other ethical issues to do with definitions of human life, the advisory committee was named (American Medical News, 1987). The first meeting did not take place until September 1988, and the date for submission of a report on fetal research had passed (Capron, 1989).

Appointing the leaders and members of this Biomedical Ethics Board and the Biomedical Ethics Advisory Committee has been fraught with delays, as has the initiation of the activities of these bodies. Adequate funding is not assured. Alexander Capron, Chairman of the Biomedical Ethics Advisory Committee notes that "one can be hopeful, but not certain that [these groups] will be able to fulfill their statutory purposes." (Capron, 1989) Whether progress in enabling embryo research to proceed will be made is open to doubt in today's political environment.

In addition to these congressional groups there has been activity in the executive arm of government. The Department of Health and Human Services intends to reactivate the EAB that expired in 1980. In September 1988, notice of a draft charter was published in the Congressional Record. The 60-day comment period has elapsed and a final revised charter is pending.

The EAB, if reactivated, will develop specific guidelines to review NIH research proposals. It seems likely that the guidelines will be influenced by the broader policy formed by the congressional groups.

Domestic and Foreign Decisions on Embryo Research

The two professional societies in the United States that represent the physicians most involved in human IVFET have considered ethical questions about the practice of IVFET and embryo research. The American College of Obstetricians and Gynecologists (ACOG) issued an opinion that IVFET is a clinically applicable procedure that can be practiced if certain standards are assured and the ethical issues are considered (American College of Obstetricians and Gynecologists, 1984). Two years later the ACOG Committee on Ethics (American College of Obstetricians and Gynecologists, 1986) issued a statement that acknowledged the ethical issues that were posed by the creation of embryos outside a uterus, focusing particularly on the dilemma posed by surplus embryos and the acceptability of research using early human embryos. The ACOG committee recommended standards to guide such research, including that human embryos could be used only if nonhuman embryos would not provide the needed knowledge. It also recommended halting research on embryos that had reached the age of 14 days. The American Fertility Society (AFS) also issued a report in 1986. This document notes eight technologies that the AFS Ethics Committee found ethically acceptable, including basic IVFET, the use of donor eggs, and the use of frozen sperm. Six procedures were found suitable for clinical experimentation, including the use of frozen eggs, and experiments on embryos up to 14 days (Fertility and Sterility, 1986). A year later, after consideration of the Vatican's Instruction for Human Life in its Origin and on the Dignity of Procreation, issued by the Congregation for the Doctrine of Faith, the AFS issued another report. This report stated that progressive degrees of respect are due with progressive development of embryos, and that experimentation can be justified and is necessary if the human condition is to be improved (Fertility and Sterility, 1988b).

Despite public debate of the issues, statements issued by religious and professional groups, and other evidence of public interest, the government of the United States, since 1979, has not followed the lead of a number of nations that have systematically examined issues related to human IVFET. However, state statutes relevant to embryo research exist. Twenty-five states restrict embryo research, and 19 of those have language that could be interpreted as prohibiting some "pre-embryo" research (Office of Technology Assessment, 1988). Internationally, some governments have issued rules or regulations to control either the clinical practice, level of research, or both. LeRoy Walters, Director of Bioethics at the Kennedy Institute of Ethics, Georgetown University, has reviewed statements on the new reproductive technologies made by committees in other countries (Walters, 1987). He notes that, since 1979, at least 85 statements have been prepared by committees representing at least 25 countries. Walters focused his analysis on the reports listed in Figure 1. His analysis of issues in human embryo research notes that four Australian committees found research on early (preimplantation) embryos to be ethically unacceptable. Eleven committees approved at least some kinds of early embryo research. Six of these accept such research only on embryos left over from clinical activities. Five committee statements (including the 1979 DHEW Ethics Advisory Board) would allow the creation of embryos for research purposes. Although the majority of committees favor limiting research on embryos to up to fourteen days, one committee allowed it only to seven days, and one only through the first cleavage (for details see Figure 1). An additional, important, position statement was issued by the Vatican in 1987. The Vatican found unacceptable IVFET, artificial insemination, and embryo research if it is not for the direct benefit of the embryo on which the procedure is performed.

In sum, numerous groups have wrestled with questions related to the ethical problems of human embryo or fetal research. Some have based their conclusions on religious tenets, some on an interpretation of scientific knowledge, some on a mixture of both. It should not be surprising that there are substantial differences in the conclusions drawn by these groups. However, to the extent that each has laid out the foundations of its arguments, the debate about the acceptability of embryo research has been advanced.

Figure 1

ISSUES IN HUMAN EMBRYO RESEARCH

1. Acceptability in principle

A HEW Ethics Advisory Board 1979	B Waller I-IV, Victoria 1982-84	C South Australia 1984	D Demack, Queensland 1984	E Council for Science and Society U.K. 1984	F Warnock, United Kingdom 1984	G Tasmania I-II 1984-85	H Ontario Law Reform Commission 1985
yes	yes	no	no	yes	yes	no[1]	yes

I Australia, Family Law Council I-II 1984-85	J Benda, German Federal Republic 1985	K Spain, Special Commission 1986	L American Fertility Society 1986	M Western Australia I-II 1984-86	N Dutch Health Council I-II 1984-86	O National Ethics Committee, France I-II 1984-86	P New South Wales I-II 1987
no	yes	yes[2]	yes[3]	yes	yes[3,4]	yes[5]	yes

[1]But a National Bio-Technology Committee should periodically review current community attitudes on the subject.
[2]But only with embryos that are nonviable and not implantable.
[3]Early embryos called "pre-embryos."
[4]"By way of exception."
[5]But only after prior approval by, and under the supervision of, the National Ethics Committee.

Source: LeRoy Walters, Unpublished Paper Presented at the Annual Meeting, American Fertility Society, Atlanta, George. October 11, 1988.

Figure 1 (cont)

ISSUES IN HUMAN EMBRYO RESEARCH

2. Sources of embryos
   a. Acceptability of using left-over embryos from clinical IVF
   b. Acceptability of creating embryos especially for research purposes

	A HEW Ethics Advisory Board 1979	B Waller I-IV, Victoria 1982-84	C South Australia 1984	D Demack, Queensland 1984	E Council for Science and Society U.K. 1984	F Warnock, United Kingdom 1984	G Tasmania I-II 1984-85	H Ontario Law Reform Commission 1985
a.	yes	yes	---	---	yes	yes	---	yes
b.	yes	no	---	---	(yes)	yes	---	yes

	I Australia, Family Law Council I-II 1984-85	J Benda, German Federal Republic 1985	K Spain, Special Commission 1986	L American Fertility Society 1986	M Western Australia I-II 1984-86	N Dutch Health Council I-II 1984-86	O National Ethics Committee, France I-II 1984-86	P New South Wales I-II 1987
a.	---	yes	yes[1]	yes	yes	yes[2]	yes[3]	yes
b.	---	no	no	yes	no	no	no	yes

[1]But only with embryos that are nonviable and not implantable.
[2]"By way of exception."
[3]But only after prior approval by, and under the supervision of, the National Ethics Committee.

Source: LeRoy Walters, Unpublished Paper Presented at the Annual Meeting, American Fertility Society, Atlanta, Georgia. October 11, 1988.

Figure 1 (cont)

ISSUES IN HUMAN EMBRYO RESEARCH

3. Permissible duration of embryo culture in vitro

A HEW Ethics Advisory Board 1979	B Waller I-IV, Victoria 1982-84	C South Australia 1984	D Demack, Queensland 1984	E Council for Science and Society U.K. 1984	F Warnock, United Kingdom 1984	G Tasmania I-II 1984-85	H Ontario Law Reform Commission 1985
14 days	14 days	---	---	2 wks.	14 days	---	14 days[1]

I Australia, Family Law Council I-II 1984-85	J Benda, German Federal Republic 1985	K Spain, Special Commission 1986	L American Fertility Society 1986	M Western Australia I-II 1984-86	N Dutch Health Council I-II 1984-86	O National Ethics Committee, France I-II 1984-86	P New South Wales I-II 1987
---	Thru the first cleavages	14 days	14 days	14 days	14 days	7 days	14 days

[1]Subject to alteration if circumstances warrant.

Source: LeRoy Walters, Unpublished Paper Presented at the Annual Meeting, American Fertility Society, Atlanta, Georgia. October 11, 1988.

Figure 1 (continued)

THE SIXTEEN EXTENDED COMMITTEE STATEMENTS
ON THE NEW REPRODUCTIVE TECHNOLOGIES: 1979-1987

A. U.S., Department of Health, Education, and Welfare (HEW), Ethics Advisory Board, HEW Support of Research Involving Human In Vitro Fertilization and Embryo Transfer (May 4, 1979)

B. Victoria, Australia, Committee to Consider the Social, Ethical and Legal Issues Arising from In Vitro Fertilization

1. Interim Report (= Waller I) (September 1982)
2. Issues Paper on Donor Gametes (= Waller II) (April 1983)
3. Report on Donor Gametes and In Vitro Fertilization (= Waller III) (August 1983)
4. Report on the Disposition of Embryo Produced by In Vitro Fertilization (= Waller IV) (August 1984)

C. South Australia, Report of the working Party on In Vitro Fertilization and Artificial Insemination by Donor (January 1984

D. Queensland, Australia, Report of the Special Committee Appointed by the Queensland Government to Enquire into the Laws Relating to Artificial Insemination, In Vitro Fertilization and Other Related Matters (= Demack Queensland, report) (March 1, 1984)

E. Council for Science and Society (United Kingdom), Working Party, Human Procreation: Ethical Aspects of the New Techniques (May 1984)

F. United Kingdom, Department of Health and Social Security, Report of the Committee of Inquiry into Human Fertilization and Embryology (= Warnock, United Kingdom) (July 1984)

G. Tasmania, Australia, Committee to Investigate Artificial Conception and Related Matters

1. Interim Report (= Tasmania I) (December 1984)
2. Final Report (= Tasmania II) (June 1985)

H. Ontario, Law Reform Commission, Report on Human Artificial Reproduction and Related Matters (tabled June 13, 1985)

I. Australia, Family Law Council

1. Interim Report (= Family Law Council I) (July 1984)
2. Creating Children: A Uniform Approach to the Law and Practice of Reproductive Technology in Australia (= Family Law Council II) (July 1985)

Page 2 (Figure 1 continued)

J. Federal Republic of Germany, Minister for Research and Technology and Justice Minister, Working Group, In Vitro Fertilization, Genome Analysis, and Gene Therapy (= Benda, German Federal Republic) (November 1985)

K. Spain, Congress of Deputies, General Secretariat, Special Commission for the Study of Human In Vitro Fertilization and Artificial Insemination, Report (April 10, 1986)

L. American Fertility Society, Ethics Committee, Ethical Considerations of the New Reproductive Technologies (September 1986)

M. Western Australia, Committee to Enquire into the Social, Legal and Ethical Issues Relating to In Vitro Fertilization and Its Supervision

1. Interim Report (= Western Australia I) (August 1984)
2. Report (= Western Australia II) (October 1986)

N. Netherlands, Health Council, Committee on In Vitro Fertilization and Artificial Insemination by Donor

1. Interim Report on In Vitro Fertilization (= Dutch Health Council I) (October 10, 1984)
2. Report on Artificial Reproduction, with Special Reference to In Vitro Fertilization, Artificial Insemination with Donor Sperm, and Surrogate Motherhood (= Dutch Health Council II) (October 16, 1986)

O. France, National Consultative Committee on Ethics

1. Report on Ethical Problems Related to Techniques of Artificial Reproduction (= National Ethics Committee I) (October 23, 1984)
2. Report on Research Involving Human Embryos In Vitro and Their Use for Medical and Scientific Purposes (= National Ethics Committee II) (December 15, 1986)

P. New South Wales, Australia, Law Reform Commission

1. Surrogate Motherhood: Australian Public Opinion (= New South Wales I) (May 1987)
2. Artificial Conception, Discussion Paper 2: In Vitro Fertilization (= New South Wales II) (July 1987)

Other Barriers to Scientific Progress

The research agenda developed by the Institute of Medicine, Board on Agriculture committee (see Chapter 4) identifies many areas in which further research would make major contributions to improvements in medically assisted conception in humans and animals. As noted in Chapter 1, improving the application of medically assisted conception would benefit society in several ways, including making possible the preservation of some endangered species, as well as providing some relief from infertility and making production of meat and milk more economical. Progress, however, has been delayed by a number of factors, most of which stem from the generally controversial nature of concepts surrounding issues in reproduction, and specific controversies related to elements important to reproductive research -- zygotes and embryos.

As a prerequisite to developing recommendations to advance the science base of medically assisted conception, the committee first examined the impediments. In addition to the ethical considerations referred to earlier, the following barriers deserve particular emphasis: 1) deficiencies in the scientific base of this area of reproductive biology; 2) the resources available for research in this area of science; 3) lack of mechanisms for communication within the reproductive research community among basic scientists, clinicians, and animal husbandry scientists; 4) fear of abuses of reproductive technologies; 5) a relative lack of sympathy and understanding of the problems being experienced by infertile couples; 6) lack of a cohesive public interest group favoring such research, in contrast to well organized opposition; 7) limited health insurance coverage of IVFET services; 8) limited sources of research materials for experiments relating to human beings and to animals; and 9) the present dilemma of our society concerning how to handle our ethical disagreements.

Almost none of these factors is independent of the others; rather a causal relationship is often found. Some of the barriers identified may be amenable to policy intervention. The following sections briefly review the impact of the major barriers to progress and how they have come to be in existence.

Deficiencies in the Science Base

The papers presented at the committee's workshop and the research agenda developed from that workshop (see Chapters 2 and 4) indicate deficiencies in the scientific underpinnings of reproductive biology, and identify many areas in which further research efforts would make major contributions to improvements in medically assisted conception. The deficiencies are on three levels: basic science knowledge; knowledge needed to improve the technologies being used for medically assisted conception, such as cryobiology; and knowledge needed to improve both human and animal clinical practice of IVFET. As explained in detail in

Chapter 2, deficiencies in all three levels limit both the quality and efficiency of the practice of IVFET.

Research directed toward improving the reproduction of food-producing animals such as cattle is often ahead of research in the area of human reproduction. As a result, the use of in vitro sperm capacitation, artificial insemination with frozen semen, transfer of frozen embryos, oocyte harvesting from dead animals (the Lazarus effect), and the splitting of the dividing cells of the early embryo into more than one individual animal, are routinely used in animal husbandry. A number of factors have allowed medically assisted conception to proceed at a faster pace in animals than in humans: research using food-producing animals has been stimulated by an expected economic return on the research investment, a larger volume of materials is available for study, and research has not been subject to as many ethical constraints. The latter two factors noted -- greater volume of material available for study and fewer ethical constraints -- derive from a difference in the esteem accorded to animals and humans by many members of society. The economic value of IVFET, which acts as a spur to progress in the animal area, is to some, although lesser extent, paralleled in human IVFET. That there is an economic value to IVFET for humans is indicated by the fact that clinicians can charge, and people are willing to pay, quite substantial fees for the procedure. This has enabled a small amount of research to be conducted in the absence of federal help.

Research Funding

Approximately $115 million annually is spent on research in human reproductive processes. The major sources of private funding (the Ford, Rockefeller, and A.W. Mellon foundations, and the Population Council) in 1985 together contributed $2.8 million to research related to reproductive processes. This includes grants to investigate male and female infertility, fertilization, zygote transport, preimplantation development and implantation, and reproductive endocrinology (National Institutes of Health, 1988). Federal agencies are the principal support for research. In 1986 they provided $109 million for research in reproductive processes (National Institutes of Health, undated).

The National Institute of Child Health and Human Development (NICHD) of the National Institutes of Health provides the major portion of federal support of human reproductive research. Approximately $100 million per year, one third of the budget of NICHD, is spent in the reproductive sciences branch of NICHD on contracts and grants to academic centers and to the NIH centers for reproductive biology. In addition, the Contraceptive Development Branch, which uses mainly contract mechanisms to support contraceptive technologies, contributes to basic reproductive research. Federal funds for research relating to agricultural animal reproduction are available from the U.S. Department of Agriculture.

Funding for basic research in human reproductive biology is undoubtedly constrained by the lack of vocal and focused advocate groups that have been formed for some diseases such as cancer and heart disease. Lacking such a constituency, a major increase in federal support is unlikely. Yet, as the research agenda in Chapter 4 suggests, additional investment in research in reproductive biology can expect to be repaid in improvements in the reproductive health of the nation.

Lack of Communication Among Researchers

Discussion with the scientists and clinicians at the committee's workshop revealed an underuse of available mechanisms for communications among the individuals involved with various aspects of research in reproductive biology -- basic, clinical, animal sciences, etc. Also revealed was a desire for greater communication to allow cross-fertilization of ideas and development of ongoing relationships among investigators pursuing similar approaches to problems.

Difficulty in establishing communications is sometimes caused by the locale of investigators and the way science is organized. Basic scientists are frequently Ph.D.s in basic science departments such as anatomy or physiology. They often use animal models for their studies and have little or no contact with patients. Clinicians have an M.D. degree with subsequent residency training and are generally housed in clinical departments such as obstetrics and gynecology. They are frequently heavily involved with patient care. S2me researchers working in animal sciences have veterinarian degrees, others have basic science degrees, and frequently work in departments related to animal sciences or agricultural practices and deal with animals of economic importance.

The excitement and stimulation experienced by the individuals attending this IOM workshop, which encouraged interaction with individuals from differing backgrounds, underscored how infrequently meaningful interactions among these communities occur, and how useful this interaction could be.

Societal Concerns

The fear of abuses of reproductive technologies, the seeming lack of societal sympathy and understanding of the problems experienced by infertile couples, and the lack of a cohesive public interest support group are to some extent related phenomena.

Fears of abuses of the new reproductive technologies are legitimate concerns in a society that has not consistently shown an ability to monitor the ethical implications of scientific progress. Nor has this society established deliberative bodies that might reassure the public that applications of reproductive science will be controlled in ways sensitive to the delicate balance between preserving ethical standards and improving the human condition.

This is not to imply that concerns about the uses of new reproductive technologies can be easily quieted by the creation of deliberative or regulatory bodies. Rather, that to date the U.S. has not grappled with the issues as some other nations have done. However, if renewed efforts to put in place the needed mechanisms are successful in dispelling some of the worst fears about research abuses and the new technologies, scientific progress may be furthered.

It is clear that concerns are deeply felt. The strength of opposition to legal abortion demonstrates concerns that result from religious conviction and concerns about social values. The strength of support for legal abortions demonstrates concern for the freedom of individuals and the social impact of restrictive reproductive policies. Advances in science are making available technologies that open up new opportunities. Some view these technologies as opportunities to enhance the human condition. Others view them as potentially damaging to the social fabric. Indeed, there already exist technologies that are used in animal husbandry that members of society will not wish to have applied to human beings. Animals can presently be cloned, and genetic alterations have been made to animal germ cell lines. These are some of the issues that deserve careful scrutiny and public input. It is therefore important that society has mechanisms to consider and to debate the application of new technological capabilities and to ultimately provide safeguards against their misuse.

Reinforcing a sense of discomfort with some of the possible uses and abuses of new reproductive technologies, and also constraining efforts to advance in this area, may be a sense that infertility is a less vital concern than, for example, life threatening illness, disability, or more generally accepted signs of ill health. Although private third-party payers reimbursed roughly 70 percent of total non-IVFET infertility expenditures in 1980, the services are usually covered only if "they are associated with medical conditions or diseases requiring diagnosis and/or treatment and not solely related to infertility and fertilization" (Office of Technology Assessment, 1988). Insurance coverage for IVFET is still limited, in part because the procedure is considered experimental. However, coverage is available for many parts of the IVFET workup (Office of Technology Assessment, 1988). If concerns about the allocation of constrained health care resources continue to be high on the nation's agenda, it seems likely that less "acute" conditions such as infertility will lose out to more generally accepted forms of medical care, and will be available only to those who can pay the price. Also, a high cost procedure such as IVFET, likely to benefit a very limited number of people, will be further compromised in a cost-conscious health services environment.

That there is no single cohesive public interest group pressing for heightened research in the area of reproductive biology might be partially explained by the complex religious and ethical issues involved with certain types of research, by the lack of a cohesive research community, by fear of abuses in the area of reproductive technology, by a low public awareness of the size of the nation's infertility problem and the complex effect of infertility on individuals and couples, or a combination of these and other factors.

Sources of Research Material

The committee's workshop provided many excellent examples of instances in which information about reproductive physiology derived from animal models has been useful in understanding human physiology. Instances, however, were also presented in which the human processes have marked differences from those elucidated by animal models, and to move forward in understanding the human physiology requires the use of human tissue. An example of this is investigation of reasons for the developmental failure of human embryos, and cryopreservation of human eggs and zygotes. Preservation of human tissue by freezing allows excess zygotes to be thawed for use at a later time. At present, we do not know all the possible negative consequences of freezing and thawing of the human tissue, nor do we know the optimum conditions under which these procedures should be done. However, cryopreservation has crept into clinical practice despite these uncertainties. This has happened in part because of the dilemma caused by excess embryos. Discarding them is unacceptable to some. Therefore cryopreservation has become an acceptable option.

In addition, although specific primates are good models for some aspects of human reproductive physiology, there are only a limited number of monkeys of desirable species in captivity and many of them are presently being used for AIDS research. To optimize the use of each primate, it would be helpful to develop noninvasive procedures for steps such as oocyte harvesting. Such a procedure would also be of great importance to patients.

The committee believes that restrictions on the amount of material available for research use is slowing the rate of progress in developing a scientific base for IVFET and the technological advances that would make it more efficient. These limitations are the result of a concern that human material be used with proper respect, and that the use of animals for research purposes be controlled to ensure that they are not abused.

There exist ways, however, of enlarging the amount of available research material while preserving a proper consideration for these issues. Making available material from organs that have been surgically removed would increase the supply of human tissue. For instance, when an organ such as the ovary is removed for medical reasons, it might be made available for harvesting oocytes and eggs.

In sum, as discussed in many places in this report, ethical, and social concerns underlie many of the barriers to progress in reproductive research and the clinical application of new technologies. Lack of mechanisms for resolving such disagreements has inhibited progress in the necessary debate that must precede the development of policies. It is encouraging that Congress has established the Biomedical Ethics Boards, and that the executive branch of government is taking steps to reactivate the Ethics Advisory Board. If these bodies become functional the nation will have taken steps to establish entities to handle difficult issues in reproductive biology.

REFERENCES

American College of Obstetricians and Gynecologists. 1986. Ethical Issues in Human In Vitro Fertilization and Embryo Placement. Committee on Ethics. ACOG Committee Opinion Number 47. Washington, D.C.

American College of Obstetricians and Gynecologists. 1984. Human In Vitro Fertilization and Embryo Placement. Committee on Gynecologic Practice. Committee Statement. Washington, D.C.

American Medical News. 1987. Ethics Panel Members Are Named. Aug 14.

Association of American Medical Colleges. 1988. Fetal Research and Fetal Tissue Research. Washington, D.C.: Association of American Medical Colleges. June.

Capron, Alexander Morgan. 1989. Bioethics on the Congressional Agenda. Hastings Center Report. March/April pp. 22-23

Department of Health, Education and Welfare. 1979. HEW Support of Research Involving Human In Vitro Fertilization and Embryo Transfer. Report and Conclusions. May 4. Washington, D.C.: U.S. Government Printing Office.

Fertility and Sterility. 1986. Ethical Considerations of the New Reproductive Technologies. The Ethics Committee of The American Fertility Society. 46(3 Suppl 1): IS-94S

Fertility and Sterility. 1988a. In vitro fertilization/embryo transfer in the United States: 1985 and 1986 results from the National IVF/ET Registry. 49(2):212-215.

Fertility and Sterility. 1988b. Ethical considerations of the new reproductive technologies. By the Ethics Committee (1986-7) of the American Fertility Society in light of Instruction on the Respect for Human Life in its Origin and on the Dignity of Procreation issued by the Congregation for the Doctrine of the Faith. Feb;49(2 Suppl 1):IS-7S

Jones, Howard W. Jr. 1989. Eastern Virginia Medical School. Personal Communication, January 23.

Journal of the American Medical Association. 1988. IVF Registry Notes More Centers, More Births, Slightly Improved Odds. 259(13):1920-1921.

National Institutes of Health. 1988. Inventory and Analysis of Private Agency Population Research. 1984 and 1985. U.S. Department of Health and Human Services, Public Health Services, National Institutes of Health. June.

National Institutes of Health. Undated. Inventory and Analysis of Federal Population Research. Fiscal year 1986. U.S. Department of Health and Human Services, Public Health Service, National Institutes of Health.

Office of Technology Assessment. 1988. Infertility: Medical and Social Choices. OTA BA BP 48. August. Washington, D.C.: Office of Technology Assessment.

Van Blerkom, Jonthan. 1989. Professor of Molecular, Cellular, and Developmental Biology, University of Colorado at Boulder. Personal Communication, January.

Walters, LeRoy. 1987. Ethics and New Reproductive Technologies: An International Review of Committee Statements. Hastings Center Report. June:3-9

CHAPTER 4

RESEARCH AGENDA AND RECOMMENDATIONS

This chapter lists the areas of research that, in the view of the committee and workshop participants, hold particular promise in advancing the knowledge base and efficacy of medically assisted conception for humans and other animals. In addition, this chapter outlines the conclusions and specific recommendations that were made by the committee following the workshop. The research agenda items are organized into three main categories. The first and largest category includes promising areas of basic scientific investigation. The second category encompasses areas in which improvements in technology are needed. The third category includes questions that can be approached by cooperative agreements among clinical centers involved in IVFET. In addition to organizing the research agenda into common areas, these categories also generally reflect the different funding mechanisms that might be required to support such investigations. Research grants to support basic science projects could be obtained through competitive grant programs at the National Institutes of Health or the National Science Foundation. Formula funds and competitive grants can be obtained from the United States Department of Agriculture. Technological development, however, could be supported by contracts from projects already supported by the NIH or other agencies. Coordination of information from IVFET clinics could be accomplished by any one of a number of agencies or professional societies.

Research Agenda

The topics listed below are areas in which further research was recommended by workshop participants and committee members. It therefore reflects the areas of investigation considered to be promising by the committee. Work in these areas is expected to increase understanding of the biology of reproduction with the hope that increased knowledge will eventually lead to improvements in practice of IVFET in humans or other animals, or to advances in the area of contraception. Research areas are listed here in summary form and apply both to lower animals and human beings unless specifically noted. The reader should refer to Chapter 2 and the papers published in Appendix A for detailed discussion of these areas.

Basic Science

Male Gametogenesis

o Definition of the role of cell adhesion molecules in interactions between Sertoli cells and developing sperm cells.

- Understanding the function of differential protein synthesis in different stages of sperm development.

- Determination of the role of paracrine factors including fibroblast growth factor, somatomedin C, epidermal growth factor, and interleukin-1 on the development and differentiation of male gametes.

- Structural analysis to identify normal and abnormal sperm and the development of markers for abnormal sperm.

- Understanding of the biochemistry of sperm capacitation.

Female Gametogenesis

- Analysis of the effects of superovulation or hormonal stimulation protocols on oocyte development and maturation. This work should also examine differences among species.

- Development of ways to mature oocytes in vitro.

- Investigation of ways to naturally stimulate oocyte and follicular development.

- Investigation into the biochemistry of meiotic arrest and the factors, such as cyclic AMP, purines, calcium, and maturation-promoting factor, that may mediate this process.

- Development of ways to produce or synthesize hormones from non-human primates to be used in ovarian stimulation.

- Definition of the role of ovarian estrogen in oocyte maturation and ovulation and the interactions between estrogen and paracrine factors including fibroblast and epidermal growth factors, insulin-like growth factor, transforming growth factor, and inhibin.

- Definition of the point at which oocytes become sensitive to factors that influence their development.

- Elucidation of the processes that underlie oocyte depletion, to determine why oocytes are lost at a predictable rate throughout life.

- Investigation into ways to augment natural hormone release.

- Investigation into the biochemistry of protein synthesis and modification in ovarian cells.

Fertilization

o Investigation into the biophysics of cell membranes as it relates to sperm and egg interactions at fertilization.

o Continued investigation to identify the genes for zona proteins in various species, especially humans.

o Further delineation of the role of zona proteins, especially ZP2 and ZP3, in sperm binding.

o Understanding of the biochemistry of the modification of zona proteins in preventing polyspermy.

o Elucidation of the molecular determinants of antibody formation to zona proteins and their possible role in contraceptive strategies.

o Definition of the biochemical mechanisms of the cortical reaction in the egg and the effects of this reaction on zona proteins.

o Determination of the physiological significance of germinal vesicle breakdown and the biochemistry of sperm chromatin decondensation.

o Definition of the molecular events associated with formation of the male and female pronuclei.

o Definition of the molecular events during zygote formation and the first cleavage.

Preimplantation Development

o Definition of the metabolic requirements of early embryos at different stages.

o Determination of embryonic gene expression.

o Assessing the potential of individual embryonic cells and defining the point at which embryonic cells are committed to particular fates.

o Identification of substances produced by early embryos that signal changes in the uterus prior to implantation.

o Improvements in embryo multiplication and embryo splitting, especially for food-producing animals.

Implantation

o Definition of the biochemical events that make the uterus permissive to implantation.

o Definition of the factors released by embryos that cause endometrial changes at the site of implanation.

o Identification of the role of embryo-released factors in suppressing the immune responses of the mother.

o Isolation and analysis of substances released by endometrial cells and their effects on embryos.

o Continued work with in vitro models of human implantation to study the biochemistry and mechanisms of embryo-endometrial interactions, especially the role of extracellular matrix proteins and the biochemistry of trophoblast invasion of the endometrium.

Technological Advances

o Improved cryopreservation techniques, including freezing and thawing protocols for eggs and embryos.

o Improved resolution of ultrasonography for localization and noninvasive harvest of oocytes, eggs, and embryos--would have particular usefulness for non-human primates and food-producing animals.

o Development of new culture media and methods for in vitro maturation of oocytes.

o Development of safe methods of biopsy of early embryos for preimplantation diagnosis of genetic diseases.

Clinical Research Opportunities

The following areas are those in which a coordinated data collection effort across IVFET clinical centers would improve the quality and success rates of IVFET nationally and, possibly, internationally.

o Evaluation of hormonal stimulation protocols in terms of number of oocytes harvested, quality of oocytes, and rate of fertilization success.

o Documentation on the incidence of abnormal implantation rates in IVFET practice and correlation of incidence with particular stimulation protocol used.

- Collection of information regarding the incidence of abnormal eggs and embryos, failed fertilization, and developmental arrest of embryos.

- Analysis of data pertaining to synchronization of embryonic stage with endometrial stage and development of methods to improve synchronization.

- Collection of information on sharing of spare oocytes and arrested embryos for research purposes.

Conclusions and Recommendations

Developing Research Policy

The lack of a mechanism for dealing with ethical disagreement over the use of embryos in research has slowed the rate of progress in research by, in effect, placing a moratorium on the use of federal funds for eight years. This has had undesirable results: the human clinical practice of IVFET is less effective than it might have been had research progressed at a faster pace; other socially desirable goals such as improved contraception, better techniques to preserve endangered species, and more cost-effective methods of producing food have developed at a pace slower than is optimal.

The recent appointment of the Biomedical Advisory Committee by the Biomedical Ethics Board, to report to Congress by November 1990 on embryo research issues, could be a step toward a solution. The committee also applauds the intention to revive the Ethics Advisory Board of the Department of Health and Human Services to rule on the ethical acceptability of research relating to human embryos, which is required before federal funding of such a research grant can be considered. However, until these groups become fully functional and show evidence of progress, their impact must remain in question.

If these groups can assume leadership roles in resolving the difficult issues of reproductive research, and develop guidelines for research that are based on information provided by science, as well as on concepts that are ethically acceptable to society, research in reproduction will be able to move forward. But if these groups become paralyzed because of political considerations or an inability to develop a framework for the resolution of differences of opinion, another organization should take over the role. The committee recommends that, <u>if the groups currently being formed fail to come to conclusions concerning embryo and fetal research, a non-governmental organization should be established to develop guidelines for embryo and fetal research based on the most advanced knowledge that science can muster, and with serious consideration of the expressed values of society. The group should be</u>

composed of individuals with expertise in the relevant scientific disciplines, representatives of the lay public, and experts in the legal, ethical and social issues. The organization should be housed in an institution that would allow it to conduct its deliberations free from undue pressures from political and special interest groups. A model for such activities can be found in the Voluntary Licensing Authority of Great Britain. This group was established after a governmental committee recommended a statutory licensing authority. Recognizing that it would be some time before legislation would be completed, the Medical Research Council and the Royal College of Obstetricians jointly sponsored the voluntary body. Five of its 13 members are lay people. The group has a mandate to undertake five major activities:

- to approve a code of practice on research related to human fertilization and embryology;
- to invite all centers, clinicians and scientists engaged in research on IVF to submit their work for approval and licensing;
- to visit each center before it is granted a license;
- to report to the sponsoring organizations; and
- to make known publicly the details of both approved and unapproved work.

Basic Science Foundations

The number and range of topics included in this chapter's research agenda indicate the exciting potential for productive scientific exploration. Funding that would allow investigation of the areas targeted in the research agenda would allow significant advances to be made in understanding reproductive processes. The committee believes that fundamental research to enhance the basic science foundations of reproductive biology should be stimulated and supported. This includes studies of human beings, laboratory animal models, and food-producing animals. The knowledge that would be generated is fundamental to an understanding of how to reverse infertility, to new approaches in the area of contraception, and to increasing the world's food supply.

This report attempts to define not only the state of knowledge in reproductive science relating to IVFET, but to assess and highlight some research opportunities. It is important that aspects of male as well as female reproductive biology be studied. It is also important that investigators make use of some opportunities that are largely ignored today. These opportunities occur as a result of clinical activities as well as research activities. For example, eggs that have failed to become fertilized can become material for studies seeking chromosomal abnormalities. Fertilized eggs that fail to develop may be used to investigate the reasons for developmental failure, and to answer questions about the natural wastage that occurs in pregnancy.

Research of this type would generally be funded by the grant mechanisms of the National Institutes of Health and by the United States Department of Agriculture. For adequate attention to a research agenda, however, administrators of NIH need mechanisms to insure that studies on a variety of these topics are being funded. For this an RFP or a contract mechanism might be necessary to insure adequate coverage of the various aspects of reproductive biology. Foundations are also encouraged to consider increased support for basic studies in reproduction. A stable funding base for reproductive research will encourage young, well-trained scientists to pursue research in reproductive biology.

The committee recommends that a vigorous program for funding of a basic science agenda in reproductive biology be maintained in a coordinated fashion by an appropriate office in the National Institutes of Health.

Applied Research

Research needs to be stimulated concerning technologies used in medically assisted conception in food producing animals and in human beings. Lack of support in these areas is leading to inadequate scientific underpinnings for safe and effective clinical practice. An example of a technique used, but not carefully evaluated for possible detrimental effects, is freezing eggs or zygotes. Further experiments should be conducted to assess the effects on safety and viability of this technology which is standard practice in many IVFET clinics. Other areas of technology that need to be developed include less invasive ways to retrieve oocytes, ways to mature oocytes in vitro, and ways to assess the quality of spermatozoa or oocytes to be used for fertilization.

Since these studies would be technology driven, it is unlikely that the research will be funded by a grant mechanism, therefore a contract mechanism should be used.

The committee recommends that applied research into technologies used in medically assisted conception be undertaken to provide a firm foundation for the safe and effective practice of in vitro fertilization and embryo transfer. Such applied research should be coordinated by the appropriate office at the National Institutes of Health.

Clinical Research Opportunities

Perhaps the most obvious missed opportunity is the failure to learn from the diverse experiences of the approximately 160 clinical programs that provide human IVFET. In addition to scientific questions, there are questions to do with the organization of clinics and the outcomes of procedures. The answers to these questions would enable practitioners to work more effectively, and enable policy makers to make decisions on the basis of the best available information. Such questions include: Who provides the quality assurance for facilities and procedures used (including the training of those providing services)? How are protocols developed? How standard are they? What are the outcomes for each protocol? What is the role of Institutional Review Boards in the establishment and maintenance of clinical facilities?

Some specific questions to be answered from data that could be collected from clinics include: What ovarian stimulation protocols are the most successful? Is chromosomal damage associated with any of the procedures used? What indicates whether a zygote will implant successfully? Much data relating to these questions already exist in the centers. Other questions may require a cooperative prospective study to be undertaken.

Clinical IVFET centers can also provide unique opportunities for important studies. For example, human eggs that fail to fertilize in vitro are material that could be used to investigate the phenomenon of failed fertilization. Improved understanding of reasons for failure has implications for reversing infertility and ensuring conception. Another area for investigation for which the IVFET centers are particularly suited is arrested zygotic development. In producing embryos by in vitro fertilization, it has been noted that certain zygotes stop dividing. The reasons for such arrested zygotic development are unknown. Have these cells died? Can one tell from studying the newly dividing zygotes which ones will be most likely to initiate a successful pregnancy?

Failed fertilization in the peri-implantation period in both the natural situation and during the procedures of IVFET is an important area to be understood. Developmental failure in the early stages of embryonic development appears to be a normal event for the human species. Whether this loss results from lethal genetic defects, chromosomal anomalies, biochemical or structural abnormalities, or technical difficulties needs further elucidation. Research that seeks to understand the basis of reproductive failure, and its relationship to hyperstimulation should be encouraged. Coordinated studies utilizing the mass of material and experience from IVFET centers could begin to answer these and other questions. In addition, the experience of centers performing IVFET for farm animals could provide valuable information.

The committee applauds the activities of the various professional societies that have issued non-binding statements about the quality of practice of IVF. The American Fertility Society has also provided a voluntary registry for centers. On the animal side, The American Embryo Transfer Society has started to establish some quality measures for commercial bovine embryo transfer. States in general get involved by licensing physicians, hospitals, and clinics, and as yet have not played a dominant role in assuring quality of care.

The committee believes that a mechanism is needed to monitor and evaluate clinical practice so that existing information that is relatively easy and inexpensive to collect can be disseminated. This would enable clinicians to build on the broadly based experience of the community and help ensure that patients have access to information about developments in IVFET and to well-informed physicians. The committee recommends that a mechanism for multi-centered data collection be established to monitor and evaluate human and veterinary practices of medically assisted conception in order to improve the safety, effectiveness, and quality of clinical practice. A cooperative group composed of the relevant professional societies should be established to fund and initiate data collection under the direction of an inter-society council composed of representatives of each participating organization.

Improving Communications

The IOM Workshop on Medically Assisted Conception brought together researchers from basic science, clinical practice, and animal sciences. The resulting interaction was viewed as extremely helpful by investigators from each of these communities. The committee recommends that a mechanism (or multiple mechanisms) be found for fostering continued communication between researchers in diverse areas of reproductive science. The initiative should come both from NIH research administrators who could sponsor additional workshop opportunities and from the professional societies, either individually or through an intersociety council.

APPENDIX A

Committee on Basic Science Foundations of Medically Assisted Conception

Papers Presented At Workshop
August 21 - 23, 1988

1. TECHNOLOGIES FOR ASSISTED CONCEPTION/EMBRYO TRANSFER IN AGRICULTURAL AND VETERINARY PRACTICE
   Neal L. First

2. OVARIAN STIMULATION IN NON-HUMAN PRIMATES
   Robert F. Williams

3. RESPONSES TO GONADOTROPINS IN DOMESTIC ANIMALS
   Joanne Fortune

4. MEMBRANE MODELS: EVOLUTION FROM THE FLUID-MOSAIC STANDBY
   Roy Hammerstedt

5. MAMMALIAN OOCYTE MATURATION: MECHANISMS FOR REGULATION AND PROSPECTS FOR PRACTICAL APPLICATION OF IN VITRO TECHNOLOGY
   John J. Eppig

6. INTRAGONADAL CONTROL OF TESTIS FUNCTION
   William W. Wright

7. PARACRINE REGULATION IN THE OVARY
   James M. Hammond

8. EXPERIMENTAL APPROACHES TO THE STUDY OF EARLY DEVELOPMENTAL FAILURE IN HUMAN REPRODUCTION
   Jonathan Van Blerkom

9. THE FERTILIZING SPERM: STRUCTURE, MATURATION AND FUNCTION
   Patricia M. Saling

10. MOLECULAR EVENTS PRE- AND POST-FERTILIZATION OF MOUSE EGGS: OOCYTE MATURATION, EGG ACTIVATION, AND POLYSPERMY BLOCK
    Richard M. Schultz

11. GLYCOLYTIC PATHWAY IN PREIMPLANTATION MAMMALIAN EMBRYOS
    John D. Biggers

12. REGULATIVE POTENTIAL OF MICROMANIPULATED EMBRYOS
    Virginia Papaioannou

13. UTERINE RECEPTIVITY, MATERNAL RECOGNITION OF PREGNANCY AND EARLY EMBRYONIC LOSS
    R. Michael Roberts

14. TOWARDS A CELLULAR AND MOLECULAR UNDERSTANDING OF IMPLANTATION IN THE HUMAN: IMPLICATIONS FOR ASSISTED REPRODUCTIVE TECHNOLOGIES
    Christos Coutifaris, M.D., Ph.D., Jerome F. Strauss, III, M.D., Ph.D., Harvey Kliman, M.D., Ph.D.

TECHNOLOGIES FOR ASSISTED CONCEPTION/EMBRYO TRANSFER IN AGRICULTURAL AND VETERINARY PRACTICE

Neal L. First

INTRODUCTION

The natural reproductive processes of domestic animals have been controlled or assisted largely for the purpose of producing a larger number of offspring from individuals of superior meat or milk production.

For some technologies a secondary benefit has also been increased reproductive efficiency. Thus, the purpose for development of methods for assisting natural reproduction has been different between animal agriculture and human medicine where enhancement of reproduction has been the focus.

The developed and developing technologies to be discussed here are artificial insemination, superovulation, embryo transfer, freezing of embryos, sexing of embryos, multiplication of embryos by bisection and by cloning, production of embryos in vitro and modification of embryos by gene transfer.

Artificial Insemination

The oldest and thus far most used biotechnology is artificial insemination. More than 60% of the 10 million United States dairy cows and nearly all West European dairy cattle are presently mated by artificial insemination (Betteridge, 1986). The strength of this tool comes from the ability to statistically identify and select the very best bulls in the country in terms of genes for milk production and the ability to extend each ejaculate to produce more than 500 inseminations. Artificial insemination has been standard practice in the dairy industry for more than 30 years. During this period its use has contributed heavily to a doubling of the milk production of each cow and reduced the number of cows consuming the nation's grain and forage resources by 50% (Reid, 1978). It has been available but little used in other species except in limited economic situations of low labor or high animal value.

Superovulation

Superovulation by injection of a follicle stimulating hormone has been utilized for many years to increase by 5- to 10-fold the ovulation rate of cattle, sheep and swine (Pineda and Bowen, 1980). While this treatment is common practice in the cattle embryo transfer industry as well as for human in vitro fertilization, better methods are needed. The ovulatory response is unpredictable ranging from 1 to 20 or more in cattle (Massey and Oden, 1984). Ovulation does not always occur (Moor et al., 1984) and an increase in frequency of oocytes with abnormal chromatin occurs for all species superovulated and studied thus far (First and Eyestone, 1987). For cattle and sheep some increase in yield of fertilizable oocytes has been achieved by use of an early luteal priming dose of follicle stimulating hormone in addition to the usual midluteal treatments (Ware et al., 1988). A more uniform superovulation and increased frequency of high quality transferable bovine embryos has also been reported from the use of recombinant derived pure FSH material (Chappel et al., 1988).

Embryo Transfer

More recently superovulation and embryo transfer have been used to multiply scarce exotic breeds of cattle and to accelerate genetic improvement by expanding the maximum number of offspring possible from genomic combinations of the best females and progeny tested sires. The embryos are collected from cows induced by the use of hormones to ovulate ~5 to 15 eggs. The embryos are then transferred nonsurgically to other recipient cows whose estrous cycles are synchronous with the donor cow (Seidel, 1981). Bovine embryos are commonly stored frozen if more embryos are harvested than the number of recipient cows available (Massip et al., 1987) or bisected to double their number if insufficient embryos are harvested (Baker and Shea, 1985; Leibo, 1988). These latter procedures are currently in use in the embryo transfer industry. The use of embryo transfer has resulted in collection of the best dairy cows into a small number of herds designed to produce and sell embryos. In 1987 this commercial embryo transfer industry performed approximately 150,000 transfers in the USA and 250,000 in the world, of which approximately 30% were with frozen embryos (Seidel, personal communication).

So far neither artificial insemination nor embryo transfer is widely practiced in domestic species other than cattle. This is either for reasons of low economic payback or technical difficulties such as low semen extension, inability to freeze semen effectively or absence of nonsurgical transfer methods. Nevertheless these technical problems are being solved. For example, the first method for nonsurgical transfer of swine embryos was recently reported (Sims and First, 1987). In spite of their past impact on the dairy industry, the tools of artificial insemination and embryo transfer as practiced are slower than desired in effecting productivity changes, and are restricted to the gene pool of existing livestock breeds. Artificial insemination allows genetic change only through the male side of the pedigree and embryo transfer results in a cohort of embryos after superovulation, mating and recovery that are no more alike than siblings. Because heritabilities are low for most desired production traits, methods are needed for producing large numbers of duplicate copies of high performance individuals or embryos.

Freezing of Embryos

Freezing embryos for storage, cryopreservation, is a valuable technology important in facilitating embryo transfer and in preserving rare genetic traits. Embryos of cattle can be frozen and, after thawing and transfer, produce pregnancies with nearly the same efficiency as fresh embryos (i.e., 45-60%; Massip et al., 1987). The embryo transfer industry often stores frozen cattle embryos to preserve surplus; in 1986, for instance, it froze 30% of the cattle embryos it later transferred (Seidel, unpublished).

Frozen cattle embryos are routinely thawed by a one-step method that allows nonsurgical embryo transfer directly from the straw, as with artificial insemination (Leibo, 1982; Renard et al., 1982; Chupin et al., 1984; Massip and van der Zwalmen, 1984). Following freezing and thawing procedures, 50-80% of livestock embryos survive, including cattle (Lehn-Jensen et al., 1981; Kennedy et al., 1983; Renard et al., 1983; Pettit, 1985), sheep (Ware and Boland, 1987), goat (Chemineau et al., 1986) and horse embryos (Slade et al., 1984, 1985; Takeda et al., 1984).

Effectiveness of embryo freezing varies with mammalian species. Pig embryonic blastomeres have a relatively high lipid content (Edidin and Petit, 1977) that impedes freezing (Polge et al., 1974). Because each species possesses a physically unique embryo, a universal cryoprotection scheme has not been possible.

Sexing of Embryos

Sexing of embryos before transfer is especially sought by the dairy cattle industry where females are the desired milk producing unit. To be useful sexing techniques must be accurate, efficient, rapid and without detrimental effects on the embryos. The three most successful approaches have been 1) cytogenetic karyotyping of the cells of the blastocyst, 2) immunological detection of a male specific antigen on male embryos and 3) use of DNA hybridization probes to identify Y (male) chromosome specific DNA.

<u>Sexing by Karyotyping</u>. Cytogenetic methods are highly accurate and also allow identification of aberrant chromosomal karyotypes. However, karyotyping requires a large number of colcemid arrested cells to insure a few readable metaphase chromosome displays (King, 1984). Embryonic biopsies of sufficient size to yield sufficient cells may be damaging to the embryo. One approach to providing sufficient cells has been to bisect the embryo followed

by cytogenetic analysis of the smaller half and transfer of the larger. In one experiment this resulted in a normal pregnancy rate after transfer of the half embryo but sex could be determined for only 60% of the half embryos subjected to cytogenetic analysis. Sex when identified was predicted with 100% accuracy (Picard et al., 1984). Recent research has resulted in methods to increase the frequency of metaphase spreads (Kent and First, 1988). Nevertheless, a large number of blastomeres, 15-20, is needed for accurate assay. Cytogenetic analysis does provide an accurate standard for quick evaluation (less than 15 hr) of other methods.

Sexing by Immunological Methods. The sex of mammalian embryos is determined by the presence or absence of the Y chromosome derived from the father since males are the heterogametic sex and can produce either X or Y bearing sperm cells while oocytes from the female contain only X chromosomes. Antigens which are coded from Y specific genes are found in and on male but not female cells. Male specific antigens are expressed as early as the 8-cell stage in mice. One such antigen has been called H-Y (Krco and Goldberg, 1976; Epstein et al., 1980; reviewed by Haseltine, 1983). Fluorescent labeled antibodies to one or more male specific antigen provide a way to recognize individuals expressing male specific genes. Except for cases of chromatin translocations which often cause intersexuality, the presence of a Y chromosome antigen specifically identifies a male.

Male specific antigens such as H-Y antigen are known to be highly conserved across species and present on cells of at least 70 species of vertebrates including mice, rats, cattle, dogs, goats, donkeys, horses, pigs and humans as well as on cells of the female, the heterogametic sex, of birds (Wachtel, 1984a). Both polyclonal (Krco and Goldberg, 1976; Epstein et al., 1980; White et al., 1982) and monoclonal (Koo and Varano, 1981) antibodies have been prepared against the serological H-Y antigen.

The H-Y antigen has been detected on preimplantation murine embryos by incubation with H-Y antiserum and complement (Krco and Goldberg, 1976; Epstein et al., 1980; White et al., 1982; reviewed by Wachtel, 1984a,b). Cell lysis occurred in approximately 50% of the embryos which were exposed to murine H-Y antiserum and complement from guinea pigs; but embryos cultured in medium with H-Y antiserum or complement alone were not damaged. Karyotypes of embryos that were unaffected by culture with antiserum and complement showed that 92% were female (Epstein et al., 1980). When unaffected embryos were transferred to pseudopregnant recipients, 86% of the pups born were female (White et al., 1983). The obvious disadvantage of this complement dependent cytolytic method for detection of male specific antigen is that survival of male embryos is reduced. When a male specific antibody and fluorescent labeled second antibody were utilized together the male specific antigen on male embryos was identified without toxicity to embryos of either sex

(White et al., 1983). Removal of complement and use of a fluorescent labeled second antibody for detection of binding of the first antibody have resulted in identification of both male and female embryos without cytotoxic death of the male embryos (White et al., 1984). These studies in mice suggest that it should be possible to sex embryos of bovine and other species by use of an antibody to one or more male specific antigens.

The accuracy of White et al. (1984, 1987a,b,c) in predicting each sex from application of this assay to several species is shown in Table 1. As indicated in Table 1 for the bovine the assay was nearly perfect for identification of female embryos (89%). However, only 80% of male embryos were identified. This reduced efficiency for males appeared to occur because dead cells of some embryos exhibited an autofluorescence thus when embryos with dead cells were female the accuracy of male identification was reduced. It is likely that this problem could be eliminated by choosing a fluorescent label of a different wave length. A perfect prediction of sex may not occur from this method since females with a fragment of Y chromosome translocated to an X chromosome may test as males by this test or by DNA hybridization. Overall, the accuracy for bovine was 86%. Using a similar method Wachtel et al. (1988) report an accuracy in bovine embryos of approximately 85%. While large scale testing of this antisera and second antibody system will be essential it would appear that with minor adjustments in the method such as indicated above, a method for determining the sex of embryos of domestic animals is at hand and ready for use. Its wide scale use will depend on the availability of antisera recognizing male cells.

Table 1. Accuracy of immunological detection of embryonic H-Y antigen in various species[a]

Species	Embryos Correctly Sexed: Male	Female	References
Murine	78%	83%	White et al., 1983
	67%	80%	Piedrahita and Anderson, 1985
Bovine	80%	89%	White et al., 1987a
Ovine	88%	82%	White et al., 1987b
Porcine	77%	86%	White et al., 1987c

[a]From White, 1988.

Sexing by Y Chromosome Specific DNA Hybridization

The third method is based on the isolation, cloning and subsequent labeling of unique and repetitive DNA sequences from the Y chromosome. Most of the unique and transcribed sequences coding for critical male specific substances such as H-Y antigens are located at or near the pseudoautosomal (X pairing region) of the Y whereas repetitive sequences such as GATA-GATA or GACA-GACA tend to be dispersed over the length of the Y[1/2]. These Y specific fragments are used as probes to locate homologous sequences present in DNA from blastomeres, trophoblasts, amniocytes and other cell types (Gosden et al., 1984; Moyzis et al., 1987; West et al., 1987). As few as three to five blastomeres can be biopsied from embryos and using an oligonucleotide polymerase chain reaction for signal amplification, embryonic sex can be determined in from 1 to 3 hr (Grey and Langlois, 1987; Ou et al., 1987). Numerous Y specific probes are currently available for sex selection in humans (Page et al., 1987; Kent and First, 1988), bovine (Leonard et al., 1987; Ellis et al., 1988; Popescu et al., 1988) and equine (Kent et al., 1988a,b).

Multiplication of Embryos

The ability to produce multiple copies of an individual or embryo is of interest not only to researchers, but also to the livestock industry. Genetic identicals provide the perfect control for experimental conditions thus reducing the genetic variation in experiments to zero. A large number of genetically identical embryos provides a means for embryo phenotypic selection wherein clonal lines descendent from one embryo are selected by progeny test for clonal multiplication to large numbers. This system approaches phenotypic selection and could permit rapid change in selected characteristics such as meat or milk production. Two methods of embryo multiplication will be discussed here. They are embryo bisection and nuclear transplantation.

Embryo bisection is a procedure whereby an early embryo (2-cell through the blastocyst stage) is bisected to yield either 2 cells as with a 2-cell embryo, or 2 or more cell masses as with a morula or blastocyst stage embryo. This procedure results in identical offspring in sheep (Willadsen, 1982), pigs (Willadsen, 1982; Rorie et al., 1985) and cattle (Ozil et al., 1982; Baker and Shea, 1985; Leibo, 1988). Since this procedure is successful it can be concluded that these cells are totipotent. However, the number of identicals produced by this method is limited. If the embryo is divided more than twice survival to offspring is reduced. This is likely due to the requirement of a minimum cell number at the time of blastulation. In the mouse this minimum is 8-16 cells. If blastulation occurs with fewer cells, a trophoblastic vesicle will form without an inner cell mass

(Tarkowski and Wroblewolsa, 1967). Therefore, the limit to the number of identicals produced by splitting is maximally four and efficiently two (Robl and First, 1985). This procedure is commonly used in the cattle embryo transfer industry and results in a pregnancy rate nearly equivalent to the whole embryo with the number of offspring nearly doubled (Leibo, 1988).

Nuclear Transfer

The second method for producing multiple copies of an embryo is by nuclear transplantation. A nuclear transplantation procedure has recently been shown successful in producing cloned embryos and offspring in cattle (Prather et al., 1987), sheep (Willadsen, 1986; Smith and Wilmut, 1988) and rabbits (Stice and Robl, 1988). This procedure is a modification of a procedure developed for the frog in 1952 (Briggs and King, 1952).

As shown in Fig. 1 the procedure involves transfer of a blastomere or nucleus from a valuable embryo of a multicellular stage into an enucleated metaphase II oocyte with subsequent development to a multiple cell stage and use as a donor in a serial recloning. This procedure is being developed in private industry as well as by the cited authors. Collectively in the USA and Canada several hundred pregnancies have been produced in cattle and recloning has been performed. Thus far the largest number of calves cloned from one embryo has been seven. These were born at Granada Genetics in 1987.

A system for cloning of embryos useful to the livestock industry depends on the ability to produce offspring from donor embryos of large cell number and the ability to reclone as the clones develop to advanced cell number or to multiply donor cells in culture.

Studies with sheep at Edinburgh, Scotland, suggest this should be possible. The frequency of development to blastocyst after use of donor cells from blastocyst inner cell mass was 56% and pregnancies resulted (Smith and Wilmut, 1988).

In Vitro Production of Embryos

The production of embryos in vitro from abattoir recovered oocytes is best developed for cattle (Lu et al., 1987; Eyestone and First, 1988); although offspring have been produced from in vitro fertilization of in vitro developed oocytes in sheep (Crozet et al., 1987) and swine (Cheng et al., 1986). There are at least three reasons for producing embryos of cattle in vitro. First, this technique provides large numbers of embryos for commercial transfer and calf production. The value of dairy calves is sufficiently low relative to beef calves in Europe and Japan that there are economic incentives for transfer of in vitro produced beef embryos into dairy cow recipients, particularly

Figure 1

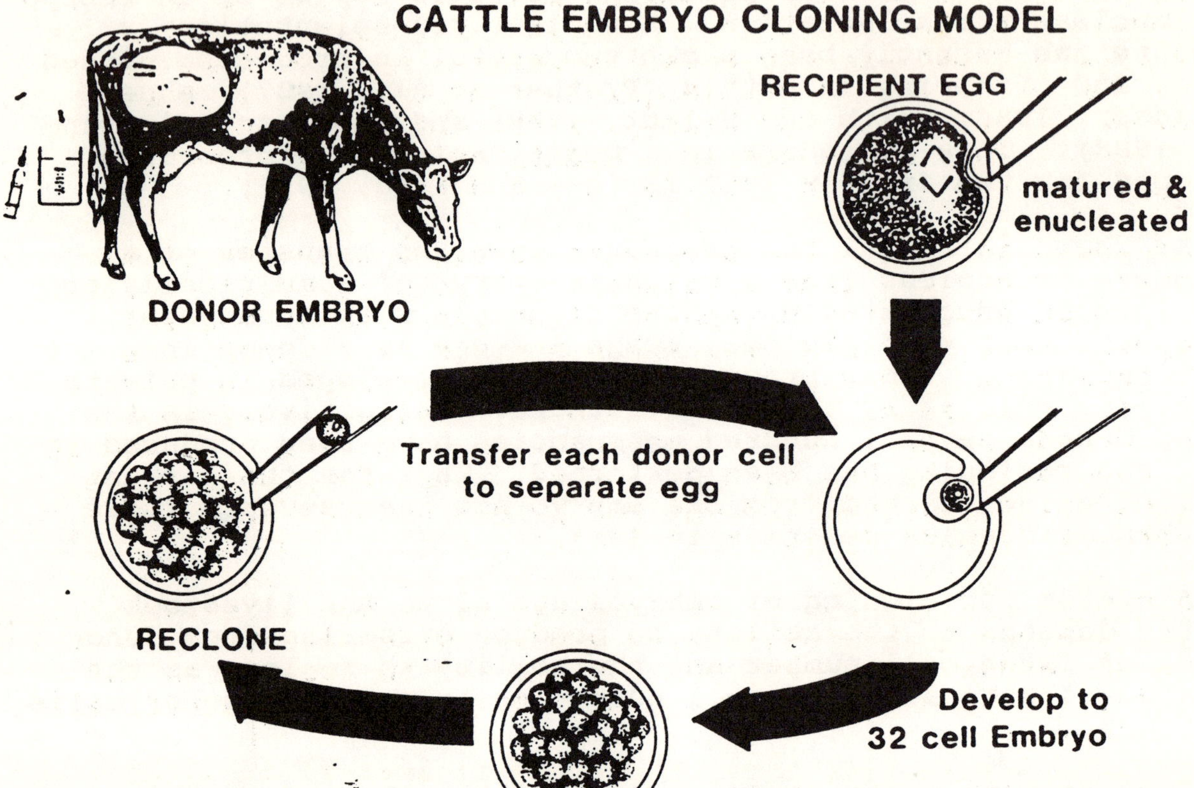
CATTLE EMBRYO CLONING MODEL
RECIPIENT EGG
matured & enucleated
DONOR EMBRYO
Transfer each donor cell to separate egg
RECLONE
Develop to 32 cell Embryo

with the goal of inducing twinning. In Ireland and Japan large commercial ventures have been established for in vitro production of cattle embryos. Second, in vitro produced embryos are highly valuable for research where large numbers or precise timing of fertilization and development are needed. Third, the economic feasibility of embryo cloning by nuclear transfer requires that the enucleated oocytes be produced in vitro from abattoir recovered ovaries and that the new zygote be developed in vitro to a stage suitable for recloning.

In vitro production of embryos requires development of technology in three areas, oocyte development and maturation, in vitro fertilization and in vitro embryo development.

Oocyte Maturation

In domestic species oocytes recovered from follicles matured in vivo either with or without superovulation can be fertilized and proceed through embryo development with good success (cattle: Leibfried-Rutledge et al., 1987; swine: Cheng et al., 1986). However, oocytes recovered from small follicles (1-5 mm) many of which have not completed growth and development produce zygotes which fail to complete embryo development (sheep: Moor and Trounson, 1977; Crosby et al., 1981; cattle: Leibfried-Rutledge et al., 1987). Development is enhanced when the underdeveloped oocytes undergo in vitro maturation in the presence of hormone stimulated granulosa cells (Staigmiller and Moor, 1984; Critser et al., 1986a; Lu et al., 1987) and to a lesser extent with cumulus cells confined in a small volume of medium (Critser et al., 1986b; Sirard et al., 1988). The embryo developmental signals developed during this co-culture or the beneficial material from the granulosa cells are unknown. At present the frequency of bovine blastocysts developing from in vitro fertilization of in vivo matured oocytes is approximately ≥45 (Brackett et al., 1982; Sirard et al., 1985, 1986; Leibfried-Rutledge et al., 1987), from immature oocytes co-cultured with granulosa cells it is approximately 23 to 63% (Critser et al., 1986a; Lu et al., 1987; Xu et al., 1987; Fukui and Ono, 1988), from culture with significant cumulus contribution per volume of medium it is 20 to 30% (Critser et al., 1986b; Goto et al., 1988; Sirard et al., 1988) and from immature oocytes cultured so a "helper cell" effect is lost ≤20% (Critser et al., 1986b and unpublished; Sirard et al., 1988).

In Vitro Fertilization

The second part of production of embryos in vitro is the sperm capacitation and fertilization system. Here numerous capacitation systems have been used including high ionic strength media and glycosaminoglycans such as heparin-sulfate and fucose sulfate, aging, pH shift, calcium ionophores and caffeine (reviewed by First and Parrish, 1987, 1988). In general any agent which causes Ca^{++} entry into the sperm acrosome and causes a pH increase within the sperm causes capacitation (reviewed by First and Parrish, 1988). From this and with incubation in serum-free medium at body temperature for a given species in vitro fertilization has been successful in cattle, sheep, swine and goats (First and Parrish, 1987).

Development of Embryos In Vitro

Embryos of each domestic species can be developed with good efficiency to the blastocyst stage or later by transfer at the 1-cell or 2-cell stage into the oviduct of the respective species. For cattle the embryos can also be successfully developed in the oviduct of the sheep or rabbit (Sirard et al., 1986; Eyestone et al., 1987). Embryos of all domestic species do not develop to morulae or blastocysts when cultured in any of the common culture media (Wright and Bondioli, 1981). Their development is blocked at the transition from maternal to zygotic control of development (Barnes, 1988; First and Barnes, 1988). The embryos remain alive but with cleavage arrested and in the resting phase (Eyestone and First, 1989b).

Recently bovine (Eyestone and First, 1987, 1988) and ovine embryos (Gandolfi and Moor, 1987) have been cultured through the period of blocked development and to the blastocyst stage with good efficiency by co-culture with oviduct epithelial cells or media conditioned by cultured oviduct cells (Eyestone and First, 1988). In the bovine the essential oviduct material is in the protein fraction but its identity is unknown. In the sheep the essential oviduct component is believed to be a protein of 92.5 Kd or its combination with a 46 Kd protein. The 92.5 Kd fucose rich glycoprotein increases greatly in the oviduct just before the block period, translocates to the zona and embryo and disappears by the blastocyst stage (Gandolfi and Moor, 1988).

Whether the bovine oviduct factor is the same is unknown. A protein fraction from trophoblast cells also has similar activity in enhancing embryo development (Heyman et al., 1987). The nature of these embryotrophic compounds needs elucidation as well as their mode of action and the way in which the blocked development relates to initiation of embryonic transcription and the transition from short to long cell cycles (Barnes, 1988; First and Barnes, 1988). In spite of these gaps in our knowledge, supplementation of embryo cultures with frozen oviduct cell conditioned media has provided an in vitro method for development of bovine embryos which has resulted in an approximately normal pregnancy rate (50%) after transfer into cows (Eyestone and First, 1989a).

Gene Transfer

The first production of transgenic mice by Gordon and Ruddle (1981) and the evidence that mice transgenic for rat (Palmiter et al., 1982) or human (Palmiter et al., 1983) growth hormone grew to nearly twice normal size greatly excited animal scientists with hopes of producing transgenic livestock. In the ensuing years more than 400 strains of transgenic mice have been produced for use in studying problems of biology, medicine and animal agriculture.

However, only a few new transgenic lines of domestic animals have been produced and in many cases the expected performance has not been achieved. The slow progress has been largely due to the low efficiency of production of transgenics by microinjection of DNA into pronuclei and to the high economic value of each egg microinjected.

At present transgenic swine have been produced in at least five different laboratories, transgenic sheep in at least three and transgenic cattle embryos, fetuses or offspring in three laboratories (Rexroad and Pursel, 1988; Murray et al., 1988).

A second problem has been failure of expression of the desired response (i.e., growth) or failure of expression at the desired time or in the desired tissue (Rexroad and Pursel, 1988). These problems are being resolved as more is learned about the promoter and enhancer sequences used with the gene of interest. Especially exciting are the possibilities for targeting gene expression exclusively to skeletal muscle for alteration of the meat product (Shani et al., 1987) or to the mammary gland for alteration of the composition of milk or for production of pharmaceutical proteins in milk (Simons et al., 1987).

Conclusions

A large array of gamete and embryo biotechnologies have been developed for use in animal agriculture. Older technologies with high economic value such as artificial insemination, embryo transfer, freezing and splitting of embryos have over time developed to high efficiency. Newer technologies such as embryo sexing, in vitro production of embryos, embryo cloning and gene transfer show promise for commercial use but require research to become more efficient. Rapid development of each technology is enhanced in a given species by availability of gametes and embryos as well as the existence of supporting technology such as nonsurgical embryo transfer and economic incentives.

REFERENCES

Baker, R.D. and B.F. Shea. 1985. Commercial splitting of bovine embryos. Theriogenology 23:3-12.

Barnes, F.L. 1988. Characterization of the onset of embryonic control and early development in the bovine embryo. Ph.D. Thesis, University of Wisconsin-Madison.

Betteridge, K.J. 1986. Increasing productivity in farm animals. In: Reproduction in Mammals: Manipulating Reproduction (Austin and Short, eds.), Cambridge University Press, Cambridge, pp. 1-47.

Brackett, B.G., D. Bousquet, M.L. Boice, W.J. Donawick, J.F. Evans and M.A. Dressel. 1982. Normal development following in vitro fertilization in the cow. Biol. Reprod. 27:147-158.

Briggs, R. and T.J. King. 1952. Transplantation of living nuclei from blastula cells into enucleated frogs' eggs. Zoology 38:455-463.

Chappel, S., C. Looney and K. Bondioli. 1988. Bovine FSH produced by recombinant DNA technology. Theriogenology 29:235a.

Chemineau, P., R. Procureur, Y. Cognie, P.C. LeFeure, A. Lacatelli and D. Chupin. 1986. Production, freezing and transfer of embryos from a blue tongue infected goat herd without blue tongue transmission. Theriogenology 26:279-290.

Cheng, W.T.K., R.M. Moor and C. Polge. 1986. In vitro fertilization of pig and sheep oocytes matured in vitro and in vivo. Theriogenology 25:146.

Chupin, D., B. Florin and R. Procureur. 1984. Comparison of two methods for one-stem in straw thawing and direct transfer of cattle blastocysts. Theriogenology 21:455-459.

Critser, E.S., M.L. Leibfried-Rutledge, W.H. Eyestone, D.L. Northey and N.L. First. 1986a. Acquisition of developmental competence during maturation in vitro. Theriogenology 25:150 (abstract).

Critser, E.S., M.L. Leibfried-Rutledge and N.L. First. 1986b. Influence of cumulus cell association during in vitro maturation of bovine oocytes on embryonic development. Biol. Reprod 34(Suppl 1):192 (abstract).

Crosby, I.M., J.C. Osborn and R.M. Moor. 1981. Follicle cell regulation of protein synthesis and developmental competence in sheep oocytes. J. Reprod. Fertil. 62:575-582.

Crozet, N., D. Huneau, V. Desmedt, M.C. Theron, D. Szollosi, S. Torries and C. Sevellec. 1987. In vitro fertilization of sheep ova. Gamete Res. 16:159-170.

DiBerardino, M.A. 1980. Genetic stability and modulation of metazoan nuclei into eggs and oocytes. Differentiation 17:17-30.

Edidin, M. and V.A. Petit. 1977. The effect of temperature on the lateral diffusion of plasma membrane proteins. In: The Freezing of Mammalian Embryos, Ciba Foundation Symp. 52, Elsevier, Excerpta Medica, North Holland, Amsterdam, pp. 155-164.

Ellis, S.B., K.R. Bondioli, M.E. Williams, J.H. Pryor and M.M. Harpold. 1988. Sex determination of bovine embryos using male-specific DNA probes. Theriogenology 29:242a.

Epstein C.J., S. Smith and B. Travis. 1980. Expression of H-Y antigen on preimplantation mouse embryos. Tissue Antigens 15:63-67.

Eyestone, W.H., J. Vignieri and N.L. First. 1987 Co-culture of early bovine embryos with oviductal epithelium. Theriogenology 27:228.

Eyestone, W.H. and N.L. First. 1988. Co-culture of early bovine embryos with oviductal tissue. Proc. 11th Intern. Congr. AI and Anim. Reprod., Dublin.

Eyestone, W.H. and N.L. First. 1989a. Co-culture of early cattle embryos to the blastocyst stage with oviductal tissue or in conditioned medium. J. Reprod. Fertil. 85:715-720.

Eyestone, W.H. and N.L. First. 1989b. Characterization of development arrest in early bovine embryos cultured in vitro. J. Reprod. Fertil. (submitted).

First, N.L. and J.J. Parrish. 1987. In vitro fertilization of ruminants. Second International Symposium on Reproductive in Domestic Ruminants. J. Reprod. Fertil. 34:151-165.

First, N.L. and W.H. Eyestone. 1987. Reproductive efficiency in domestic animals. Proceedings Vth World Congress of In Vitro Fertilization and Embryo Transfer. Norfolk, VA.

First, N.L. and J.J. Parrish. 1988. Sperm maturation and in vitro fertilization. 11th International Congress on Animal Reproduction and Artificial Insemination, Dublin.

First, N.L. and F.L. Barnes. 1988. Development of preimplantation mammalian embryos. Symposium on Development of Preimplantation Embryos and Their Environment. Kyoto, Japan, July 14-16, 1988.

Fukui, Y. and H. Ono. 1988. In vitro development to blastocyst of in vitro matured and fertilised bovine oocytes. Vet. Rec. 122:282.

Gandolfi, F. and R. M. Moor. 1987. Stimulation of early embryonic development in the sheep by co-culture with oviduct epithelial cells. J. Reprod. Fertil. 81:23-28.

Gandolfi, F. and R. M. Moor. 1988. Interactions between somatic and germinal cells during early development. Proc. 22th Intern. Congr. AI and Anim. Reprod., Dublin.

Gordon, J.W. and F.H. Ruddle. 1981. Integration and stable germline transmission of genes injected into mouse pronuclei. Science 214:1244-1246.

Gosden, J.R., C.M. Gosden, S. Christie, H.J. Cooke, J.M. Morsman and C.H. Roedeck. 1984. The use of cloned Y chromosome-specific DNA probes for fetal sex determination in first trimester prenatal diagnosis. Hum. Gen. 66:348-351.

Goto, K., Y. Kajihara, S. Kosaka, M. Koba, Y. Nakanishi and K. Ogawa. 1988. Pregnancies after co-culture of cumulus cells with bovine embryos derived from in vitro fertilization of in vitro matured follicular oocytes. J. Reprod. Fertil. 83:753-758.

Grey, J.W. and R.G. Langlois. 1986. Chromosome classification and purification using flow cytometry and sorting. Ann. Rev. Biophys. Chem. 15:195-235.

Haseltine, F.P. 1983. Sex differentiation, current concepts. In: The Biological Basis of Reproductive and Developmental Medicine (J.B. Warshaw, ed.), Elsevier Biomedical, Amsterdam.

Heyman, Y., Y. Menezo, P. Chesne, S. Camous and V. Garnier. 1987. In vitro cleavage of bovine and ovine early embryos: improved development using co-culture with trophoblastic vesicles. Theriogenology 27:59-68.

Jones, K.W. and L. Singh. 1982. Conserved sex-associated repeated DNA sequences in vertebrates. In: Dover GA, Flavell RB (eds) Genome Evolution, Academic Press, London.

Kennedy, L.G., M.P. Boland and I. Gordon. 1983. The effect of embryo quality at freezing on subsequent development of thawed cow embryos. Theriogenology 19:823-832.

Kent, M.G. and N.L. First. 1988. Nonantigenic methods of prenatal sex determination. Amer. Fertil. Soc. Workshop, Hands-on IVF, Cryopreservation, and Micromanipulation, April 25-29, 1988, Madison, WI.

Kent, M.G., H. E. Schneller, K.S. Guise, K.O. Elliston, K. Jones and S.S. Wachtel. 1988a. Hybridization of sex specific Bkm DNA in XY sex-reversed females of the horse (in press).

Kent, M.G., K.O. Elliston, W. Schroeder, K. Guise and S. Wachtel. 1988b. Conserved repetitive DNA sequences (Bkm) in normal males and sex reversed famales of the horse detected by in situ hybridization. Cytogenet. Cell Genet. (in press).

King, W.A. 1984. Sexing embryos by cytological methods. Theriogenology 21:7-17.

Koo, G.C. and A. Varano. 1981. Inhibition of H-Y cell mediated cytolysis by monoclonal H-Y specific antibody. Immunogenetics 14:183.

Krco, CJ and E.H. Goldberg. 1976. Detection H-Y (male) antigen on 8-cell mouse embryos. Science 193:1134-1135.

Lehn-Jensen, H., T. Greve and A.D. Navas. 1981. Two-step freezing of cow embryos in 1.4 m glycerol. Theriogenology 15:427-432.

Leibo, S.P. 1982. A one-step method for direct nonsurgical transfer of frozen thawed bovine embryos. In: Intern. Congr. Embryo Transfer in Mammals, France:97.

Leibo, S.P. 1988. Cryopreservation of ova and embryos. Amer. Fertil. Soc. Workshop, Hands-on IVF, Cryopreservation, and Micromanipulation, April 25-29, 1988, Madison, WI.

Leibfried-Rutledge, M.L., E.S. Critser, W.H. Eyestone, D.L. Northey and N.L. First. 1987. Development potential of bovine oocytes matured in vitro or in vivo. Biol. Reprod. 36:376-383.

Leonard, M., M. Kirszenbaum, C. Cotinot, P. Chesne, Y. Heyman, M.G. Stinnakre, C. Bishop, C. Delouis, M. Vaiman and M. Fellous. 1987. Sexing bovine embryos using Y chromosome specific DNA probe. Theriogenology 27:248.

Lu, K.H., I. Gordon, M. Gallagher and H. McGovern. 1987. Pregnancy established in cattle by transfer of embryos derived from in vitro fertilisation of oocytes matured in vitro. Vet. Rec. 121:259-260.

Massey J.M. and A.J. Oden. 1984. No seasonal effect on embryo donor performance in the southwest region of the USA. Theriogenology 21:196-217.

Massip, A.P. and P. van der Zwalmen. 1984. Direct transfer of frozen cow embryos in glycerol-sucrose. Vet. Rec. 115:327-328.

Massip, A.P., P. van der Zwalmen and F. Ectors. 1987. Recent progress in cryopreservation of cattle embryos. Theriogenology 27:69-79.

Moor, R.M. and A.O. Trounson. 1977. Hormonal and follicular factors affecting maturation of sheep oocytes in vitro and their subsequent developmental capacity. J. Reprod. Fertil. 49:101-109.

Moor, R.M., Th.A.M. Kruip and D. Green. 1984. Intraovarian control of folliculogenesis: limits to superovulation? Theriogenology 21:103-116.

Moyzis, R.K., L.S. Albright, L.S. Bartholdi, L.L. Cram, J. Draven, T. Hildebrand, J.L. Lonmire, J. Meyne and T. Schruarzacher-Robinson. 1987. Chromosome-specific DNA sequences: Novel markers for genetic analysis. Chromosoma 95:375-386.

Murray, J.D., C.D. Nancarrow and K.A. Ward. 1988. Techniques for the transfer of foreign genes into animals. Proc. 11th Intern. Congr. Animal Reprod. & AI, Dublin, pp. 19-27.

Ou, C.Y., S.W. Kwok, D.W. Mitchell, D.H. Mack, J.J. Snisky, J.W. Krebs, P. Feorino, D. Warfield and G. Schochetman. 1987. DNA amplification for direct detection of HIV-1 in DNA of peripheral blood mononuclear cells. Science 239:295-298.

Ozil, J.P., Y. Heyman and J.P. Renard. 1982. Production of monozygotic twins by micromanipulation and cervical transfer in the cow. Vet. Rec. 110:126-127.

Page, D.C., R. Mosher, E.M. Simpson, E. Fisher, G. Mardon, J. Pollack, G. McGillivray, A. de la Chapelle and L. Brown. 1987. The sex-determining region of the human Y chromosome encodes a finger protein. Cell 51:1091-1104.

Palmiter, R. D., R. L. Brinster, R. E. Hammer, M. E. Trumbauer, M. G. Rosenfeld, N. C. Birnberg and R. M. Evans. 1982. Dramatic growth of mice that develop from eggs microinjected with metallothionein-growth hormone fusion genes. Nature 300:611-615.

Palmiter, R. D., G. Norstedt, R. E. Gellnas, R. E. Hammer and R. L. Brinster. 1983. Metallothionein-human GH fusion genes stimulate growth of mice. Science 222:809-814.

Pettit, W.H., Jr. 1985. Commercial freezing of bovine embryos in glass ampules. Theriogenology 23:13-16.

Piedrahita, J.A. and G.B. Anderson. 1985. Investigation of sperm cytotoxicity as an indicator of ability of antisera to detect male-specific antigen on preimplantation mouse embryos. J. Reprod. Fertil. 74:637-644.

Pineda, M.H. and R.A. Bowen. 1980. Embryo transfer in domestic animals. In: Veterinary Endocrinology and Reproduction (McDonald, ed.), Lea & Febiger, Philadelphia, pp. 476-502.

Polge, C., I. Wilmut and L.E.A. Rowson. 1974. The low temperature preservation of cow, sheep and pig embryos. Cryobiology 11:560.

Popescu, C.P., C. Cotinot, J. Boscher and M. Kirszenbaum. 1988. Chromosomal localization of a bovine male specific probe. Annales De Genetique 31(1):39-42.

Prather, R.S., F.L. Barnes, M.L. Sims, J.M. Robl, W.H. Eyestone and N.L. First. 1987. Nuclear transplantation in the bovine: assessment of donor nuclei and recipient oocyte. Biol. Reprod. 37:859-866.

Reid, J.T. 1978. In: Agricultural and Food Chemistry: Past, Present, Future (R. Teranishi, ed.), AVI Press, Westport, Connecticut, pp. 120-135.

Renard, J.-P., Y. Heyman and J.-P. Ozil. 1982. Congelation de lembryon bovine: une nouvelle methode decongelation pour le transfert cervical d'embryons conditionnes une seule fois en aillettes. Ann. Med. Vet. 126:23-32.

Renard, J.-P., Y. Heyman, P. Leymanie and J.-C. Plat. 1983. Sucrose dilution: a technique for field transfer of bovine embryos frozen in the straw. Theriogenology 19:145.

Rexroad, C.E. and V.G. Pursel. 1988. Status of gene transfer in domestic animals. Proc. 11th Intern. Congr. Animal Prod. and A.I. Dublin, pp. 28-35.

Robl, J.M. and N.L. First. 1985. Manipulation of gametes and embryos in the pig. J. Reprod. Fertil. 33:101-114.

Rorie, R.W., S.A. Voelkel, C.W. McFarland, L.L. Southern and R.A. Godke. 1985. Micromanipulation of day-6 porcine embryos to produce split-embryo piglets. Theriogenology 23:225.

Seidel, G.E., Jr. 1981. Superovulation and embryo transfer in cattle. Science 211:351-358.

Shani, M., D. Shinar, P. Einat, D. Yaffe and Y. Dekel. 1987. EMBO Workshop: Germline Manipulation of Animals, Scotland, abstract.

Simons, J.P., I. Wilmut, A.J. Clark, A.L. Archibald, J.O. Bishop and R. Lathe. 1987. EMBO Workshop: Germline Manipulation of Animals, Scotland, abstract.

Sims, M.M. and N.L. First. 1987. Nonsurgical embryo transfer in swine. Amer. Soc. Anim. Sci., Logan, Utah (abstract).

Sirard, M.A., R.D. Lambert, D.P. Menard and M. Bedoya. 1985. Pregnancies after in vitro fertilization of cow follicular oocytes, their incubation in rabbit oviduct and transfer to the cow uterus. J. Reprod. Fertil. 75:551-556.

Sirard, M.A., R.D. Lambert, D.P. Menard and M. Bedoya. 1986. Pregnancies after in vitro fertilization of cow follicular oocytes, their incubation in rabbit oviduct and their transfer to the cow uterus. J. Reprod. Fertil. 75:551-56.

Sirard, M.A., J.J. Parrish, C.B. Ware, M.L. Leibfried-Rutledge and N.L. First. 1988. The culture of bovine oocytes to obtain developmentally competent embryos. Biol. Reprod. 39:546-552.

Slade, N.P., T. Takeda, E.L. Squires and R.P. Elsden. 1984. Development and viability of frozen thawed equine embryos. Theriogenology 21:263.

Slade, N.P., T. Takeda, E.L. Squires, R.P. Elsden and G.E. Seidel, Jr. 1985. A new procedure for the cryopreservation of equine embryos. Theriogenology 24:45-58.

Smith, L.C. and T. Wilmut. 1988. Factors influencing nuclear transplantation in sheep embryos. J. Reprod. Fertil 1:10 (abstract).

Staigmiller, R.B. and R.M. Moor. 1984. Effect of follicle cells on the maturation and developmental competence of ovine oocytes matured outside the follicle. Gam. Res. 9:221-229.

Stice, S.L. and J.M. Robl. 1988. Nuclear reprogramming in nuclear transplant rabbit embryos. Biol. Reprod. 36:657-664.

Takeda, T., R.P. Elsden and E.L. Squires. 1984. In vitro and in vivo development of frozen-thawed equine embryos. In: 10th Intern. Congr. on Anim. Reprod. and A.I., Urbana-Champaign, IL:246.

Tarkowski, A.K. and J. Wroblewolsa. 1967. Development of blastomeres of mouse eggs isolated at the 4- and 8-cell stage. J. Embryol. exp. Morph. 36:155-180.

Wachtel, S.S. 1984a. H-Y Antigen and the Biology of Sex Determination. Grune Stratton, NY.

Wachtel, S.S. 1984b. H-Y antigen in the study of sex determination and control of sex ratio. Theriogenology 21:18-28.

Wachtel, S.S., M.G. Kent, D. Nakamura. 1988. Embryonic sex selection in bovine. J. Reprod. Fertil. (in press).

Ware, C.B. and M.P. Boland. 1987. Effect of varying glycerol and sucrose concentration combinations on embryo survival rate in a one step cryoprotectant removal from frozen-thawed ovine embryos. Theriogenology 27:721-728.

Ware, C.B., D.L. Northey, M.P. Boland and N.L. First. 1988. Early cycle FSH-p priming as a prelude to superovulating gonadotropin administration in ewes and heifers. Anim. Reprod. Sci. 16:97-105.

West, J.D., R.R. Angell, S.S. Thatcher, J.R. Gasden, N.D. Hartie and A.F. Glasier. 1987. Sexing the human pre-embryo by DNA-DNA in situ hybridization. Lancet 1:1345-1347.

White, K.L., G.M. Lindner, G.B. Anderson and R.H. Bon Durant. 1982. Survival after transfer of sexed mouse embryos exposed to H-Y antisera. Theriogenology 18:655-662.

White, K.L., G.M. Lindner, G.B. Anderson and R.H. Bon Durant. 1983. Cytolytic and fluorescent detection of H-Y antigen on preimplantation mouse embryos. Theriogenology 19:701.

White, K.L., M.W. Bradbury, G.B. Anderson and R.H. Bon Durant. 1984. Immunofluorescent detection of a male specific factor on preimplantation bovine embryos. Theriogenology 21:275.

White, K.L., G.B. Anderson and R.H. BonDurant. 1987a. Expression of a male-specific factor on various stages of preimplantation bovine embryos. Biol. Reprod. 37:867-873.

White, K.L., G.B. Anderson, R.L. Pashen and R.H. BonDurant. 1987b. Detection of histocompatibility-Y antigen: identification of sex of pre-implantation ovine embryos. J. Reprod. Immunol. 10:27-32.

White, K.L., G.B. Anderson, T.J. Berger, R.H. BonDurant and R.L. Pashen. 1987c. Identification of a male-specific histocompatibility protein on preimplantation porcine embryos. Gam. Res. 17:107-113.

White, K.L. 1988. Identification of embryonic sex by detection of male specific antigens. Amer. Fertil. Soc. Workshop, Hands-on IVF, Cryopreservation, and Micromanipulation, April 25-29, 1988, Madison, WI.

Willadsen, S.M. 1982. Micromanipulation of embryos of the large domestic species. In: Mammalian Egg Transfer (C.E. Adams, ed.) CRC Press, Boca Raton, Florida, pp. 185-210.

Willadsen, S.M. 1986. Nuclear transplantation in sheep embryos. Nature 320:63-65.

Wright, R.W. and K.R. Bondioli. 1981. Aspects of in vitro fertilization and embryo culture in domestic animals. J. Anim. Sci. 53:702-729.

Xu, K.P., T. Greve, H. Calleson and P. Hyttel. 1987. Pregnancy resulting from cattle oocytes matured and fertilized in vitro. J. Reprod. Fertil. 81:501-504.

Ovarian Stimulation in Non-Human Primates

Robert F. Williams, Ph.D.

INTRODUCTION

This article will review methods of ovarian stimulation in nonhuman primates. Since the majority of reproductive biology research with non-human primates is undertaken with macaques (_M._ _mulatta_ - rhesus and _M._ _fasicularis_ - cynomolgus), the focus of this article will be on the normal and stimulated folliculogenic process in the genus _Macaca_. Presumably, mechanisms regulating folliculogenesis, ovulation, luteal function and the menstrual cycle are similar in other "old world" primates. In contrast, "new world" primates have reproductive cycles shorter than those of the old world species (1). For reproductive biology, the most extensively utilized "new world" species are _Saimiri sciureus_ (squirrel monkeys) and _Callithrix jacchus_ (marmosets). Though the following explanation of the menstrual cycle does not apply to new world primates, the problems associated with ovarian stimulation are applicable.

The Menstrual Cycle

The menstrual cycle of macaques is approximately four weeks in length, being about equally divided into follicular and luteal phases. During the second week of the follicular phase, a single follicle becomes grossly evident on one of the two ovaries; it continues maturation until mid-cycle, when ovulation and transformation into a corpus luteum occurs. The major endocrine events (see Figure 1) of the menstrual cycle are a geometric increase in peripheral serum concentrations of estradiol, emanating from the maturing follicle, midcycle surges of LH and FSH, induced by the rising estradiol levels, and two weeks of progesterone secretion from the corpus luteum, a differentiated structure derived from the ovulatory follicle. With diminished function of the corpus lutem, progesterone levels decline and menstruation occurs. Additionally, a significant increase in basal FSH concentrations occurs in the last week of one cycle and continues through the first week of the follicular phase of the next cycle. Some reports have attributed the initiation of follicle growth for the next cycle to this rise in FSH concentrations. (2).

The minimum interval for folliculogenesis to occur in macaques is two weeks. If the periovulatory follicle is lyzed (3), the corpus luteum removed (3), or the menstrual cycle suppressed by progesterone (4), the interval for new follicle growth is approximately two weeks in duration, equal in length to the follicular phase. However, it remains unknown from what follicular stage follicle growth is initiated at the beginning of this two week interval - primordial, primary, secondary, or tertiary (antral) follicles. Since an increase in medium-sized antral follicles (0.5-1.0mm) occurs at the transition from one cycle to another, it is presumed that the ovulatory follicle develops from one of these medium-sized antral follicles. The two week follicular phase has been divided into the intervals of recruitment, selection and dominance. Recruitment is the process whereby a cohort of follicles begins to mature, leading to a follicle capable of ovulation. The size of the cohort during this first week of the menstrual cycle remains unknown. During the process of selection (approximately day 5-7 of the menstrual cycle) the single follicle, which may lead to ovulation, becomes biochemically distinct from all other follicles in both ovaries. Finally, the process of dominance is the mechanism by which the selected follicle and its successor, the corpus luteum, dictates the course of events in the hypothalamus, the pituitary and ovaries. The expression of dominance by the preovulatory follicle has been hypothesized to be achieved by the active secretion of an inhibitor of the development of other follicles (5), or, coincident with a diminished need for FSH support by the dominant follicle (2), the inhibition of FSH levels by estradiol secretion from this follicle, thus diminishing available FSH for follicles requiring greater support than the dominant follicle. Irrespective of the mechanism, it is evident that the dominant follicle precludes the development of other follicles during the second week of the follicular phase. Thus, by this second week, an asymmetrical ovarian function occurs in macaques; one ovary has the primary gametogenic role during the interval of the dominant follicle and the other ovary is functioning as an endocrine organ (5). In fact, differential negative feedback on FSH secretion occurs between the two ovaries, with the ovary without the dominant follicle having a specific inhibition of FSH secretion. These concepts of recruitment, selection and dominance must be considered in planning treatments for the induction of folliculogenesis. Such treatments must be initiated prior to the expression of dominance by the preovulatory follicle.

Regulation of Folliculogenesis

Both LH and FSH are required for successful folliculogenesis. Numerous studies of episodic secretion of LH secretion have demonstrated the association of hourly pulses of LH with successful folliculogenesis culminating in ovulation and the absence of such secretory patterns with anovulation (6). Administration of GnRH and resultant induction of such patterns of LH secretion in monkeys with either arcuate nucleus lesions (7) or juvenile status (8) results in the induction of ovulatory menstrual cycles. Thus, the episodic secretion of LH appears to be an integral component of the processes regulating folliculogenesis. The patterns of FSH secretion required for folliculogenesis are less clear. Because of the longer circulating half-life of FSH, spontaneous or GnRH induced episodic patterns of FSH secretion are not as evident as those of LH. The late luteal/early follicular phase increase in basal FSH concentrations has been proposed as the determinant of folliculogenesis for the next ovulation (2). Though FSH secretion during this interval is critical for normal folliculogenesis culminating in ovulation and normal luteal function, the specific increase in FSH which occurs during this transition from one cycle to another is not essential. Such a rise does not occur prior to the first ovulatory cycle during sexual maturation, the first ovulatory cycle postpartum, nor during the follicular phase following repeated follicular destructions by cautery (9). Thus, this elevation of FSH secretion is not essential for normal folliculogenesis culminating in ovulation. However, some degree of FSH secretion during this interval is critical for normal folliculogenesis. Both inadequate and short luteal phases have been described as spontaneously occurring in macaques (10). Presumably, these luteal dysfunctions reflect aberrant folliculogenesis (11). In monkeys having these spontaneous luteal defects, FSH concentrations in the early follicular phase are below the normal range. Monkeys treated during the first week of the menstrual cycle with porcine follicular fluid had a transient inhibition of FSH concentrations; ovulation was followed by aberrant luteal function, presumably induced by the decreased FSH concentrations in the early follicular phase (12). Thus, during the interval of normally elevated FSH secretion in the transition from one cycle to the next, FSH plays a significant role in the maturation of the developing follicle, but the elevation in FSH concentrations is not essential.

Ovarian Stimulation

Any treatment regimen designed to induce multiple follicular development must override a physiologic system operating to permit only a single follicle to mature to ovulation. Thus, treatment must be initiated prior to the emergence of the dominant follicle. Additionally, the stimulation must not only promote follicular development but must also provide for normal development of granulosa and thecal cells so that luteal function is normal.

Four gonadotropic preparations in combination with human chorionic gonadotropin are each successful in inducing multiple follicular development: human menopausal gonadotropin (see Figure 2; 13), human follicle stimulating hormone (14), pregnant mare's serum gonadotropin (15), and porcine FSH (16). Treatment with these gonadotropin preparations in combination with hCG results in multiple follicular development and resumption of meiosis by the oocyte. Oocytes may be aspirated from the follicles prior to 36 hours post hCG or flushed from the oviduct after 36 hours post hCG treatment. Oocyte maturation is variable and will range from immature (prophase I) through mature (metaphase II). The hCG is typically administered after serum estradiol concentrations have achieved at least 600 pg/ml. Though follicular development and oocyte maturation occurs with these treatment regimens, luteal function is not the same as in normal menstrual cycles. Because of the multiple follicular development and correspondingly increased amount of luteal tissue following hCG treatment, progesterone concentrations are markedly increased in the luteal phase of stimulated cycles relative to the spontaneous cycle (13). However, despite increased progesterone concentrations, the function of the luteal tissue is diminished relative to normal tissue. Following gonadotropin-induced ovarian stimulation, LH binding sites and LH-binding site coupling to the adenylate cyclase system are reduced (17).

Typically, though not uniformly, LH surges do not occur in gonadotropin-stimulated cycles, despite serum concentrations of estradiol normally associated with the induction of LH/FSH surges. Apparently, inhibitory products of the stimulated follicles block the positive-feedback action of estradiol on LH/FSH secretion (18). The ovarian factor which blocks the positive feedback action of estradiol has been referred to as gonadotropin-surge-inhibiting factor (GnSIF); whether GnSIF activity is due to a compound distinct from inhibin remains to be determined (19).

A proportion of monkeys treated with exogenous gonadotropins do have a spontaneous LH surge in response to the rising elevation of estradiol. Since the increasing estrogen concentrations are due to the collective production of estrogens, by a multitude of follicles, the degree of follicular maturation which has normally occurred when positive-estrogen feedback occurs in the spontaneous cycles has not yet occurred in monkey with stimulated folliculogenesis. Thus, when a spontaneous gonadotropin surge occurs in a gonadotropin stimulated monkey, the surge is not synchronized to follicular maturation, resulting in premature luteinization of the granulosa with attendant increased progesterone production (13). The precocious secretion of progesterone precludes normal follicular development and oocyte maturation.

Even among those subjects which do not have a premature LH surge, there is a heterogeneous follicular response to gonadotropic stimulation. Typically, based on estrogen response to gonadotropic stimulation, the females have retrospectively been classified as "low", "medium" and "high" responders to gonadotropic stimulation (20). No method has been developed for predicting the response of a given subject to gonadotropin treatment.

The latter two problems (i.e., premature LH surge and heterogeneous response to gonadotropins) can be ameliorated by adjunctive therapy with a gonadotropin releasing hormone (GnRH) analog for the inhibition of endogenous gonadotropin secretion (20). Agonists of GnRH, though initially promoting pituitary gonadotropin secretion will induce "down-regulation" of pituitary function and inhibition of gonadotropin release, when administered at a sufficient frequency and duration. Without the initial stimulatory phase, GnRH antagonists will immediately inhibit gonadotropin secretion. Simultaneous treatment with GnRH antagonist and exogenous gonadotropin will prevent the premature LH surge and will increase the response to exogenous gonadotropin. The estradiol response of "medium responders" has been observed to be greater with adjunctive antagonist treatment than without such treatment.

Another sequela of ovarian stimulation with exogenous gonadotropins is the development of hyperprolactinemia during the luteal phase of the stimulated cycle. Though not present in all stimulated cycles, this transient hyperprolactinemia is prevalent when estrogen concentrations have been profoundly elevated during the follicular phase and continue elevated into the luteal phase (13). Though estrogen dependent, the hyperprolactinemia is not induced by the estrogen milieu, but rather by the synergistic interaction of estrogen and progestin (21). Apparently at least one week of estradiol concentrations in excess of 250 pg/ml is required for priming of the pituitary, followed by sustained estradiol concentrations coincident with progesterone levels at physiologic concentrations between 2 and 4 ng/ml. In this model, hyperprolactinemia develops coincident with the progesterone elevation; estradiol or progesterone alone do not induce this hyperprolactinemia. In the stimulated menstrual cycles, these conditional patterns of estradiol and progesterone will occasionally occur with attendant hyperprolactinemia.

Problems with Stimulated Ovarian Function

As described above, the response to exogenous gonadotropin therapy may be heterogeneous, with varying degrees of follicular development and variable oocyte maturation. Part of this heterogeneity is due to the endocrine milieu of the subject and/or the occurrence of a premature LH surge. Additional variability may be introduced by the human chorionic gonadotropin therapy administered to induce final follicular development and oocyte maturation. HCG has a long half-life relative to LH. Following the injection of hCG in monkeys, if follicles are not aspirated, but are permitted to ovulate, final follicular maturation continues for days and ovulations will occur from 48 to 120 hours after hCG treatment (22). This response is probable due to hCG acting on a heterogenous population of follicles, with the least mature follicles at the time of hCG treatment being those which ovulate last.

Despite the problems with inducing follicular and oocyte maturation and management of the endocrine milieu, the greatest difficulty limiting research in primate folliculogenesis and oogenesis is the availability of research subjects. Non-human primates are of a much more limited supply than laboratory or domestic species. This problem could be overcome by reutilization of research subjects for ovarian stimulation protocols; however, the gonadotropin preparations utilized at this time are of human or non-primate origin, thus the treated individuals develop neutralizing antibodies to the gonadotropin preparations (23, 24), precluding reutilization of the subject with the same gonadotropin preparation. Even if the immune response to gonadotropins did not occur, or could be blocked, ethical considerations of how many times should a non-human primate be utilized in research protocols and how many surgical procedures should be performed would become limiting factors. Currently, oocytes must be recovered from non-human primates by either laparotomy or laparoscopy; ethical considerations encourage restriction of the performance of multiple surgical procedures on animals utilized in research. Thus, the limited availability of non-human primates, the unavailability of non-human primate gonadotropins, and the limited reutilization of research subjects are greater problems in the investigation of folliculogenesis and oogenesis than are the biologic problems associated with inducing follicular growth and oocyte maturation.

Solutions for Problems with Stimulated Ovarian Function

The variable ovarian response to gonadotropic stimulation may be understood and overcome by additional research into the basic mechanisms regulating folliculogenesis in non-human primates. Numerous questions remain unanswered relative to primate folliculogenesis:

a) What is the time course for development of a primordial follicle into a periovulatory follicle?

b) What stages of folliculogenesis are gonadotropin dependent? What are the specific roles of LH and FSH?

c) Is folliculogenesis a continuous process, or can it be arrested without atresia occurring?

d) During recruitment, what is the size of the cohort of follicles from which the dominant follicle is selected?

e) By what mechanism does "selection" of the dominant follicle occur and how is dominance of the periovulatory follicle sustained?

f) When ovarian function is stimulated by exogenous gonadotropins, what pool of follicles is responsive to this treatment?

g) How are oocyte and follicular maturation interlinked?

h) Can mechanisms of folliculogenesis be extrapolated from nonprimate species to primates? Estrogen receptors have been demonstrated in the granulosa cells of rats (25); *in vivo*, rat granulosa cell proliferate following DES treatment (26). In macaques, estradiol receptors have not been demonstrated in granulosa cells (27), and monkeys do not exhibit a granulosa cell proliferation when treated with DES (28). Thus, are mechanisms of folliculogenesis different in primates then in non-primate species?

Determination of answers to these questions will significantly improve our understanding of folliculogenesis and should allow for greater control of the follicular response to stimulation regimens with exogenous gonadotropins and more homogenous follicular and oocyte responses. Additionally, basic investigations into oocyte maturation should provide critical insights needed for the development of methods for the *in vitro* maturation of oocytes. Thus, when a heterogeneous distribution of oocytes is obtained, the immature oocytes can be matured *in vitro* to a stage of fertilizability.

The interlinked problems of no non-human primate gonadotropins for ovarian stimulation protocols, immune responses to available gonadotropins, and limited supply of non-human primates present a scenario which is difficult to resolve. There will never be sufficient quantities of non-human primate pituitaries or urine to allow for the purification of gonadotropins by convential technologies. Cell lines are being prepared by recombinant DNA technology to produce non-human primate gonadotropins. However, it would seem highly unlikely that quantities of gonadotropin available or cost of production would permit the formulation of preparations for routine ovarian stimulation. An alternative to the use of exogenous gonadotropins, is the augmentation of endogenous gonadotropin secretion. This can currently be accomplished with the pulsatile administration of GnRH or clomiphene citrate treatment, but the degree of ovarian stimulation is not as extensive as occurs with exogenous gonadotropins, and the pulsatile administration of GnRH is problematic, and thus not practical for routine ovarian stimulation. Possibly, activin (29), a member of the inhibin family which induces FSH secretion, may be useful for ovarian stimulation. In the future, when its actions are understood and it is available, it or an agonistic analog may be useful in promoting ovarian stimulation. Of course, if the structure of activin is heterogeneous between species, then immune response of recipient animals will be a problem. Evolving national norms will dictate the appropriateness of reutilization of research subjects in multiple stimulation protocols. Certainly, the need for surgical recovery of oocytes can be eliminated in the future by the continual improvements in the resolution of sonography equipment. When resolution has been significantly improved, oocytes can be recovered non-surgically by sonographically guided procedures. Then the ethical consideration will not be the performance of multiple surgical procedures, but rather the reutilization of nonhuman primates in multiple ovarian stimulation protocols.

In summary, relative to the utilization of non-human primates in research on medically assisted conception, the major difficulties associated with the stimulation of folliculogenesis are not biological problems associated with the function of the hypothalamic-pituitary-ovarian axis, but rather are managerial and logistical problems caused by the limited availability of non-human primates, the unavailability of non-human primate gonadotropins for ovarian stimulation, and the ethical considerations of research subject reutilization. Thus, the majority of basic science research on medically assisted conception will need to be undertaken in laboratory or domesticated species and the non-human primate utilized for only selected studies.

ACKNOWLEDGEMENT

The assistance of Martha Forrester, Dara Leary, and Patricia Reese in the preparation of graphics and the manuscript is sincerely appreciated.

Schematic representation of the endocrine profiles for a normal menstrual cycle in macaques.

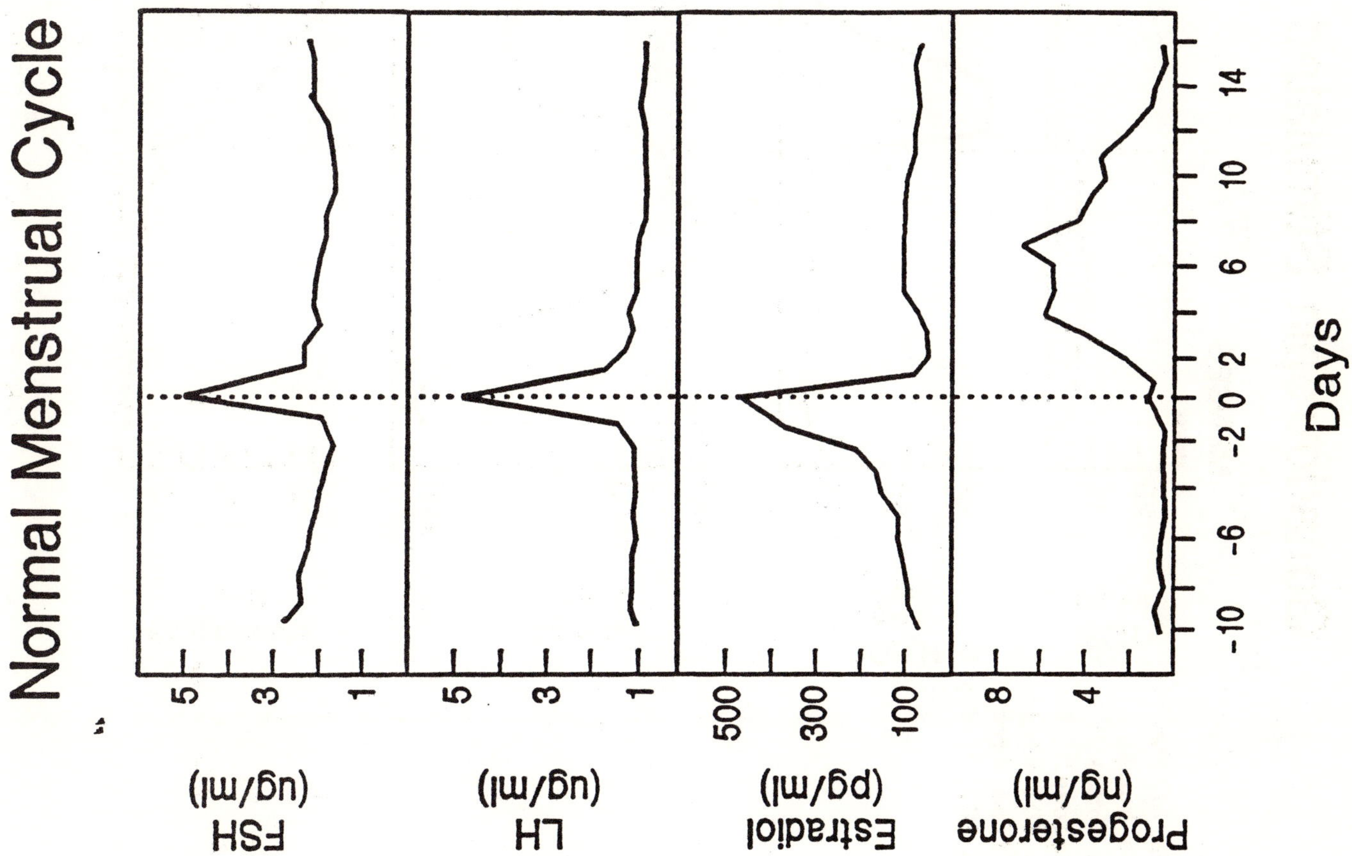

Figure 1

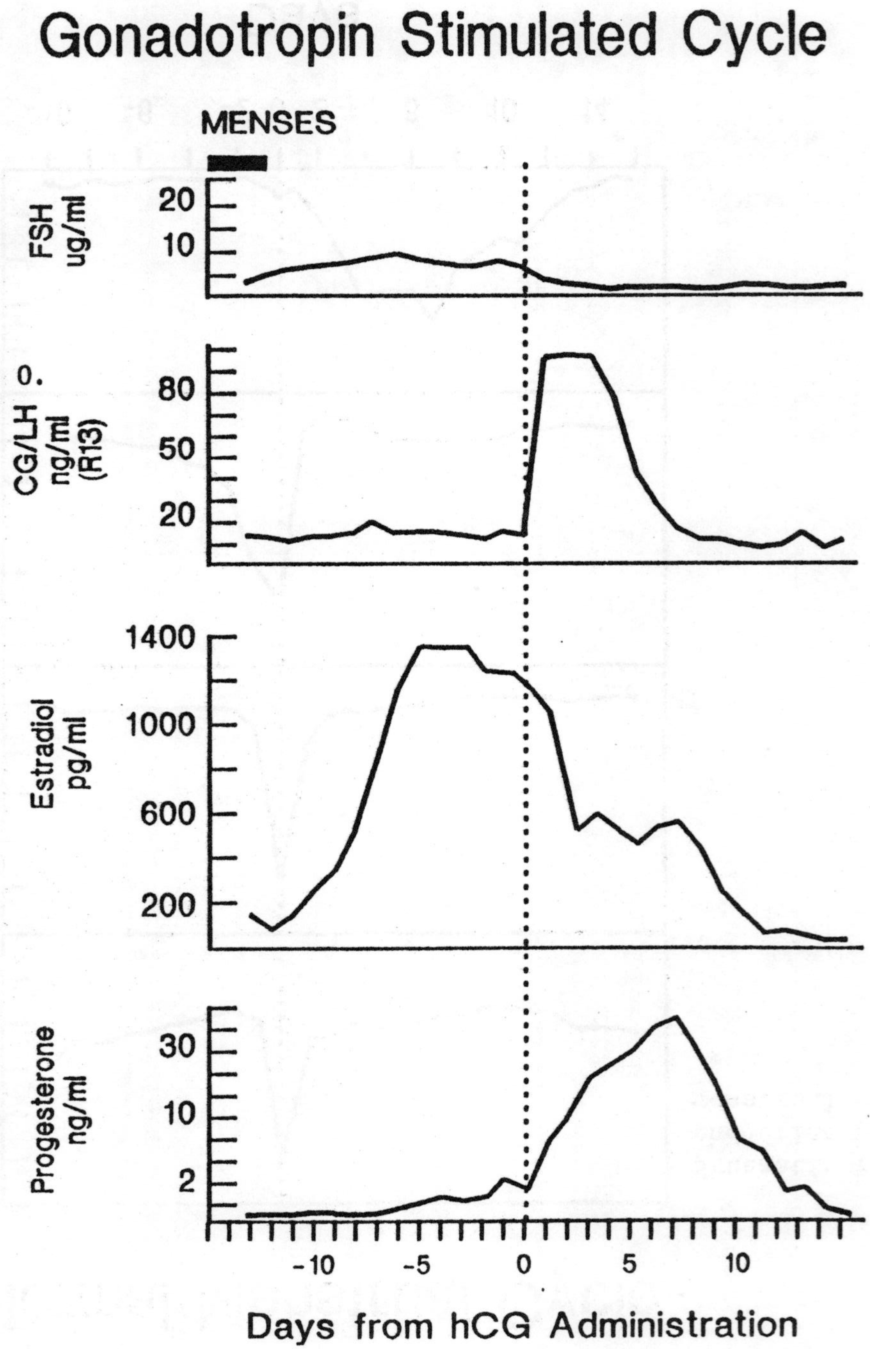

Schematic representation of the endocrine profiles for a human menopausal gonadotropin (hMG) stimulated cycle in macaques. In such a cycle, hMG (37.5 IU) would have been administered from days -10 to -1 and hCG (1000 IU) on day 0.

Figure 2

REFERENCES

1. Koering MJ (1986) Ovarian architecture during follicle maturation. In: Dukelow, WR and J Erwin (eds) Reproduction and Development Volume 3 of Comparative Primate Biology, Alan R. Liss, Inc, New York, p 215.

2. Zeleznik AJ and SG Hillier (1984) The role of gonadotropins in the selection of the preovulatory follicle. Clinical Obstetrics and Gynecology 27:927.

3. Goodman AL, WE Nixon, DK Johnson, and GD Hodgen (1977) Regulation of folliculogenesis in the cycling rhesus monkey: Selection of the dominant follicle. Endocrinology 100:155.

4 Goodman AL and GD Hodgen (1977) Systemic versus intraovarian progesterone replacement after luteectomy in rhesus monkeys: Differential patterns of gonadotropins and follicle growth. J Clin Endocrinol Metab 45:837.

5. Goodman AL and GD Hodgen (1983) The ovarian triad of the primate menstrual cycle. Recent Prog Horm Res 39:1.

6. Crowley WF, M Filicori, D Spratt, and NF Santoro (1985) The physiology of gonadotropin-releasing hormone (GnRH) secretion in men and women. Recent Prog Horm Res 41:473.

7. Knobil E, TM Plant, L Wildt, PE Belchetz, and G Marshall (1980) Neuroendocrine control of the rhesus monkey menstrual cycle: Permissive role of the hypothalamic gonadotropin-releasing hormone. Science 207:

8. Wildt L, G Marshall, and E Knobil (1980) Experimental induction of puberty in the infantile female rhesus monkey. Science 207:1373.

9. diZerega GS, WE Nixon, and GD Hodgen (1980) Intercycle serum follicle-stimulating hormone elevations: Significance in recruitment and selection of the dominant follicle and assessment of corpus luteum normalcy. J Clin Endocrinol Metab 50:1046.

10. Williams RF and GD Hodgen (1982) The reproductive cycle in female macaques. Amer J Primatol Suppl 1:181.

11. diZerega, GS and GD Hodgen (1981) Luteal phase dysfunction infertility: A sequel to aberrant folliculogenesis. Fertil Steril 35:489.

12. Stouffer RL and GD Hodgen (1980) Induction of luteal phase defects in rhesus monkeys by follicular fluid administration at the onset of the menstrual cycle. J Clin Endocrinol Metab 51:669.

13. Collins RL, RF Williams, and GD Hodgen (1984) Endocrine consequences of prolonged ovarian hyperstimulation: Hyperprolactinemia, follicular atresia, and premature luteinization. Fertil Steril 42:436.

14. Schenken RS, RF Williams, and GD Hodgen (1984) Ovulation induction using "pure" follicle-stimulating hormone in monkeys. Fertil Steril 41:629.

15. Bavister, BD, BE Boatman, K Collins, DJ Dierschke, and SG Eisele (1984) Birth of rhesus monkey infant after in vitro fertilization and nonsurgical embryo transfer. Proc Natl Acad Sci USA 81:2218.

16. Yano J and KG Gould (1985) Induction of follicular growth in the squirrel monkey (*Saimiri sciureus*) with human urinary follicle stimulating hormone (Metrodin). Fertil Steril 43:799.

17. Stouffer RL, GD Hodgen, PE Graves, DR Danforth, KM Eyster, and JS Ottobre (1986) Characterization of corpora lutea in monkeys after superovulation with human menopausal gonadotropin or follicle stimulating hormone. J Clin Endocrinol Metab 62:833.

18. Schenken RS and GD Hodgen (1983) Follicle-stimulating hormone induced ovarian hyperstimulation in monkeys: Blockade of the luteinizing hormone surge. J Clin Endocrinol Metab 57:50.

19. Danforth DR, MJ Sinosich, TL Anderson, CY Cheng, CW Bardin, and GD Hodgen (1987) Identification of gonadotropin surge-inhibiting factor (GnSIF) in follicular fluid and its differentiation from inhibin. Biol Reprod 37:1075.

20. Hodgen GD, D Kenigsberg, RL Collins, and RS Schenken (1985) Selection of the dominant follicle and hormonal enhancement of the natural cycle. Ann NY Aca Sci 442:23.

21. Williams, RF, JG Gianfortoni, and GD Hodgen (1985) Hyperprolactinemia induced by an estrogen-progesterone synergy: Quantitative and temporal effects of estrogen priming in monkeys. J Clin Endocrinol Metab 60:126.

22. Abbasi R, D Kenigsberg, D Danforth, RJ Falk, and GD Hodgen (1987) Cumulative ovulation rate in human menopausal/ human chorionic gonadotropin-treated monkeys: "Step-up" versus "Step-down" dose regimens. Fertil Steril 47:1019.

23. Platia MP, G Bloomquist, RF Williams, and GD Hodgen (1984) Refractoriness to gonadotropin therapy: How to distinguish ovarian failure versus pseudoovarian resistance caused by neutralizing antibodies. Fertil Steril 42:779.

24. Bavister BD, C Dees, and RD Schultz (1986) Refractoriness of rhesus monkeys to repeated ovarian stimulation by exogenous gonadotropins is caused by nonprecipitating antibodies. Am J Reprod Immun Microbiol 11:11.

25. Richards JS, JJ Ireland, MC Rao, GA Bernath, AR Midgley, Jr., and LE Reichert, Jr. (1986) Ovarian follicular development in the rat: Hormone receptor regulation by estradiol, follicle stimulating hormone and luteinizing hormone. Endocrinology 99:1562.

26. Goldenberg RL, JL Vaitukaitis, and GT Ross (1972) Estrogen and follicle stimulating hormone interactions on follicle growth in rats. Endocrinology 90:1492.

27. Hild-Petito S, RL Stouffer, NB West, and RM Brenner (1988) Localization of progesterone, but not estradiol receptors in follicular and luteal tissues during the menstrual cycle. 70th Annual Meeting of the Endocrine Society, Abstract #1121.

28. Koering MJ (1987) Follicle maturation and atresia: Morphological correlates. In: Stouffer RL (ed) The Primate Ovary, Plenum Press, New York., P.3.

29. Ling NP (1988) Isolation and characterization of gonadal polypeptides that regulate the secretion of follicle stimulating hormone. In: Hodgen GD, Z Rosenwaks, and JM Spieler, (eds) Nonsteroidal Gonadal Factors: Physiological Roles and Possibilities in Contraceptive Development, Jones Institute Press, Norfolk, USA.

RESPONSES TO GONADOTROPINS IN DOMESTIC ANIMALS

J.E. Fortune

INTRODUCTION

Superovulation and embryo transfer in domestic animals and human in vitro fertilization (IVF) programs have in common the use of large doses of gonadotropins to "persuade" additional follicles to grow to preovulatory size and to produce fertilizable oocytes that will develop into normal young. It has been known since the 1940's that both pituitary gonadotropins, LH and FSH, interact with the developing preovulatory follicle and are essential to its growth and differentiation (Fevold, 1941; Greep et al., 1942; Fraenkel-Conrat et al., 1943). Hence, protocols used for superovulation of animals and hyperstimulation of women in IVF programs are based on the idea that abnormally high doses of gonadotropins will overcome the normal restraints imposed by each species on the number of ovulations, and will recruit and select "extra" follicles for ovulation. In the first part of this manuscript I will review the effects of ovarian hyperstimulation with exogenous gonadotropins (or "superovulation") in cattle, sheep and goats, horses, and pigs. In the second part I will make suggestions for future directions for research with animal models, directions that may generate new information to improve protocols for superovulation in both domestic animals and women.

Superovulation In Domestic Animals

Much of the impetus for the use of gonadotropins to stimulate superovulation in domestic animals comes from the desire to exploit the genetic potential of superior females. Artificial insemination, especially in the cattle industry, has made possible a very wide dissemination of the genes of superior males. It is estimated that popular bulls may sire around 50,000 calves per year (Seidel, 1981). Since the use of frozen semen allows reproduction by males even after death, the genes of a highly desirable male can be disseminated widely through a population. However, the genetic improvement of species or breeds is still limited by the small number of offspring that an individual female can produce in her lifetime. Repeated superovulation of superior females and the collection and transfer of their embryos to less valuable recipients greatly enhances the reproductive potential of these females and makes it possible to disseminate their genes more widely than could be achieved through natural pregnancies and more directly than through "natural" male offspring. Among the domestic animals, superovulation and embryo transfer have had the greatest impact on the cattle industry. As techniques for ovarian stimulation and for non-surgical embryo collection and transfer were developed and improved this industry has boomed. Therefore, I will concentrate the discussion of superovulation in domestic animals primarily on cattle; other domestic species will be discussed more briefly in subsequent sections.

Superovulation of cattle

<u>a) How are cattle superovulated?</u> The bovine estrous cycle averages 21 days in length (Day 0 = day of estrus behavior). Ovulation occurs on Day 1 and is followed by a luteal phase of about 18 days and then a follicular phase of 2-4 days. The first experimental induction of superovulation in farm animals was reported by Casida et al. (1943) who used pituitary gonadotropins to superovulate cattle (for a review of the early experimental work see Willet, 1953). Current superovulatory protocols involve the injection of exogenous gonadotropin during mid-luteal phase (Day 8-14) to recruit extra follicles near the end of the luteal phase for ovulation following the next follicular phase. Several gonadotropin preparations have been used in commercial superovulation programs. In the 1970's pregnant mare's serum gonadotropin (PMSG, also called equine chorionic gonadotropin or eCG) was widely used for superovulating cattle. This gonadotropin is produced by the equine conceptus and can be obtained from the blood of pregnant mares. Although it has only LH-like action in horses, PMSG has the interesting feature of binding to both LH and FSH receptors in other mammalian species (Stewart and Allen, 1979) and exhibiting both LH-like and FSH-like activity in bioassays (Papkoff, 1981). Therefore, its effects as a superovulating treatment are probably due to its capacity to mimic the actions of both pituitary gonadotropins. The advantage of PMSG in superovulation of domestic animals is that it can be given as a single injection to produce multiple ovulations several days later. This is because the higher sialic acid content of PMSG confers on it a much longer half-life (the half-life appears to have two components - a shorter one of 40-50 h and a longer of 118-123 h; Schams et al., 1978) than the pituitary gonadotropins. However, the long half-life of PMSG is also a disadvantage, since the gonadotropin continues to stimulate the ovaries over a long period of time and this can lead to asynchronous ovulations and large unovulated follicles on the ovaries after ovulation.

Although attempts have been made to attenuate PMSG's long half-life by injecting anti-PMSG just before ovulation (see below), the use of pituitary gonadotropins has gradually displaced PMSG for superovulation of cattle. The most widely used are commercially available preparations of FSH extracted from pig (or other domestic animal) pituitaries, referred to as "FSH-P". These preparations usually contain high and variable levels of contaminating LH and are typically given as a series of 8 injections of either equal or decreasing strength for 4 days, beginning around Day 10 of the cycle. Superovulation with pituitary gonadotropins was reported to produce more transferrable embryos (Critser et al., 1980) and pregnancies (Elsden et al., 1978) than superovulation with PMSG. In a few studies human menopausal gonadotropin (HMG) has been used successfully to produce multiple ovulations in cattle (Lauria et al., 1982a,b), but it is not used commercially.

Early in the history of bovine superovulation, it became apparent that the variable length of the estrous cycle in cattle makes it difficult to control the application of gonadotropins relative to the time of luteolysis. Normal estrous cycles can vary in length from around 18-26 days and, in contrast to the situation in humans, it is the luteal phase that is the most variable. Therefore, current superovulatory regimens with PMSG or pituitary gonadotropins include the injection of prostaglandin F2a (PGF_2a) to regress the corpus luteum (CL) and reduce the variability in the interval from gonadotropin injection to estrus (Betteridge, 1977; Elsden et al., 1978; Seidel, 1981). The use of PGF_{2a} to regress the CL also makes superovulation more economical, since it allows the cycles of recipients to be synchronized with that of the donor female. Many more potential recipients are needed if their cycles are not synchronized with that of the donor. With recently developed techniques for freezing embryos, synchronized recipients are not quite as critical.

Superovulatory regimens for cattle do not usually include the injection of hCG or any other gonadotropin to trigger ovulation. With the protocols used ovulation occurs spontaneously and injection of hCG does not seem to improve the outcome.

<u>b. How successful are superovulatory treatments in cattle?</u> The protocols currently used to stimulate bovine follicular development - i.e., a single dose of PMSG or multiple doses of porcine pituitary gonadotropins, with regression of the CL by PGF_{2a} - are successful in producing multiple ovulations and transferrable embryos (see Table 1). The endpoints that have been used to look at the success of superovulation have varied - number of corpora lutea, number of ova/embryos recovered, number of "good" or "transferrable" embryos recovered, or number of young born to recipients. It is generally agreed that, since

numbers of CLs or ova/embryos recovered can give misleading ideas of success if the ova/embryos are either not recovered or not developing properly, the important endpoint is number of transferrable embryos. Hence, it is important to look at success in terms of transferrable embryos, rather than numbers of ovulations. The success of superovulation depends partially on when the embryos are collected. It was found that collection of embryos on Day 6-8 after estrus is optimal, since they can be recovered non-surgically, and in good condition, from the uterus at that time (Hasler et al., 1983, 1987; Seidel, 1981)

The major point that should be made about the success of superovulation in cattle is that it is highly variable. While it may be very successful, yielding many good embryos from one donor, it may fail miserably with others. A number of recent reviews on superovulation in cattle have emphasized this lack of consistency as a major problem (Bindon et al., 1986; Foote and Ellington, 1988; Monniaux et al., 1983; Moor et al., 1984; Murphy et al., 1984; Seidel, 1981). There is variability both among animals and also within the same animal from one treatment cycle to the next. The variability that one finds in surveying the literature has a number of sources. Season, breed, gonadotropin preparation, number of previous superovulations, and dose and timing of gonadotropin preparations may all contribute to variability of response. The literature on these sources of variation will be reviewed briefly.

It has been reported that different breeds of cattle treated with the superovulatory regimens respond differently (Betteridge, 1977; Saumande, et at., 1978). There are conflicting reports on effects of season of the year on the superovulatory response, with some authors finding differences with season (Hasler et al., 1983) and others reporting no difference (Critser et al., 1980; Hasler et al., 1987; see Betteridge, 1977 for review of earlier literature). The potential existence of variation due to breed and/or season makes comparisons among studies much more difficult.

Since the goal of bovine superovulation is to reproduce a valuable female as many times as possible, a critical question is the number of times a female can be successfully superovulated and whether repeated superovulation leads to refractoriness or a lowered response. It seems possible that animals treated repeatedly with heterologous gonadotropins could develop antibodies towards them that would lessen or prevent the response. Hasler et al. (1983) have reported that ovulation rate remained the same over the course of ten superovulations, but the number of fertilized eggs declined. Donaldson and Perry (1983) also found that transferrable embryos decreased with repeated superovulations. In contrast a decline in the number of viable embryos was not observed in studies in which animals were superovulated fewer times (Moor et al., 1984; Lubbadeh et al., 1980).

Since the timing of injections of gonadotropins, relative to luteolysis and relative to changes in follicular populations, may be critical to the success of superovulation, several investigators have examined the effects of injecting gonadotropins at different times during the cycle. Donaldson (1984) reported that there was no difference in the number of embryos or transferrable embryos when FSH-P was administered on Day 9, 10, 11, 12, or 13 of the cycle (with PGF_{2a} given on the third day of treatment). Using a similar protocol, Hasler et al. (1983) reported similar responses to treatment begun on Days 8-13, except that the responses of infertile cows were lower on Days 8 and 13. However, Lerner et al.(1986) observed that greater numbers of embryos plus ova were recovered when treatment began on Day 10 or 11, as compared with Day 7-9 or Day 12-14.

The effects of varying the dose of gonadotropin on the success of superovulation have been examined. With older donors increasing doses of FSH-P were associated with increases in the recovery of ova plus embryos, whereas increasing doses of FSH had the opposite effect on younger donors (Lerner et al., 1986).

Variability among gonadotropin preparations, even from the same supplier, may explain part of the variability observed in different superovulation trials. Murphy et al (1984) reported that the amount of LH vs. FSH immuno- or bio-activity varies considerably among different preparations of PMSG and FSH-P. They found that preparations with lower ratios of FSH/LH are less successful in producing ovulations. Donaldson et al. (1986) removed most of the contaminating LH from FSH-P and found that their FSH preparation, which had no detectable LH activity, did not increase the total number of embryos, but did increase the number and percent transferrable embryos. A subsequent field study further confirmed the efficacy of the purified FSH preparation (Donaldson and Ward, 1986).

Even when the factors mentioned above are controlled for - i.e. when animals of the same breed are given the same dose of the same gonadotropin preparation at the same time of the cycle and the same time of year, and number of previous superovulations is controlled for, there is still tremendous variability in the response among, and within, animals. Although part of this variability may be explained by the treatment of donors inadvertently at inappropriate times of the cycle (Britt and Holt, 1988), it also seems clear that there is much unexplained variability. It seems logical to conclude that this unexplained variability comes from individual genetic differences among animals that determine how they respond to supra-physiological doses of gonadotropins. The next section addresses the question of what is currently being done to try to increase the uniformity of the response.

c. Current strategies to improve the success of superovulation. One goal is to determine an optimal gonadotropin treatment. This has not been easy because of the variability of the preparations that have been marketed. Several investigators have concluded that preparations with low ratios of FSH to LH give poorer superovulatory responses. Several investigators have tested this idea by using highly purified FSH and adding known amounts of LH to it in attempts to determine the optimum FSH/LH ratio (Chupin et al., 1987; Donaldson and Ward, 1986; Murphy et al., 1984). Another recent suggestion is that it may be helpful to prime the animals with injections of FSH early in the cycle to recruit more small follicles to grow prior to the superovulatory treatment in mid-cycle (Rajamahendran et al., 1987; Ware et al., 1988). PMSG is attractive as a superovulatory gonadotropin because it need only be injected once, as opposed to 8 injections of FSH-P. Therefore, several groups have tried to improve the response to PMSG by injecting an antiserum to PMSG around the time of ovulation to effectively shorten its biologically active life in the animals and to limit the effects of PMSG on the ovary to the follicular phase (Dieleman et al., 1987; Saumande et al., 1984; Wang et al., 1987).

Some investigators have attempted to analyze what may be going wrong when superovulation fails by comparing endocrine profiles and follicular characteristics of superovulated cattle with normal cattle. As would be expected, estradiol is higher prior to ovulation and progesterone is higher following ovulation in superovulated cattle as compared with controls (Henricks et al., 1973). Plasma levels of estradiol before ovulation and progesterone after ovulation are positively correlated with the number of ovulations (Saumande, 1980; Saumande and Batra, 1985). Very high levels of estradiol after superovulation with PMSG are probably caused by the presence of large, unovulated follicles (Booth et al., 1975). Although the magnitude of the LH and FSH surges appears to be normal in superovulated animals, in some cases their timing is abnormal and they may not be coincident (Saumande, 1980; Donaldson, 1985). Further evidence of lack of normal coordination of endocrine events in superovulated animals comes from studies of individual follicles. Profiles of steroids in the follicular fluid varied from those seen in control animals at the same time relative to LH surge and there was considerable variability among superovulated animals (Fortune and Hansel, 1985). Callesen et al. (1986) found that in superovulated animals relatively normal plasma levels of progesterone and LH were associated with normal follicular steroid levels, but that animals with deviating patterns of LH and progesterone during the pre- and peri-ovulatory periods had abnormal progesterone: estradiol ratios in follicular fluid and prematurely activated or meiotically arrested oocytes. Moor et al. (1984) concluded that premature activation of oocytes during superovulation protocols causes significant loss of potential embryos.

d. Summary of superovulation in cattle. Superovulation of cattle using a combination of gonadotropin treatment and PGF_{2a} is a multi-million dollar industry in the United States. Yet it is clear that the variability of the ovarian response in terms of numbers of ovulations, but more importantly in terms of numbers of good embryos, is a definite limitation to its successful use. There are no obvious solutions to this problem and the approaches being taken now represent a fine-tuning of the system that may lead to only slight improvements, rather than more radical approaches that might lead to more substantive improvements. Recommendations for research that might provide new insight into this problem are discussed in the last section of this paper.

Superovulation of Sheep and Goats

Sheep and goats have also been successfully superovulated for embryo transfer. Induction of multiple pregnancies with gonadotropin treatment was reported as early as 1944 (Casida et al., 1944). A number of studies since that time have demonstrated that ewes can be synchronized by treatment with progestagens, and multiple ovulations readily induced by a single injection of PMSG given one day before the end of progestagen treatment or by multiple injections of pituitary extracts (see Wright et al., 1981 for references). Australian scientists have been especially active in this area of research (see Shelton et al., 1982 for reviews). Since most breeds of sheep are seasonal breeders, the induction of ovulation during the anestrous season is of interest. The use of exogenous gonadotropins for this purpose has been explored, beginning with the successful induction of ovulation by Cole and Miller (1933).

An interesting aspect of research on superovulation in sheep is that some approaches have bypassed gonadotropin treatment entirely, using more indirect means to raise endogenous gonadotropin levels and thereby recruit excess follicles. It is believed that inhibin made by the ovulatory follicle suppresses basal levels of FSH in the blood during the follicular phase and thereby helps ensure the dominance of the dominant (ovulatory) follicle. It is known that one of the prolific sheep breeds, the Booroola, which typically ovulate about five oocytes, has lower levels of inhibin than less prolific breeds of sheep (Bindon et al, 1986). Injections of antibodies against inhibin have the potential of increasing FSH levels and recruiting additional follicles to ovulate (Bindon et al., 1986). Another approach to increasing ovulation rate is immunization against steroids. Scaramuzzi et al. (1977) first reported an increase in ovulation rate from 1.1 to 2.1 in sheep immunized against androstenedione.

Induction of superovulation with gonadotropin preparations has been particularly successful in goats (Armstrong and Evans, 1983). The ability to reproduce genetically superior animals is especially important for valuable breeds, such as Angora goats. A purebred Angora female is very expensive, but through superovulation and embryo transfer, she can potentially produce many embryos that can be transferred to non-Angora recipients. Results of experiments performed in Australia and Canada by Armstrong et al. (1983a,b) showed that injection of porcine FSH in decreasing doses from day 12-15 of the estrous cycle resulted in a higher mean ovulation rate (17.6 ± 5.5/ewe) than injection of PMSG on day 12 (10.1 ± 3.0) of the cycle in feral goats (Armstrong et al., 1983a). Application of these two gonadotropin treatments to Angora females in a second study gave almost identical results (Armstrong et al., 1983b). In the second study transfer of the embryos to recipients resulted in equal rates of survival for embryos from FSH vs. PMSG-treated females. However, these authors found that survival was better when recipients had 2 or 3 corpora lutea (CL) as compared with 1 CL (63 and 75% vs. 52%) and that transfer of two embryos to recipients increased their chances of survival (65%) as compared with transfers of single embryos (65% vs. 48% survival).

Effects of Exogenous Gonadotropins on Horses

There have also been attempts to superovulate mares, with a view towards embryo transfer. Horses are seasonally polyestrus; their estrous cycles average 21-22 days in length and consist of a period of diestrus for about 15 days following ovulation, followed by an estrous period that averages about 7 days (Ginther, 1979). The mare is unique among the domestic animals in having an estrous period that is long and variable in length, lasting 2-12 days (Ginther, 1979). LH rises progressively during the estrous period, but does not peak until after ovulation. In contrast, FSH is low during most of the estrous period, but there appears to be a surge of FSH coincident with the LH peak following ovulation and a second rise in FSH at about Day 10 of the cycle (reviewed by Irvine, 1981).

In contrast to cattle, where the effect of exogenous gonadotropins is to dramatically increase the number of ovulatory follicles, injections of gonadotropins have not produced large numbers of synchronous ovulations in mares. PMSG, even in large doses, is not effective in inducing multiple ovulations (Ginther, 1979). Its lack of effectiveness in horses, contrary to the marked response observed when cattle are injected with PMSG, is probably due to the fact that PMSG is an LH-like hormone in horses. It binds with only about 4% the affinity of LH to equine LH receptors and binds not at all to equine FSH receptors (Stewart and Allen, 1979). This is in contrast to the actions of PMSG in other species, in which it binds to both LH and FSH receptors. It appears that its lack of FSH action in mares makes PMSG ineffective and unsuitable for inducing multiple ovulation in this species.

Treatment of mares with other gonadotropic preparations has been effective in inducing ovulation in seasonally anestrus mares and in increasing the ovulation rate in mares treated during the breeding season. In these studies pony mares or horse mares have been treated with either a crude extract of equine pituitary glands, an acetone extract of equine pituitary glands, or a commercially available preparation of porcine FSH (FSH-P, which probably also contains some LH). In some studies hCG has been injected at or towards the end of the treatment with FSH. Such treatments have been effective in inducing a high percentage of treated mares to ovulate outside the breeding season (>90%) and a number of these mares had multiple ovulations (overall average for mares ovulating = 2.1; Douglas et al., 1974; Lapin and Ginther, 1977; Woods et al., 1982). Treatment with these gonadotropin preparations for 5 to 7 days during the breeding season has also induced multiple ovulations. Treatment with FSH-P in two studies resulted in ovulation rates of 1.6 and 1.7 ovulations per mare (Irvine, 1981; Squires et al., 1986), whereas ovulation rates of 2-3 ovulations per mare were observed when equine pituitary extract was injected (Lapin and Ginther, 1977; Douglas, 1979; Woods and Ginther, 1983a; Squires et al., 1986). Ovulation rates for controls were always one, or very close to one, per mare. The inclusion of an injection of hCG near the end of treatment with FSH produced ovulations that were more synchronous and treatments that began on Day 11-15 produced more ovulations than treatments beginning on Day 19 (Woods and Ginther, 1983a).

In summary, the use of gonadotropins to stimulate ovulation in mares has progressed much less than with cattle and sheep and there is currently no commercially available preparation for superovulating horses. There appear to be several reasons for lack of progress with this species. First, the ovaries of mares seem much more resistant than ovine and bovine ovaries to superovulatory treatments. There may be tighter intrinsic control over ovulation rate in horses than in species like cattle, sheep, and humans. Although natural double ovulations are not uncommon (Woods and Ginther, 1983b), the birth of twins is rare in horses, indicating that there are mechanisms for preventing the development of twins when two follicles ovulate (Ginther et al., 1982; Woods and Ginther, 1983b; Ginther 1984). Second, the variable length of estrus makes embryo transfer logistically more difficult than in other species, since it is harder to synchronize the donor and recipients. Finally, many purebred organizations are conservative and stud-book regulations prohibit most artificial techniques for reproduction. However, there is a need to understand better the regulation of follicular development in mares in order to develop better treatments for infertility and because the development of methods for superovulating mares could be of particular benefit in preserving and increasing numbers of individuals in endangered equine species, through transfer of embryos to recipients of similar but non-endangered species.

Ovulation Induction in Pigs

The commercial applications of superovulation and embryo transfer techniques are more limited for swine than for the other domestic species discussed above. Since sows give birth to litters and have a relatively short gestation period (about 110 days), each fertile female is naturally capable of producing numerous progeny. However, superovulation and embryo transfer could still be used, theoretically, to increase the rate of reproduction of genetically superior animals. In the 1960's several research groups successfully ovulated or superovulated pigs with exogenous gonadotropins (Day et al., 1967; Pope et al., 1968; Bazer et al., 1969). James and Reeser (1979) showed that ovulation could be induced and embryos collected (surgically) repeatedly in the same sows. In their study follicular development was ensured by the injection of PMSG on Day 16 of the estrous cycle, at 15-20 days of lactation, or at weaning 21 to 35 days post-partum. Ovulation was timed by injecting hCG 72-96 h after PMSG injection. This treatment was applied and embryos were recovered an average of 4.8 times per donor with an average recovery rate of 15.6 transferable embryos per surgery. It should be noted that the gonadotropin treatments were used more to time follicle development and ovulation, than to "superovulate" the animals, since pigs would be expected normally to ovulate around 15 oocytes. However, these results were encouraging because they showed that embryos could be surgically collected repeatedly at 3 to 6 week intervals without the formation of extensive adhesions or any other apparent impairments of reproductive function. Other studies have shown that females can be superovulated with a single dose of PMSG on Day 15 or 16 of the cycle. Doses of 500, 750, 1000, 1250, and 1500 i.u. resulted in mean numbers of CLs of 15, 25, 26, 25, and 38, respectively (Hunter, 1980). Despite these positive results, gonadotropin treatments and embryo transfer are not much used to increase the genetic potential of superior female gilts or sows. In 1981 only 8 of 13,000 purebred seedstock producers in the U.S. used embryo transfer primarily to exploit genetically superior females (Martin, 1983). One reason for this is the difficulty of identifying superior sows. Therefore, the use of superovulation and/or embryo transfer for genetic improvement is not a high priority in the swine industry.

Embryo transfer in swine has advantages other than potential improvement of the gene pool. Disease is a major concern in the swine industry. Herds have increased in size and they are frequently confined to environmentally controlled buildings. Both these conditions encourage the spread of disease and some herds that are disease-free try to maintain that condition by becoming "closed" (i.e, not allowing new animals to be added to the herd). Embryo transfer offers a means by which new genetic material can be introduced into closed herds, without the risk of also introducing disease vectors (Martin, 1983). Embryo transfer also provides a means of repopulating diseased herds and of exporting animals to other countries. Finally, embryo transfer has proven to be a useful tool in research on pigs, especially in studies of maternal-fetal interactions (e.g., Pope et al., 1972; Rampacek et al., 1975).

In summary, embryo transfer in swine has not been exploited to increase the reproductive potential of superior females, as it has in cattle, sheep, and goats. The applications of this technique to swine center more on prevention of the spread of disease and on research applications. Although gonadotropin treatments have been used to ensure and time follicular development and ovulation, they have not been used to significantly enhance the already high ovulation rate of pigs.

Recommendations For Future Directions In Research

It is clear that while superovulation of domestic animals, especially cattle, has produced some spectacular successes, the current techniques have the important limitation of variable and unpredictable response. Superovulation is currently the weak link in the reproduction of genetically superior animals through superovulation and embryo transfer. Techniques for non-surgical collection and transfer of embryos have been well developed and major advances have been made in the cryopreservation, splitting, and sexing of embryos. However, in the last decade there has been little improvement in the success of superovulation in cattle. What kinds of research might lead to substantive improvements in the regulation of female reproduction in cattle? I believe that several approaches should be supported simultaneously.

Basic Research on Bovine Ovarian Function. Current techniques for superovulating domestic animals are based on a very minimal understanding of the regulation of follicular growth and differentiation. They depend on the knowledge that LH and FSH are important for the development of ovulatory follicles and that PGF_{2a} will cause regression of the corpus luteum. Although we do know something about the effects of gonadotropins on cultured theca and granulosa from the preovulatory follicle (McNatty et al., 1984a,b; Fortune, 1986, Fortune and Quirk, 1988; Fortune et al., 1988), we know little about the specific requirements for gonadotropin support during the course of development of ovulatory follicles. Hence, doses and times of injection of gonadotropins in superovulatory regimes have been developed empirically.

Basic research on the regulation of luteal function led to the discovery that PGF_{2a} is involved in the regression of the corpus luteum. This led to the use of PGF_{2a} and its analogues to synchronize the estrous cycles of cattle, a development that has had an enormous impact on superovulation and embryo transfer. I do not believe that more than small improvements in the success of superovulation will be made until our knowledge of the mechanisms that regulate the development and regression of follicles during normal estrous cycles is increased. One exciting and promising recent development is the use of intrarectal ultrasonography to visualize bovine ovaries. This technique was pioneered by Pierson and Ginther (1984) who showed that it was possible to see and count follicles on bovine ovaries. My laboratory has developed the use of ultrasonography to follow the development and/or regression of individual follicles over time by videotaping daily ultrasound exams (Quirk et al., 1986). The tapes can be reviewed and analyzed to reproduce the patterns of growth and regression of individual follicles. When we used these techniques to determine the pattern of follicular development through the course of the estrous cycle, the results were surprising in that they showed in all ten heifers examined a regular pattern of waves of follicular development that begin about every seven days (Figure 1; Sirois and Fortune, 1988), with one follicle in each wave growing larger, while the others regress. The regularity of these waves is exciting because it suggests a number of ways in which the development and regression of follicles can be studied experimentally. We can ask questions about what governs the timing of these waves. Why do the dominant follicles of the first and second waves not ovulate and what would stimulate them to ovulate? Knowledge of the pattern of these waves may also be directly applicable to superovulation protocols. For example, is there any relationship between the number of follicles in the waves and the superovulatory responses of individual animals?

Can gonadotropin injections be better timed to take advantage of the waves? When would the application of gonadotropins encourage more follicles to join a wave or overcome the mechanisms by which the largest follicle exerts dominance?

Through research such as outlined above we can gain insight into why superovulation so frequently fails to produce the desired large numbers of embryos. However, the number of offspring produced by a species has evolved after millions of years of selection. We are trying to subvert biological mechanisms that are very important to a species when we attempt to superovulate. With cattle there seems to be much variability among animals in terms of the extent to which these biological control mechanisms can be overcome. It may not be possible in the near future to reliably superovulate all, or even most, individuals in a domestic species. Therefore, it might make more sense to develop subtler approaches to hyper-stimulation - i.e., through better knowledge of critical times for gonadotropin action during follicular development, a protocol might be developed that could reliably produce 5-6 good embryos per treatment. Currently, large doses of gonadotropins are given in the hope that large numbers of ovulations will result. The animals must then be allowed several cycles to recover from the perturbations that result. If gentler, but more specific treatments were applied, animals might be superovulated for many cycles consecutively, leading to higher yields of embryos over a given period of time. A better understanding of follicular dynamics and their regulation by gonadotropins might also lead to an ability to fine-tune superovulation to the extent that animals would reliably ovulate two oocytes, which they could then carry to term to produce twins. This could increase the reproductive capacity of less valuable animals that were not candidates for superovulation. Although twin births cause more problems at parturition and half of female twins would be freemartins, there would still be advantages to twinning, especially in the beef industry.

<u>Other Approaches To Increasing Reproductive Capacity</u>. In the paragraphs above I have discussed potentially fruitful directions for basic research that could lead to improvements in the use of gonadotropins to superovulate domestic animals. At the same time, I believe that other approaches that do not involve gonadotropins should be explored. First, the recently isolated gonadal hormone inhibin has great potential as a regulator of fertility. This protein hormone is believed to be secreted by granulosa cells and feeds back on the pituitary to regulate selectively the secretion of FSH. It has been proposed that one follicle becomes dominant (ovulatory) because it is at

the right stage of its development to benefit from increases in plasma FSH. It then begins secreting estradiol and inhibin which feed back on the hypothalamus-pituitary to suppress FSH secretion, thereby depriving slightly smaller follicles of the critical amount of FSH needed to continue development towards ovulation. One approach, discussed above, is to identify critical times when the application of extra FSH will lead to development of additional follicles that are fairly synchronous with the ovulatory follicle. Another approach would be passive or active immunization against inhibin. It has been found that the Booroola breed of sheep, a natural superovulator, has lower levels of inhibin than less prolific breeds (Bindon et al., 1986). Although inhibin injections selectively suppress FSH, the "rebound" of FSH after the injections can cause an increase in ovulation rate (McNeilly and Wallace, 1987). Research with inhibin is now hampered by lack of availability of purified inhibin to most researchers. Availability of purified inhibin would greatly stimulate research on this potential regulator of fertility.

Second, studies on the phenomenon of follicular dominance should be encouraged. We know little about the mechanisms by which one follicle suppresses others in its cohort, yet it is these mechanisms that we try to subvert in superovulation. The production of inhibin is one potential mechanism, but there is also evidence for the production of other substances by dominant follicles, such as follicle growth inhibitor (Bindon et al., 1986) and follicle regulatory protein (Ono et al., 1986) that may act more directly to suppress the development of other follicles.

Third, approaches that do not involve attempts to manipulate the endocrine milieu should be explored. Especially promising is the idea of maturing and fertilizing bovine oocytes in vitro for transfer to foster mothers. Bovine ovaries contain around 150,000 primordial follicles at birth and thousands of them are maintained in a healthy state until about 4-6 years of age (Erickson, 1966), even though only a small number ever ovulate. Most of these oocytes reside in preantral follicles and have not acquired the ability to undergo spontaneous maturation if released from their follicles (meiotic competence). However, John Eppig's group at the Jackson Laboratory has been successful in isolating oocytes in preantral mouse follicles, growing them in vitro to meiotic competence, and producing live young (Eppig, personal communication). Adaptation of these techniques to cattle ovaries could be very exciting. For example, one or both ovaries could be removed from a valuable heifer calf and preantral follicles obtained, grown to meiotic competence, and fertilized in vitro. The embryos could then be frozen and transferred to recipients as desired. In this way information could be gained about the characteristics of a female's progeny before she herself had even reached puberty. A fourth potentially productive area is research on genetic selection for twinning in cattle or superfecundity in other domestic species (Bindon et al.,1986).

SUMMARY

The use of commercially available gonadotropin preparations, coupled with the use of PGF_{2a} to control the time of regression of the corpus luteum, has resulted in the ability to produce excess ovulations in cattle, sheep, pigs, and mares. These techniques have been most widely and successfully used with cattle and are key to the success of the wider dissemination of genes of valuable bovine females through embryo transfer. However, the variability in the response to exogenous gonadotropins is a major stumbling block to improvements in the success of superovulation in cattle and other domestic species. I have suggested a number of approaches to solving this problem. These include: 1) basic research on follicular dynamics and the roles of pituitary gonadotropins in the regulation of follicular growth and differentiation, 2) basic research on the role of inhibin in ovarian regulation in cattle and other species, 3) continued research on other mechanisms by which follicles may exert dominance, 4) the adaptation of techniques for growing mouse oocytes to meiotic competence in vitro to domestic species, 5) continuation of studies on genetic selection for multiple ovulation. I believe that we need a broad range of approaches for several reasons. First, this field is at a stalemate right now and it is hard to predict where the next big breakthrough might come. Second, it seems desirable to think in terms of different approaches for different types of animals - meiotic maturation and in vitro fertilization of thousands of oocytes from the most valuable cows, better or more reliable techniques for using gonadotropins to superovulate valuable cows, and endocrine or genetic methods for producing twins in less valuable animals. The cattle industry would benefit from having more available options. Also, a wider range of options increases the chance that some of them would also be useful in developing countries, where farmers may be unable to benefit from some techniques that work well for American producers. It seems well to remember that a high percentage of the world's domestic animals live in developing countries and that increases in their reproduction are very crucial to health and nutrition world-wide. Finally, cattle are an excellent model for humans and a wide variety of approaches to studying the regulation of follicle development in cattle could well generate new ideas and approaches for regulating fertility in women.

REFERENCES

Armstrong, D. T., and G. Evans. 1983. Factors influencing success of embryo transfer in sheep and goats. Theriogenology 19:31-42.

Armstrong, D. T., A. P. Pfitzner, G. M. Warnes, M. M. Ralph, and R. F. Seamark. 1983a. Endocrine responses of goats after induction of superovulation with PMSG and FSH. J. Reprod. Fert. 67:395-401.

Armstrong, D. T., A. P. Pfitzner, G. M. Warnes, and R. F. Seamark. 1983b. Superovulation treatments and embryo transfer in Angora goats. J. Reprod. Fert. 67:403-410.

Bazer, F. W., O. W. Robison, A. J. Clawson, and L. C. Ulberg. 1969. Uterine capacity at two stages of gestation in gilts following embryo superinduction. J. An. Sci. 29:30-38.

Betteridge, K. J. 1977. Superovulation. pp. 1-9 in Embryo Transfer in Farm Animals, K. J. Betteridge, ed., Can. Dept. Agric. Res. Branch Monograph #16.

Bindon, B. M., L. R. Piper, L. P. Cahill, M. A. Driancourt and T. O. Shea.1986. Genetic and hormonal factors affecting superovulation. Theriogenology 25:53-70.

Booth, W. D., R. Newcomb, H. Strange, L. E. Rowson, and H. B. Sacher. 1975. Vet. Record 97:366-369.

Britt, J. H. and L. C. Holt. 1988. Endocrinological screening of embryo donors and embryo transfer recipients: a review of research with cattle. Theriogenology 29:189-202.

Callesen, H., T. Greve and P. Hyttel. 1986. Preovulatory endocrinology and oocyte maturation in superovulated cattle. Theriogenology 25:71-86.

Casida, L. E., R. K. Meyer, W. H. McShan, and W. Wisnicky. 1943. Effects of pituitary gonadotropin on the ovaries and on the induction of superovulation in cattle. Amer. J. Vet. Res. 4: 76-94.

Casida, L. E., E. J. Warwick, and R. K. Myer. 1944. Survival of multiple pregnancies induced in the ewe following treatment with pituitary gonadotropins. J. Anim. Sci. 3:22-28.

Chupin, D., Y. Cognie, Y. Combarnous, R. Procureur, and J. Saumande. 1987. Effect of purified LH and FSH on ovulation in the cow and ewe. pp. 63-71 in Follicular Growth and Ovulation Rate in Farm Animals, J. F. Roche and D. O'Callaghan, eds. Dordrecht: Martinus Nijhoff Publishers.

Cole, H. H. and R. F. Miller. 1933. Artificial induction of ovulation and oestrum in the ewe during anoestrum. Amer. J. Physiol. 104:165.

Critser, J. K., R. F. Rowe, M. R. Del Campo and O. J. Ginther. 1980. Embryo transfer in cattle: factors affecting superovulatory response, number of transferable embryos, and length of post-treatment estrous cycles. Theriogenology 13:397-406.

Day, B. N., D. E. Longenecker, S. C. Jaffe, E. W. Gibson, and J. F. Lasley. 1967. Fertility of swine following superovulation. J. An. Sci. 26:777- 786.

Dieleman, S. J., M. M. Bevers, and J. Th. Gieleu. 1987. Increase of the number of ovulations in PMSG/PG treated cows by administration of monoclonal anti-PMSG shortly after the endogenous LH peak. Theriogenology 27:222.

Donaldson, L. E. 1984. The day of the estrous cycle that FSH is started and superovulation in cattle. Theriogenology 22:97-99.

Donaldson, L. E. 1985. LH and FSH profiles at superovulation and embryo production in the cow. Theriogenology 23: 441-447.

Donaldson, L. E., and B. Perry. 1983. Embryo production by repeated superovulation of commercial donor cows. Theriogenology 20:163-168.

Donaldson, L. E., and D. N. Ward. 1986. Effects of luteinizing hormone on embryo production in superovulated cows. Veterinary Record 119:625-626.

Donaldson, L. E., D. N. Ward, and S. D. Glenn. 1986. Use of porcine follicle stimulating hormone after chromatographic purification in superovulation of cattle. Theriogenology 25:747-757.

Douglas, R. H. 1979. Review of induction of superovulation and embryo transfer in the equine. Theriogenology 11:33-46.

Douglas, R. H., L. Nuti, and O. J. Ginther. 1974. Induction of ovulation and multiple ovulation in seasonally anovulatory mares with equine pituitary fractions. Theriogenology 2:133-142.

Elsden, R. P., L. D. Nelson, and G. E. Seidel, Jr. 1978. Superovulating cows with follicle stimulating hormone and pregnant mare's serum gonadotrophin. Theriogenology 9:17-26.

Erickson, B. H., 1966. Development and senescence of the postnatal bovine ovary. J. An. Sci. 25:800-805.

Fevold, H. L. 1941. Synergism of the follicle stimulating and luteinizing hormones in producing estrogen secretion. Endocrinology 28:33-41.

Foote, R. H., and J. E. Ellington. 1988. Is a superovulated oocyte normal? Theriogenology 29:111-123.

Fortune, J. E. 1986. Bovine theca and granulosa cells interact to promote androgen production. Biology of Reproduction 35: 292-299.

Fortune, J. E. and S. M. Quirk. 1988. Regulation of steroidogenesis in bovine preovulatory follicles. Journal of Animal Science 66 (Suppl. 2):1-8.

Fortune, J. E., J. Sirois, and S. M. Quirk. 1988. The growth and differentiation of ovarian follicles during the bovine estrous cycle. Theriogenology 29:95-110.

Fortune, J. E. and W. Hansel. 1985. Concentrations of steroids and gonadotropins in follicular fluid from normal heifers and heifers primed for superovulation. Biol. Reprod. 32: 1069-1079.

Fraenkel-Conrat, H., C. H. Li, and M. E. Simpson. 1943. Pituitary Gonadotropins. In Essays in Biology, p. 183-191. Berkeley: University of California Press.

Ginther, O. J. 1979. Reproductive Biology of the Mare: Basic and Applied Aspects. Ann Arbor, Michigan: McNaughton and Gunn, Inc.

Ginther, O. J. 1984. Postfixation embryo reduction in unilateral and bilateral twins in mares. Theriogenology 22:213-223.

Ginther, O. J., R. H. Douglas, and G. L. Woods. 1982. A biological embryo-reduction mechanism for the elimination of excess embryos in mares. Theriogenology 18:475-485.

Greep, R. O., A. B. van Dyke, and B. F. Chow. 1942. Gonadotropins of the swine pituitary. I. Various biological effects of purified thylakentrin (FSH) and pure metakentrin (ICSH). Endocrinology 30:635-649.

Hasler, J. F., A. D. McCauley, E. C. Schermerhorn and R. H. Foote. 1983. Superovulatory responses of Holstein cows. Theriogenology 19:83-99.

Hasler, J. F., A. D. McCauley, W. F. Lathrop and R. H. Foote. 1987. Effect of donor-embryo-recipient interactions on pregnancy rate in a large scale embryo transfer program. Theriogenology 27:139-168.

Henricks, D. M., J. R. Hill, Jr., J. F. Dickey, and D. R. Lamond. 1973. Plasma hormone levels in beef cows with induced multiple ovulations. J. Reprod. Fertil. 35:225-233.

Hunter, R. H. F. 1980. Physiology and Technology of Reproduction in Female Domestic Animals. New York: Academic Press.

Irvine, C. H. G. 1981. Endocrinology of the estrous cycle of the mare: applications to embryo transfer. Theriogenology 15:85-

James, J. E., and P. D. Reeser. 1979. Embryo recovery in swine. Theriogenology 11:47-50.

Lapin, D. R., and O. J. Ginther. 1977. Induction of ovulation and multiple ovulations in seasonally anovulatory and ovulatory mares with an equine pituitary extract. J. Anim. Sci. 44:834-842.

Lauria, A., A. Genazzani, O. Olivia, P. Inaudi, F. Cremonesi, C. Monittola, and G. Aureli. 1982a. Clinical and endocrinological investigations on superovulation induced in heifers by human menopausal gonadotrophin. J. Reprod. Fert. 66:219-225.

Lauria, A., O. Olivia, A. R. Genazzani, F. Cremonesi, S. Crotti, and M. Barbetti. 1982b. Improved method to induce superovulation in cattle using human menopausal gonadotropin (HMG). Theriogenology 18:357-364.

Lerner, S. P., W. V. Thayne, R. D. Baker, T. Henschen, S. Meredith, E. K. Inskeep, R. A. Dailey, P. E. Lewis, and R. L. Butcher. 1986. Age, dose of FSH and other factors affecting superovulation in holstein cows. J. Anim. Sci. 63:176-183.

Lubbadeh, W. F., C. N. Graves, and S. L. Spahr. 1980. Effect of repeated superovulation on ovulatory response of dairy cows. J. Anim. Sci. 50:124-127.

Martin, P. A. 1983. Commercial embryo transfer in swine: who is interested in it and why. Theriogenology 19:43-48.

McNatty, K. P., D. A. Heath, S. Lun, J. M. Fannin, J. M. McDiarmid and K.M. Henderson. 1984a. Steroidogenesis by bovine theca interna in an in vitro perifusion system. Biol. Reprod. 30:159-170.

McNatty, K. P., D. A. Heath, K. M. Henderson, S. Lun, P. R. Hurst, L.M. Ellis, G.W. Montgomery, L. Morrison and D.C. Thurley. 1984b. Some aspects of thecal and granulosa cell function during follicular development in the bovine ovary. J. Reprod. Fertil. 72:39-53.

McNeilly, A. S. and J. M. Wallace. 1987. The effect of follicular fluid on ovulation rate in the ewe. pp. 119-127 in Follicular Growth and Ovulation Rate in Farm Animals, J. F. Roche and D. O'Callaghan, eds. Dordrecht: Martinus Nijhoff Publishers.

Monniaux, D., D. Chupin, and J. Saumande. 1983. Superovulatory responses of cattle. Theriogenology 19:55-81.

Moor, R. M., Th. A. M. Kruip, and D. Green. 1984. Intraovarian control of folliculogenesis: limits to superovulation? Theriogenology 21:103-116.

Murphy, B. D., R. J. Mapletoft, J. Manns, and W. D. Humphrey. 1984. Variability in gonadotropin preparations as a factor in the superovulatory response. Theriogenology 21:117-125.

Ono, T, J. D. Campeau, E. A. Holmberg, R. N. Nakamura, E. L. Ujita, D. L. Devereaux, S. A. Tonetta, R. DeVinna, M. Ulgalde, and G. S. DiZerega. 1986. Biochemical and physiologic characterization of follicle regulatory protein: a paracrine regulator of folliculogenesis. Amer. J. Obstet. Gynecol. 154:709-716.

Papkoff, H. 1981. Variations in the properties of equine chorionic gonadotropin. Theriogenology 15:1-11.

Pierson, R. A. and O. J. Ginther. 1984. Ultrasonography of the bovine ovary. Theriogenology 21:495-505.

Pope, C. E., R. K. Christenson, V. A. Zimmerman-Pope, and B. N. Day. 1972. Effect of number of embryos on embryonic survival in recipient gilts. J. An. Sci. 35:805-814.

Pope, C. E., C. K. Vincent, and D. M. Thrasher. 1968. Effect of I.C.I. 33828 and PMS on reproduction in gilts. J. An. Sci. 27:303-314.

Quirk, S. M., G. J. Hickey, and J. E. Fortune. 1986. Growth and regression of ovarian follicles during the follicular phase of the oestrous cycle in heifers undergoing spontaneous and PGF_{2a}-induced luteolysis. Journal of Reproduction and Fertility 77:211-219.

Rajamahendran, R., R. S. Canseco, and C. J. Denbow. 1987. Effect of low dose of FSH given at the beginning of the estrous cycle and subsequent superovulatory response in holstein cows. Theriogenology 28:59-65.

Rampacek, G. R., O. W. Robison, and L. C. Ulberg. 1975. Uterine capacity and progestin levels in superinducted gilts. J. An. Sci. 41:564-567.

Saumande, J. 1980. Concentrations of luteinizing hormone, oestradiol-17b and progesterone in the plasma of heifers treated to produce superovulation. J. Endocrinol. 84:425-437.

Saumande, J. and S. K. Batra. 1985. Superovulation in the cow: comparison of estradiol-17b and progesterone patterns in plasma and milk of cows induced to superovulate; relationships with ovarian responses. J. Endocrinol. 107:259-264.

Saumande, J., D. Chupin, J. C. Mariana, R. Ortavant, and P. Mauleon. 1978. Factors affecting the variability of ovulation rates after PMSG stimulation. pp. 195-223 in Control of Reproduction in the Cow, J. M. Sreenan, ed. Martinus Nijhoff Publishers, The Hague.

Saumande, J., R. Procureur, and D. Chupin. 1984. Effect of injection time of anti-PMSG antiserum on ovulation rate and quality of embryos in superovulated cows. Theriogenology 21:727-731.

Scaramuzzi, R. J., W. G. Davidson, and P. F. A. Van Look. 1977. Increasing ovulation rate in sheep by active immunization against an ovarian steroid androstenedione. Nature 269:817-818.

Schams, D., Ch. Menzer, E. Schallenberger, B. Hoffman, J. Hahn, and R. Hahn. 1978. Some studies on pregnant mare serum gonadotropin (PMSG) and on endogenous responses after application for superovulation in cattle. pp. 122-144 in Control of Reproduction in the Cow, J. M. Sreenan, ed. Martinus Nijhoff Publishers, The Hague.

Seidel, G. E., Jr. 1981. Superovulation and embryo transfer in cattle. Science 211:351-358.

Shelton, J. N., A. O. Trounson, N. W. Moore, and J. W. James (eds.). 1982. Embryo Transfer in Cattle, Sheep and Goats. Aust. Soc. Reproductive Biology Publication. Treasurer, ASRB, University of Sydney Farms, Camden, Australia, 2570.

Sirois, J. and J. E. Fortune. 1988. Follicular dynamics during the estrous cycle in heifers monitored by real-time ultrasonography. Biology of Reproduction 39:308-317.

Squires, E. L., R. H. Garcia, O. J. Ginther, J. L. Voss, and G. E. Seidel, Jr. 1986. Comparison of equine pituitary extract and follicle stimulating hormone for superovulating mares. Theriogenology 26:661-670.

Stewart, F. and W. R. Allen. 1979. The binding of FSH, LH and PMSG to equine gonadal tissues. J. Reprod. Fert. Suppl. 27:431-440.

Wang, H., M. Wu, K. Xu, W. C. Hagele, and R. J. Mapletoft. 1987. Control of superovulation in the cow with a PMSG antiserum. Theriogenology 27:291.

Ware, C. B., D. L. Northey, M. P. Boland, and N. L. First. 1988. Early cycle FSH-p priming as a prelude to superovulating gonadotropin administration in ewes and heifers. Animal Reproduction Science 16:97-105.

Willet, E. L. 1953. Egg transfer and superovulation in farm animals. Iowa State College Journal of Science 28:83-100.

Woods, G. L., and O. J. Ginther. 1983a. Induction of multiple ovulations during the ovulatory season in mares. Theriogenology 20:347-355.

Woods, G. L., and O. J. Ginther. 1983b. Recent studies relating to the collection of multiple embryos in mares. Theriogenology 19:101-108.

Woods, G. L., S. T. Scraba and O. J. Ginther. 1982. Prospects for induction of multiple ovulations and collection of multiple embryos in the mare. Theriogenology 17:61-72.

Wright, R. W., Jr., K. Bondioli, J. Grammer, F. Kuzan, and A. Menino, Jr. 1981. FSH or FSH plus LH superovulation in ewes following estrus synchronization with medoxyprogesterone acetate pessaries. J. Anim. Sci. 52:115-118.

TABLE 1. Superovulation in Cattle: Results from Two Commercial Embryo Transfer Programs

	Hasler et al., 1983[a]	Donaldson, 1984[b]
# Animals	984	1,263
Mean Ova/Donor	8.9	10.1 (0-70)
Mean Embryos/Donor	5.1 (0-38)	--
"Good" Embryos/Donor	4.6	4.5 (0-37)
No Embryos (% Donors)	36%	15% (32.4% - no good embryos)

[a]Reported in Theriogenology 19: 83-99

[b]Reported in Theriogenology 21: 517-524

TABLE 2. Distribution of pregnancies from donor cows. Pregnancy was confirmed by palpation through the rectal wall between days 90 and 100 of gestation. (Seidel, Science 211: 354, 1981.)

Donors		Pregnancies from Superovulation (no.)		
Age (years)	Breed	First	Second	Third
1	Simmental	4	7	8
1	Simmental	5	0	9
4	Limousin	0	0	1
2	Hereford	5	1	4
2	Pinzgauer	6	1	0
2	Pinzgauer	4	8	20
13	Angus	4	2	8
10	Hereford	2	0	5
9	Hereford	9	3	7
1	Simmental	7	0	0
1	Simmental	2	1	0
4	Charolais	7	9	0
7	Brangus	2	0	1
2	Simmental	5	11	4
6	Hereford	1	11	5
8	Hereford	2	0	1
14	Hereford	12	0	1
12	Hereford	0	3	2
2	Hereford	3	1	4
4	Angus	3	0	2
7	Hereford	4	7	2
5	Angus	4	2	0
5	Angus	4	3	0
5	Angus	4	9	0
4	Simmental	1	1	0
5	Angus	0	7	2
4	Hereford	0	4	8
10	Angus	0	6	0
1	Simmental	0	2	0
5	Angus	3	6	0
3	Simmental	3	4	14
	Average	3.4	3.5	3.4

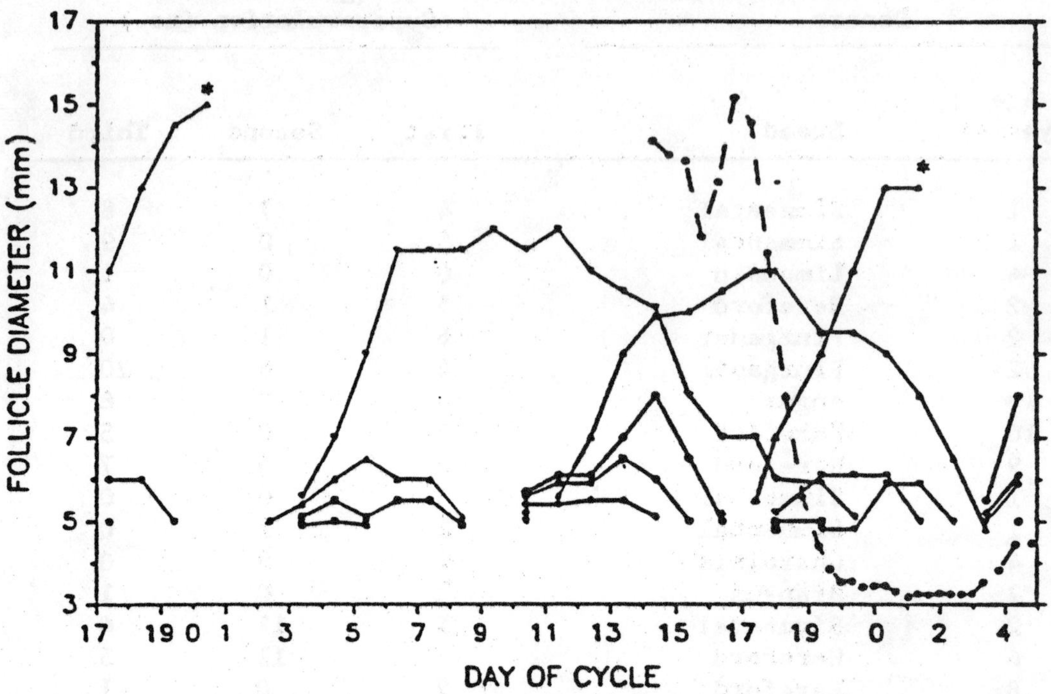

Figure 1. Pattern of growth and regression of follicles (solid lines) during a complete estrous cycle. The asterisk indicates the last day on which the ovulatory follicle was observed and the dashed line shows progesterone concentrations around the time of luteolysis.

MEMBRANE MODELS: EVOLUTION FROM THE FLUID-MOSAIC STANDBY

ROY HAMMERSTEDT

INTRODUCTION

Postulation of the fluid-mosaic model for membrane structure by Singer and Nicholson (1972) represented a thoughtful and reasoned synthesis of information obtained from observations extending over 140 years. These many contributions, summarized by Robertson (1981), began with the implicit recognition by Schleiden and Schwann that a membrane structure must bound the cell, and includes contributions from physicists, chemists and morphologists. Other key elements were: use of evolving techniques to highlight new aspects of the complex structure (eg., electron microscopy); careful quantitative analyses of the sizes and numbers of the postulated elements of the membrane (protein, sterol and phospholipids); and characterization of the physical features of membranes, either as isolated or after resolution into individual components. The resulting model has served as an framework upon which a host of experiments have been conducted to further illuminate the role of the interface between the cell interior and its environment or between compartments within the cell. Since the presentation of the fluid-mosaic model, continuous application of pre-existing plus new techniques has resulted in additional quantitative details that must be integrated into the model. Careful rereading of this classic paper, as opposed to simple inspection of the much imitated schematic Figures (see Figure 1, left panel), reveals that the authors clearly understood the limitations to their proposal and anticipated some of the modifications proposed later by others (Figure 1, right panel).

<u>Figure 1. Schematic representation of bilayer membrane</u>. The progressive evolution of our concepts of the structure of the lipid bilayer is illustrated by the initial presentation (Singer and Niclolson, 1972) on the left and an "updated" version (Cullis and Hope, 1985) on the right. Figures reprinted with permission of the publishers.

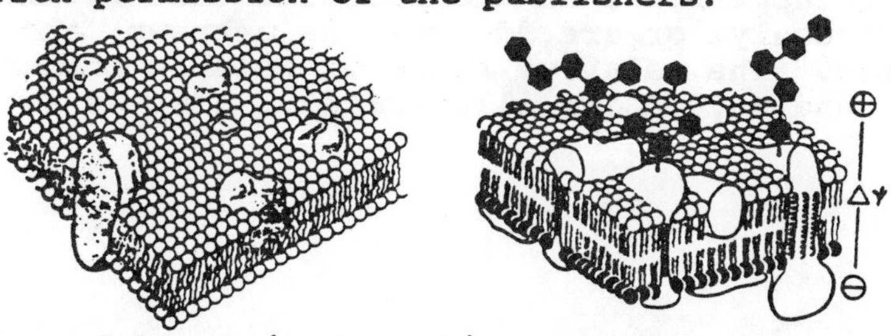

The fluid mosaic model retained earlier concepts (eg., a lipid bilayer by Danielle and Davson) while providing a new way of interpreting the distribution of protein. Proteins were considered to penetrate into the lipid (in an iceberg fashion) leaving most of the inner and outer surfaces as naked lipid. This mosaic pattern provided a portion of the name. The second aspect, fluidity, encompassed the observations of Frye and Edidin (1970) on the ability of fluorescent dyes to move freely in the membrane. These two aspects (fluidity and mosaic) remain dominant features of the evolving membrane model.

Contributions of others will be reviewed herein to illustrate: an increased awareness of the complexity of the outer surface of the plasma membrane; the asymmetric distribution of lipid components across the bilayer, and the effect that these distributions have on the physical properties of the membrane; and the lateral organization (domain structure) of the membrane. These revisions are especially important for understanding of membrane events on the surface, or plasma, membrane.

Final comments will be directed to the design and interpretation of experiments related to reproduction and will emphasize the dynamics of time frame and spatial relationships.

COMPLEXITY OF THE CELL SURFACE

The concept of integral (intrinsic) and peripheral (extrinsic) proteins found in the original fluid-mosaic model illustrates an awareness that the external face of the membrane was complex, and the differential staining of the outer surface of cells (Revel, et al, 1960) provided strong evidence for the specialized distribution of carbohydrate on the cell surface. The growing array of biophysical techniques available to study the properties of cell surfaces (pp 31-48 of Rao, 1987) provide a large body of evidence that the surface membranes of cells differ, and that "layers" of unique molecules extend from the bilayer into the surrounding media. Molecules in this succession of layers (Figure 2) "shield" the bilayer from the surrounding solution. Exposure to external phospholipases resulted in greatly different extents of hydrolysis of phospholipids of erythrocytes and erythrocyte ghosts (Zwall and Roelofson (1976) and Ottolenghe (1973)). This was because of shielding, as illustrated in Figure 2.

Figure 2. Schematic representation of the layers found on the cell surface. This scale drawing illustrates the types of molecules, and their spacial relationships, that can exist on the surface of a cell. The frequency of appearance of the proteins that extend from the bilayer to form the glycocalyx or are adsorbed is unknown, and probably differs among cell types. The relative sizes of soluble components of molecular weights of 200 and 60,000 are illustrated. Adapted from Susko-Parrish, et al., 1985.

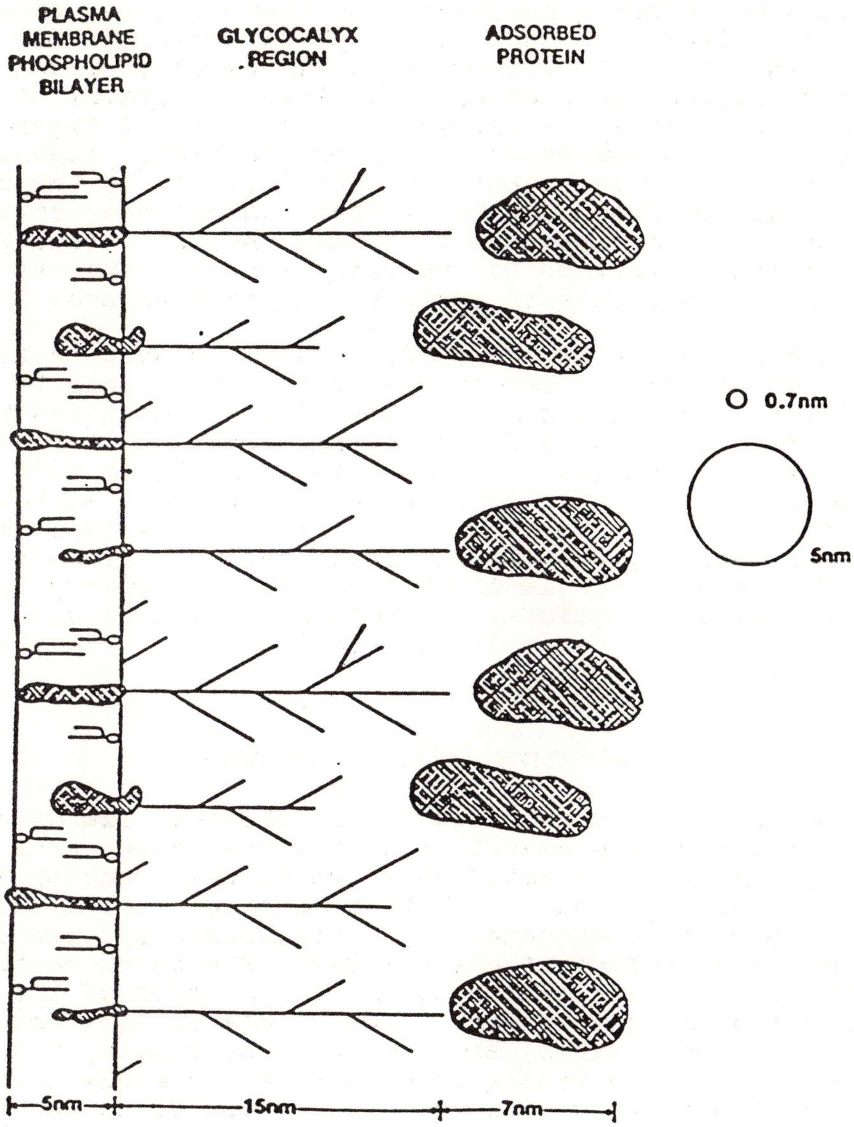
PLASMA
MEMBRANE
PHOSPHOLIPID
BILAYER
GLYCOCALYX
REGION
ADSORBED
PROTEIN
0.7nm
5nm
5nm
15nm
7nm

This succession of layers undoubtedly exists on the surface of many-to-most cells, but the distribution of the components and their density on the surface is difficult to establish. Adsorbed protein serves as a general descriptor for many of these macromolecules, and the necessity of washing cells before analysis will remove loosely bound components before the analysis begins. These complicating factors make interpretation of data obtained from such suspensions very difficult.

Hormone-receptor binding analyses often reveal alterations in the number of receptor sites as a function of physiological state. In many cases, but certainly not all, subsequent analyses established that biosynthetic and degradative processes resulted in removal of (or addition to) receptors from the bilayer. Inspection of Figure 2, which is drawn to scale, shows how access of an extracellular ligand (the large sphere represents a molecule of about 60,000 daltons) to the bilayer is dependent on the disposition of other surface components exterior to the bilayer. In contrast, the approach of a small molecule (eg. glucose; small sphere) is less likely to be impeded. Such hinderances often are not considered in the analysis of surface binding phenomena.

If the ligand for study has only one class of binding sites, the binding properties can be accurately assessed. Unfortunately, multiple classes of binding sites introduce complexities unrecognized by most investigators. The reader should consult the articles of Klotz (1982, 1983) for examples and details. One example of multiple sites is the analysis of lectin binding sites on the sperm surface. Susko-Parish et al. (1985) tested the use of iodinated lectin and classical Skatchard analysis for this purpose; no reliable quantitative data were obtained. Subsequently, Magargee et al (1988) utilized flow cytometry, under carefully standardized conditions, for cell-by-cell analysis of the same phenomenon and were able to quantitate relative binding to the sperm surface for a variety of lectins.

ASYMMETRY OF CELL SURFACES

This concept also was considered in the original fluid-mosaic model, where the pictorial displays clearly indicate that components in the membrane were unequally distributed relative to the exterior and interior faces (Figure 1). Further, based on staining properties mentioned previously, carbohydrate components (eg., gangliosides) were selectively enriched on the exterior face of the bilayer. The major contribution of the past decade was the delineation of the orientation of the phospholipids of the bilayer. These concepts are lucidly described in a review by Cullis and Hope (1985) and the texts by Houslay and Stanley (1982) and Jain (1987); the reader is encouraged to study these contributions for the primary references and exposure to the limitations to the general statements presented below.

Analysis of the lipid components in membranes involves: isolation of the cell(s) of interest free of "significant" contaminants; removal and purification of membranes, including providing proof of their point of origin in the cell; extraction of the lipids from the recovered membrane vesicles; and analysis of the lipid classes. Methods to accomplish these interdependent steps have been developed, and the resulting data base describes the major lipids for a variety of cell and membrane types. The large variety specific molecular forms (eg., phospholipid types with a variety of acyl substituents) has been bewildering, in that no general structural solution to the common biological problem of assembly of a stable bilayer was apparent. Major contributions of the last decade have been in identifying common features of these components to allow general statements about membrane structure.

The phospholipid asymmetry of plasma membranes for five cell types is presented in Figure 3. For a membrane of the dimensions of a cell, a stable bilayer will have the phospholipids equally divided between the outer and inner surfaces (monolayers or leaflets). However, the individual phospholipid types need not be equally distributed toward the inner and outer monolayers. This certainly is true for the plasma membrane, where phosphatidylcholine (PC) and sphingomyelin (SPM) are preferentially oriented "out" with phosphatidylethanolamine (PE), phosphatidylserine (PS) and phosphatidylinositol (PI) preferentially oriented "in". Since the phospholipid classes differ in the general types of acyl substituents, with PC and SPM having the more saturated members and PE, PS and PI having the more unsaturated components, this asymmetry in lipid by head group also introduces asymmetry in the hydrocarbon portion of the bilayer. To a first approximation, this asymmetry between outer and inner leaflets could result in differences in flexibility gradient (freedom of motion proceeding from the bilayer face into the interior) for the exterior and interior faces. Model system studies also establish that cholesterol (CHOL) preferentially interacts with SPM and PC relative to PE, PS and PI, thereby introducing the possibility of a asymmetric distribution of the sterols. Definitive proof of the asymmetric distribution of cholesterol in cells has been difficult because of a dearth of methods to establish sterol localization, but the data for model systems argue strongly for this feature.

<u>Figure 3: Phospholipid distributions for surface membranes</u>. The percentage of the phospholipids that are oriented to the outer vs the inner faces of the plasma membrane are shown for: A, human erythrocyte; B, rat liver sinusoidal surface; C, rat liver continuous surface; D, pig platelet; and E, BHK-21 cells. Abbreviations for the lipids are presented in the text. Taken from Cullis and Hope (1985) with permission of the publisher.

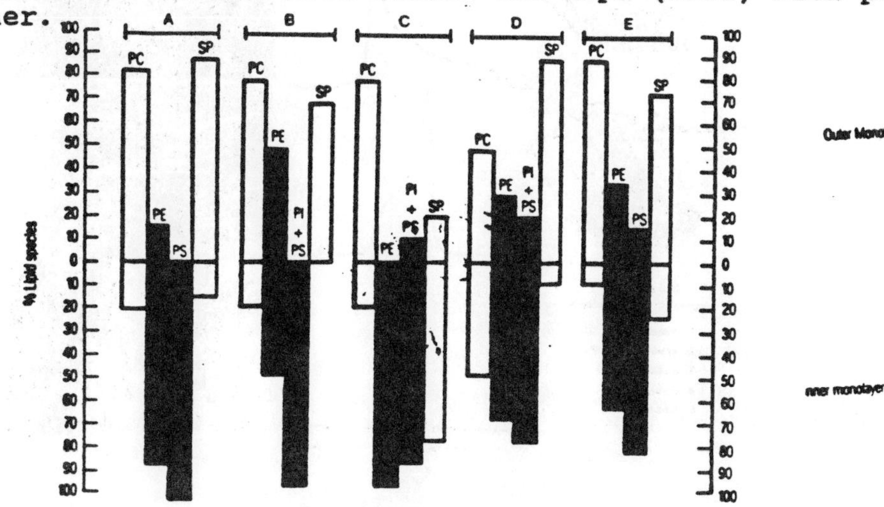

Detailed studies of model lipid dispersions have proven most useful to the resolution of these difficulties, in that precise spectroscopic and microscopic methods can be applied to the analysis of lipid assembly into phases. A clear representation of several possibilities (Figure 4) considers two general cases. First and most predominant, a situation where molecules <u>cannot</u> exhibit total freedom of motion in all dimensions is classified as an anisotropic systems (bilayer and hexagonal (H_{II})). The second, where complete freedom of motion can occur, is an isotropic mode. Specific examples include a bilayer (orientation of headgroup out, and respective acyl side chains toward each other and intercalated), the hexagonal phase (polar head groups in and acyl side chains out) and any of several orientations (vesicles, micellar, etc.) where the members have isotropic motion. Each is recognizable by virtue of its nuclear magnetic resonance (NMR) spectroscopic features and the images gathered by freeze fracture electron microscopy (EM) analysis. Detailed studies of many types of lipids have established that each has a preference for the phase it will adopt when placed in an aqueous environment. Some lipids (PC, SPM, PS (at pH > 3), PI) prefer the bilayer while others (PE, PS (at pH < 3), CHOL) prefer the hex_{II} orientation. If a membrane was composed of only one lipid, reasonably accurate predictions of the phase of the membrane formed could be made. Even then, predictive ability is limited as evident from observations that addition of secondary components (cholesterol or long-chain unsaturated fatty acids) to lipids preferring the bilayer form will induce a transition to a hexagonal orientation.

<u>Figure 4. Unique NMR spectra and freeze-fracture patterns for various phospholipid phases</u>. Preparation of lipid dispersions to yield putative bilayer, hexanonal and isotropic phases, followed by analyses by NMR (center panel) and electron microscopy (right panel) established the unique features of each phase. Taken from Cullis and Hope (1988) with permission of the publisher.

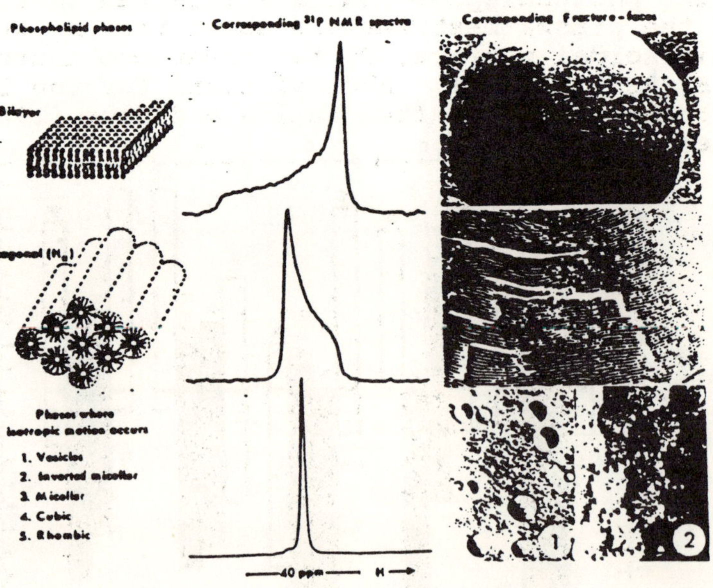

Considerable attention has been given to the changes in the degree of molecular ordering observed when temperature of the lipid phase is altered. At low temperatures the bilayer exists in a crystalline state, where the acyl chains are ordered. Raising the temperature then results in an alteration of state (to the liquid-crystalline array) where the acyl chains are disordered. Much attention has been focused on these transitions in phase state, most often characterized using techniques such as differential scanning calorimetry (Figure 5). Abrupt changes in molecular order of the acyl groups with increasing temperature are observed for pure components, but addition of other lipids (eg., cholesterol) eliminate these sharp transitions. The net effect for actual cellular membranes is that no such sharp transition occurs, and crystalline domains are unlikely to exist.

<u>Figure 5. Relationship between lipid composition and phase transition</u>. Careful heating of lipid preparations during analysis by differential scanning calorimetry tests for the temperature of conversion from the crystalline to the liquid-crystalline state. Pure lipids (DPPC, diphosphatidylphosphatidylcholine) undergo this transition at an unique temperature, but introduction of increasing amounts of a second lipid (cholesterol) results in a much broader range in the transition temperature. Adapted from a presentation by Houslay and Stanley (1982).

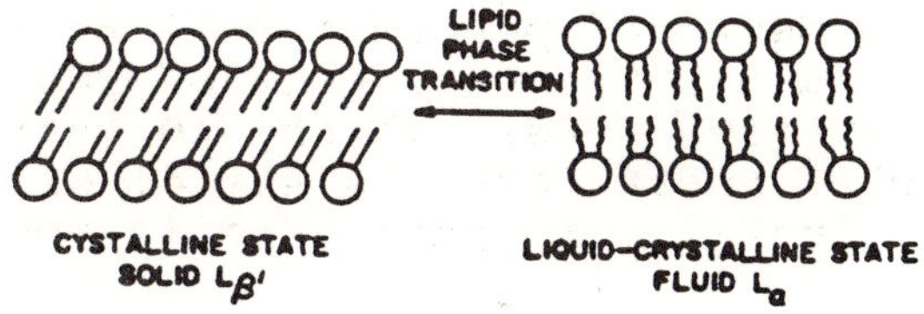

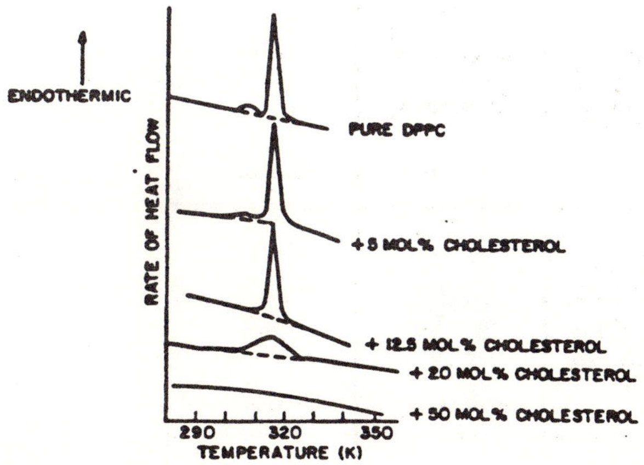

To recapitulate, detailed biophysical studies have established that quite dissimilar lipids, when place in water and allowed to assemble into aggregates, assume one of a limited number of overall orientations. Since the hexagonal phase would not provide the essential permeability barrier for demarcating the cell boundary, this form cannot provide the major form in the cell membrane. The bilayer form is preferred. Recent studies, however, are consistent with the conclusion that localized and transitory formation of hexagonal forms may be important in the function of the intact membrane.

These studies of model systems have uncovered another feature of lipids that may further simplify analysis of the interaction of lipids in the membranes. This is the concept that lipid interactions, and therefore patterns of assembly into large scale structures, can be interpreted in terms of the geometric shape of individual lipid molecules. The primary presentations (Israelachvili, et al., 1980, Kuypers, et al., 1984 and Gruner, 1985) establish that the dominant features are the molecular volume, head group area, and depth of the hydrocarbon side chains. From these considerations, the lipids can be grouped into molecular shapes and ultimately to predict the phase into which they will assemble (Figure 6). A bilayer could be envisioned as a collection of shapes, some with a predisposition to assume one of the other phase forms (Figure 7).

Figure 6. Types of lipid packing shapes assumed by various lipids. Lipids can be grouped according to molecular parameters of molecular volume, surface area and depth of penetration into the bilayer. These considerations yield a limiting number of geometric shapes (cylinders, cones and inverted cones) which, in turn, can be related to the type of polymorphic phase form the system will assume when placed into water. See text for further details and references.

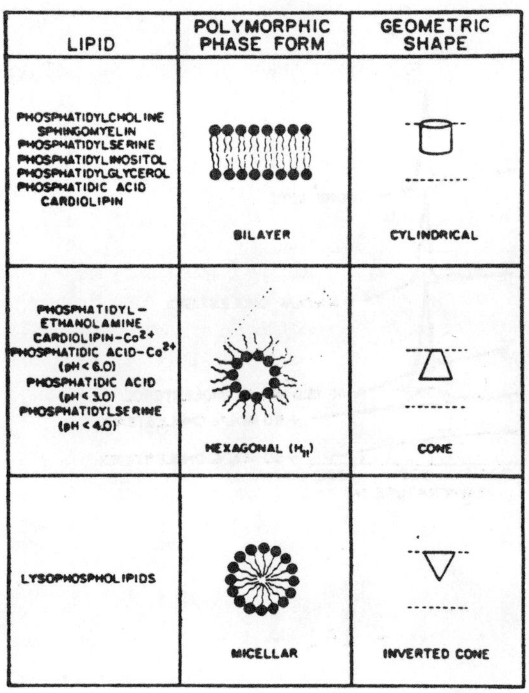

Figure 7. Schematic representation of the assembly of lipid into the bilayer. A collection of lipid shapes, selected from those presented in Figure 6, have been assembled to form a bilayer. Such a description using rigid geometrical forms is imperfect in that gaps between lipids would exist (see outer leaflet). These gaps could be accomodated by the flexibility of the lipids as well as by the insertion of proteins into the membrane.

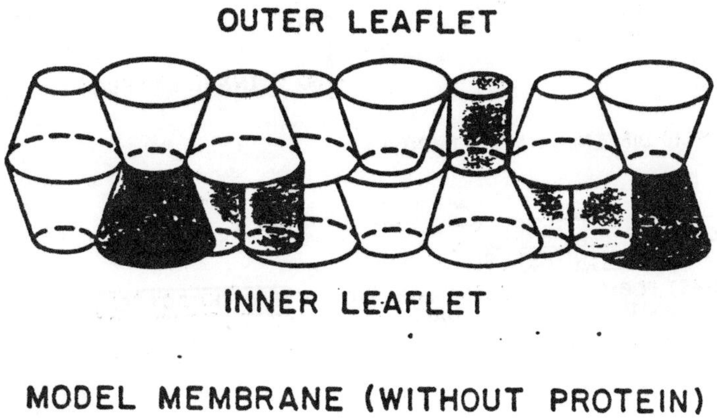

Results of recent studies on differentiation of stem cells to erythrocytes (Rayler et al., 1985) reveal changes in the "inner vs outer" display of the polar headgroups of phospholipids (Figure 8). Such changes provide one mechanism to greatly alter the surface during cellular differentiation. A second striking example (Kuypers, et al. 1984; Op den Kamp et al., 1985) is the in vitro incubation of erythrocytes, under conditions where surface lipids were exchanged without transbilayer movement from inner to outer bilayer leaflet, and a resulting gross alteration of the shape of the erythrocyte. This alteration can be explained by exchanging lipid molecules residing in the outer leaflet of the bilayer. Such a substitution would result in an alteration of the surface area of that leaflet, and force a change in the overall shape of the cell (Figure 9).

Figure 8. Schematic representation of the changes in the orientation of phospholipids during mouse erythroid differentiation. Differentiation of stem cells into erythrocytes has been intensively studied, with Friend cells considered by some investigators as a model for an intermediate stage in the process. Data provided by Rayler, et al. (1985) establish that the cell types differ in their orientation of phospholipids toward the outer and inner leaflets.

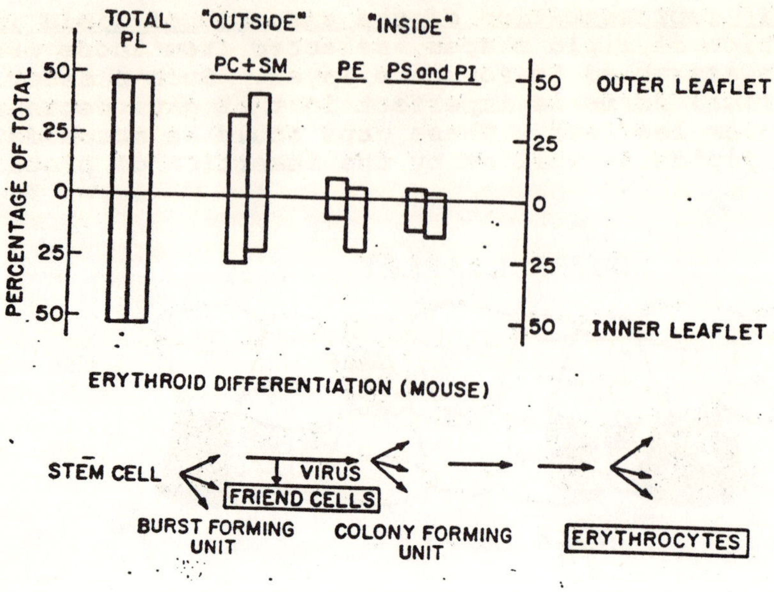

<u>Figure 9. Effect of substitution of lipids of the outer leaflet of erythrocytes on the overall shape of the cell</u>. Incubation of erythrocytes with donor vesicles of containing unique phosphatidylcholine molecules (differing in the acyl substituents, 16:0 vs 18:2) and phospholipid exchange proteins allows selective alteration of the outer surface of the bilayer (provided the time of the experiment is limited to less that that required for transbilayer movement). Such treatment converts the normal erythrocyte (center, bottom) to distinct forms. Taken from Op den Kamp, et al. (1985) with permission of the publisher.

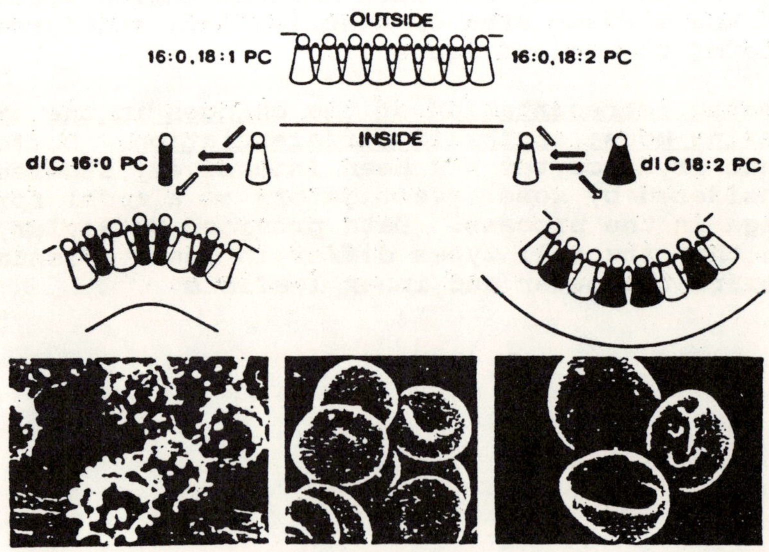

Controlled fusion of membranes represents a common membrane event, often associated with fundamental processes critical for reproduction. Careful studies of model systems are providing suggestions for the mechanism(s) by which these processes occur; no single mechanism has yet emerged from these studies (see Ohki et al., 1987). One potential mechanism involves inverted micelle forms, transiently produced by relocation of selected lipid components within the bilayer, to provide those essential features (localized points of stress, integration and relief of stress) necessary for formation of the new membrane forms (Figure 10).

Figure 10. Model for the role of minor lipids in the fusion of lipid vesicles. Lipid vesicles (panel a) can come together (panel b) and bring their lipid bilayers into molecular contact. Minor lipids, with a propensity for formation of hex II structures, could play a key role in the interaction of the membrane bilayers (panel c) before separation or fusion (panel d). Taken with Cullis and Hope (1988) with permission of the publisher.

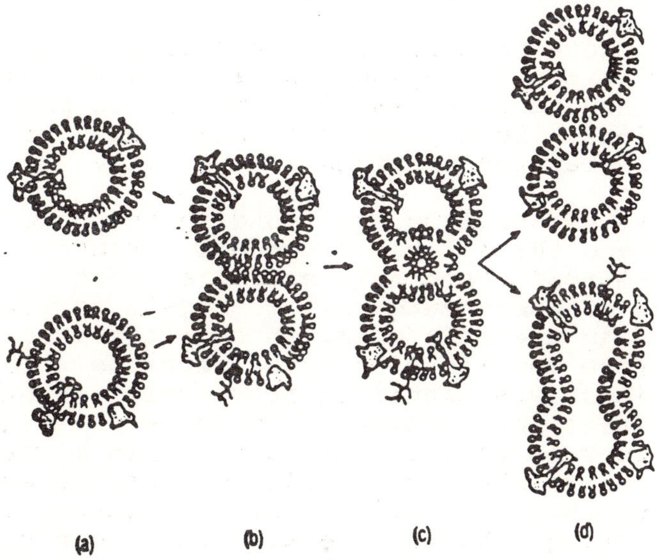

Existing data clearly establish that phospholipid asymmetry exists on cellular membranes and that this condition can persist over the lifetime of the cell. Such an unequal distribution carries with it the implication of energy expenditure to maintain disequilibrium. This could be accomplished by processes such as direct enzymatic action (phospholipid exchange enzymes moving lipids preferentially to one side of the bilayer) or synthesis and retention of proteins possessing unique binding specificities for the phospholipid head groups. Data support both concepts, but no evidence uniquely supports any single mechanism.

To summarize this section, asymmetric orientation of phospholipids on the separate bilayer leaflets exists. Their tendencies to form unique molecular aggregates can result in atlered cell functions and shape. These distributions phospholipids provides an excellent framework for the design and interpretation of experiments dealing with a variety of membrane related events.

LIPID DOMAINS

This concept, also anticipated by Singer and Nicholson, focuses on the extent to which the membrane is considered to be "fluid". Recent data clearly establish that the term must be used carefully. Fluidity is the inverse of viscosity. The implicit assumption, derived from the dictionary definition, is that the membrane would have an equivalent bulk viscosity in all dimensions. Limitations to use of the term for membrane structures have been discussed (see Lands, 1980). The predominant restriction is that freedom of motion in the membrane is very dimension-dependent in that lateral (two-dimensional on the surface), translational (across the bilayer) and rotational (around the axis of the acyl groups) motions each have their own unique features. This results in situations where molecules (eg., phospholipids of the bilayer boundary) can have great freedom of lateral motion but highly restricted translational motion. The general texts cited in the preceding section, as well as the review by Wolf (1988) and sections of the book by Loew (1988), provide an excellent introduction to the topic.

Even after acceptance of these limitations, it now is apparent that more stringent restrictions to movement of individual components within the membrane exist. They include restrictions due to phase segregation, lipid-lipid interactions, lipid-protein interactions, or morphologically recognized demarcation between portions of the cell.

Examples of phase separations are found in the data gathered from model systems of two and three lipid component mixtures. The predictions from the Gibbs' phase rule establish that at a given temperature a mixture of components will segregate to form unique gel phases (each enriched in specific components of the original mixture). These could tend to form domains within the membrane. The second form, lipid-lipid interactions, was mentioned previously; an example would be the apparent preferential interaction of cholesterol with selected phospholipids that occur as part of the total lipid mixture. Specific lipid-protein interactions also occur, although direct experimental proof of the extent of the interactions and the factors responsible for the interaction is lacking. The last feature is recognition of morphologically distinct portions of the membrane, where points of restriction (eg., tight junctions) preclude movement from one section of the surface to the other.

Direct proof of these aspects, much less the molecular basis for the restrictions, has been difficult to accumulate. However, the current implication that a variety of important cellular processes (membrane bound enzyme activation, sperm capacitation, receptor movements) are regulated in part by these membrane features assures that the topic will be intensively studied in the near future.

RELATIVE SPEED OF MEMBRANE EVENTS

Progress in the study membrane events related to reproductive processes undoubtably will proceed as a form of derivative science, where data from other areas (eg., erythrocyte and lymphocyte membrane structure-function; model liposome characterization) will serve as examples for experimental design. This is desirable. Progress with blood cells is easier because these cells have simplified membrane features and they are actively studied by a relative abundance of experienced investigators. The next two sections outline precautionary statements regarding this approach.

First, what is the most appropriate display of data to establish cause-and-effect relationships (Figure 11)? The tendency is to array data in terms of the maximum total response, leading to the assignment of order of events of A before B before C (in the left panel of Figure 11). This conclusion implies a known stoichiometry of interaction between the successive elements in the series and that no "overshoot" in the production of enzyme products occurs. While this may be true for soluble enzymes, the factors that regulate the activity of membrane-bound enzymes are less well known. Indeed, since they operate in a mileau where conformational adjustments may be slowed by surrounding molecules, it is highly probable that production of product beyond that needed for initiation of the next step will occur. The essential relationship is to supply "enough" product to satisfy the binding needs for the next step in the pathway. One example involves the sequence of events that follow activation of adenylcyclase, where absolute amount of cAMP produced in the first step is not important once a concentration sufficient to saturate the binding proteins involved in the next step of the sequence is achieved. The power of retrospective analysis, possible after detailed study of the sequence, reveals the quantitative coupling between individual steps. This provides the revised interpretation of data in the right panel of Figure 11. Thus, hypotheses should not prematurely die because of our lack of understanding of the stoichiometry of membrane related events.

The next point of consideration is the effect of the time axis on analysis of data (Figure 11). Testing of a hypothesis requires that experimental observations be made with sufficient rapidity to clearly establish the kinetic relationships between factors of interest. The elegant hypotheses related to events of reproduction often are tested by evaluating features related to changes in membrane permeability, movement of molecules in/on the membrane or modification of surfaces. A listing of such events (Figure 12), when scaled to a common axis (logarithmic, in seconds), is embarassingly informative. It alerts us to the extreme rapidity of membrane events described in other portions of this paper and to the fact that all too often the true scale of measurements is inappropriate for the events to be studied.

Figure 11. Effect of selection of axis dimensions on the the assignment of sequence in a metabolic process. For a simulated system, data for response of three enzymes (evaluated by assay of products of each) after addition of a stimulent are given (left panel). For each enzyme, the data were normalized to the maximum amount of product detected over the total time of the experiment. Precursor-product assumptions would suggest the relationship of A before B before C. It is possible that a detailed examination of the stoichiometry of the individual steps relative to overall pathway needs (see text for examples) would reveal that C is produced in vast excess, and A in moderate excess relative to B. Such considerations would be consistent with the sequence C before A before B.

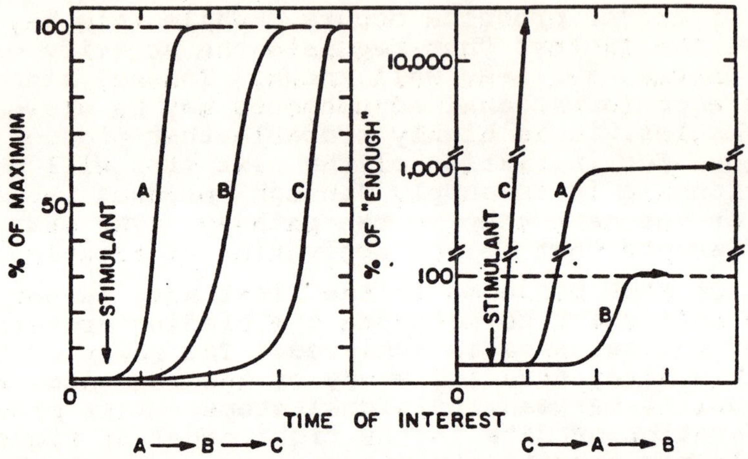

__Figure 12. Relationship of the time scale of events occuring on membranes to those of reproduction__. Events in/on membranes differ greatly in time, ranging from picoseconds (10^{-12}) to months (10^{7} seconds) while molecular and organismal events of reproduction are confined to a much smaller range of time. The effects of such considerations on the design of experiments are discussed in the text. Adapted from a presentation by Jain (1987).

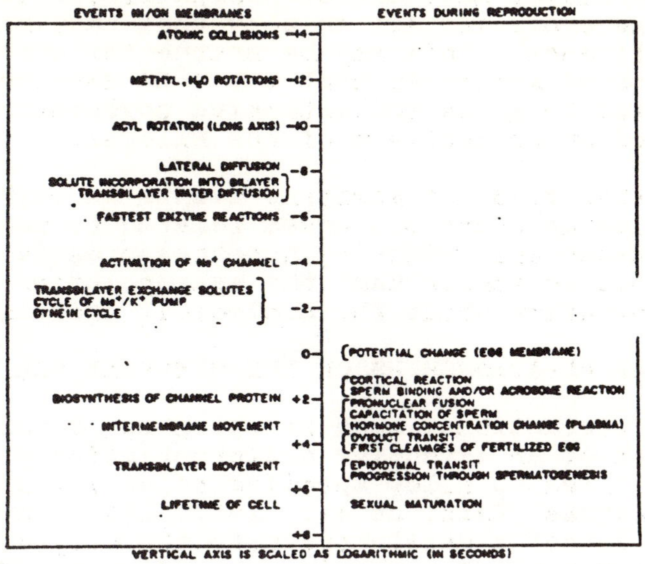

The comparison in Figure 11 will be used for two illustrations. In the first, very unique experimental designs are essential for any direct test of a postulated relationship of regulation of a cellular event involved in reproduction via a mechanism such as alteration of Na^+

movement through channels. Since these processes differ in speed by at least four orders of magnitude, no single incubation will test the desired relationship. Since direct test is impossible, elimination of alternative possibilities is most difficult. The required experimental design, use of several experiments featuring overlapping analyses of several cellular events to provide the desired temporal relationship, is rarely used. The second illustrates an opposite effect. Review of many discussions of transbilayer movement of phospholipids leaves the impression that these processes are so slow that they may be of minimal physiological significance. This may be untrue for events such as the epididymal maturation of sperm, in that the 7-14 days required for the process are sufficient to allow for extensive movement of phospholipids between the inner and outer leaflets of the bilayer.

It is evident that study of membrane events is very immature, especially for complex membrane processes related to reproduction. Great care must be taken when adapting hypotheses derived from studies on more simple systems to assure that the extrapolation is consistent with additional information about the biological process.

SPATIAL RELATIONSHIPS ON THE MEMBRANE SURFACE

Cell-to-cell interactions, such as sperm-egg binding and fusion, reflect the end point of an extensive differentiation process whereby the surfaces are modified to allow specific molecular interactions. Katz, Cone and colleagues (Katz, et al., 1987; 1988) have calculated the forces necessary to tether a motile sperm to the oocyte, and provide an order-of-magnitude estimate of a requirement of up to 10 noncovalent bonds between the respective surfaces. This serves as a minimum estimate, since mouse sperm binding to the zona pellucida may ultimately involve 10-50 x 10μ molecules per sperm head (see review by Wassarman, 1987). Since it is possible to induce specific membrane vesiculation processes, related to the acrosome reaction, using liposome surfaces that are much smaller than those of the egg (Graham et al, 1986; Graham and Hammerstedt, 1988), the following questions merit consideration. What are the spatial requirements for a specific interaction, with special emphasis on relating the body of microscopic data with the emerging biochemical description of the cell surface? Given the large difference in size of a liposome and an oocyte, are comparisons of events occurring on each surface directly comparable?

The scale drawings in Figure 13 illustrate both comparisons. On the left is a model of bovine sperm binding to an oocyte, while in the center (at 10X increase in scale) is binding of a liposome to a bovine sperm. At this level, direct comparison of events between the sperm surface and an oocyte vs a liposome appear inappropriate. However, when viewed from the perspective of the bonds to be formed (right panel, at a further 100X increase in scale) it is clear that significant interactions of the sperm surface can occur with either surface. The 10 bonds needed for tethering will come from portions of the surface that are indistinguishable using microscopic techniques. These illustrative examples serve to show the difficulty of direct comparisons of data gathered using diverse techniques ranging from microscopy to biochemical and biophysical analyses.

Figure 13. Size relationships of the cells and structures involved in sperm egg fusion. In panel A a bovine sperm is shown superimposed on a bovine egg, in panel B a liposome is showm bound to the sperm head and in panel C the liposome is shown relative to the bilayer of sperm surface (see Figure 2). The marker bars illustrate the 1000X increase in magnification. The effects of these considerations on the design and interpretation of experiments is presented in the text.

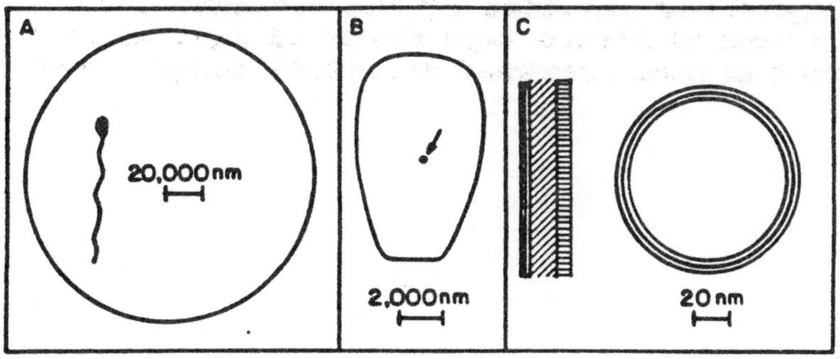

SUMMARY

Introduction of the fluid-mosaic model of a membrane in 1972 remains a landmark contribution to cell biology. As to be expected, subsequent contributions have refined the model by reemphasizing and clearly depicting the modes of molecular heterogeneity possible on any given surface. These unique orientations and motional properties of the molecules ultimately will allow discrimination among superficially similar, but uniquely regulated, membrane events.

Data from model systems aid understanding the potential characteristics of membrane surfaces involved in fundamental processes of reproduction. However, the challenge to those interested in reproduction is to make appropriate extrapolations from a model system(s) to the complex cell-cell interactions of interest.

The objective of this overview is to introduce recent contributions that provide a startpoint for interpretation of data on changes in membrane composition. As an example, Parks and Hammerstedt (1985) described changes, associated with epididymal transit, in lipid composition of the plasma membrane overlying the acrosome of ram sperm. Concomitant with this passage, sperm acquire the capacity to undergo the acrosome reaction (Williams, et al., 1987). The challenge is to devise an appropriate means to test for a relationship between changes in structure (lipid and protein) and changes in function (ability for specific vesiculation). Unique interpretations have not been provided, since if the reported lipid changes are considered in terms of the concepts of membrane fluidity, a prediction of an increase, decrease or no change in this parameter can be made! The conclusion drawn depends on the model system accepted for comparison. The alteration in lipid composition should introduce a change in the asymmetry of lipids across the bilayer. This could have extreme effects on the interaction of this sperm surface with other surfaces. Distinctions among the various possibilities requires careful extrapolation from model studies, with a greater emphasis on direct evaluation of the physical features of the membranes of interest.

ACKNOWLEDGEMENTS

Financial support was provided by the National Institutes of Health (HD 13099) and the United States Department of Agriculture (GAM 8502535). Dr. James Graham provided essential suggestions for Figures 12 and 13.

REFERENCES

Cullis, P. R. and M. Hope 1985 Physical properties and functional roles of lipids in membranes. pp 25-72 in Biochemistry of Lipids and Membranes, D. Vance and J. Vance eds. Menlo Park, CA: Benjamin/Cummings Publishing.

Cullis, P. R. and M. J. Hope 1988 Lipid polymorphism, lipid asymmetry and membrane fusion. in Molecular Mechanisms of Membrane Fusion, S. Ohki, D. Doyle, T. D. Flanagan, S. W. Hui, and E. Mayhew eds. Plenum Press New York pp 37-51.

Frey, L. D. and M. Edidin, 1970 The rapid intermixing of cell surface antigens after formation of mouse-human heterokaryons. J. Cell Sci. 7: 319-335.

Graham, J. K., R H. Foote and J. J. Parrish 1986 Effect of dilaurylphosphatidylcholine on the acrosome reaction and subsequent penetration of bull spermatozoa into zong-free hamster eggs. Biol Reprod 35: 413-427.

Graham, J. K. and R. H. Hammerstedt 1988 Induction of motility, the acrosome reaction and egg penetration in epididymal ram sperm by liposomes. Society Study Reproduction, 37 (Suppl 1) Abst 309.
Gruner, S. M. 1985 Intrisic curvature hypothesis for biomembrane lipid composition: A role for nonbilayer lipids. Proc. Natl. Acad. Sci. (USA) 82: 3665-3669.

Houslay, M. D. and Stanley, K. K. 1982 Dynamics of Biological Membranes New York: John Wiley and Sons.

Israelachvili, J. N., S. Marcelja, and R. G. Horn 1980 Physical principles of membrane organization. Quart. Rev. Biophysics 13: 121-200.
Jain, M. K. 1987 Introduction To Biological Membranes, 2nd Edition, John Wiley and Sons, New York.

Katz, D. F., E. Z. Drobnis, G. N. Cherr, J. Baltz, A. I. Yudin, R. A. Cone, and L. Y. Cheng 1987 The biophysics of sperm penetration of the cumulus and zona pellucida. Pp 275-285. in New Horizons in Sperm Cell Research, H. Mohri ed. Tokyo: Japan Sci. Soc. Press.

Katz, D. F., E. Z. Drobnis, J. Baltz, and N. DeMestre 1988 Analysis of forces generated by sperm against the zona pellucida. Society Study Reproduction 38 (Suppl 1) Abstract 285

Klotz, I.M 1982 Number of receptor sites for Scatchard graphs: facts and fantasies. Science (USA) 217:1247-1249.

Klotz, I. M. 1983 Number of receptor sites from Scatchard and Klotz graphs: a constructive critique. Science (USA) 220: 981.

Kuypers, F. A., B. Roelofsen, W. Berendsen, J. A. F. Op den Kamp, and L. L. M. van Deenen 1984 Shape changes in human erythrocytes induced by replacement of the native phosphatidylcholine with species containing various fatty acids J. Cell Biol. 99: 2260-2267.

Lands, W. E. M. 1980 Fluidity of Membrane Lipids Pp 69-73 in Membrane Fluidity eds M. Kates and A. Kukses Humana Press, Clifton NJ

Lowe, L. M. 1988 Spectroscopic Membrane Probes. Vol 1 Boca Raton FL: CRC Press.

Magargee, S. F., M. E. Kunze and R. H. Hammerstedt 1988 Changes in lectin-binding features of ram sperm surfaces associated with epididymal maturation and ejaculation. Biol. Reprod. 38: 667-685.

Ohki, S., D. Doyle, T. D. Flanagan, S. W. Hui, and E. Mayhew 1987 Molecular mechanisms of membrane fusion. New York: Plenum Press.

Op den Kamp, J. A. F., B. Roelofsen, and L. L.M. van Deeden Structural and dynamic aspects of phosphatidylcholine in the human erythrocyte membrane. Trends Biochem. Sci. 10: 320-323.

Ottolenghi, A. 1973 Preparation and characterization of mouse intestinal phospholipase. Lipids 8: 415-425.

Parks, J. E. and R. H. Hammerstedt 1985 Developmental changes occurring in the lipids of ram epididymal spermatozoa plasma membrane. Biol. Reprod. 32: 653-668.

Rao, K. V. 1987 Role of cell surface in development. Vol 1 Boca Raton FL: CRC Press.

Rayler, A., P. H. van der Schaft, B. Roelofsen, J. A. F. Op den Kamp 1985 Phospholipid localization in the plasma membrane of Friend erythroleukemic cells and mouse erythrocytes. Biochem. 24: 1777-1783.

Revel, J. P., L. Napolitano and D. W. Fawcett 1960 Identification of glycogen in electron micrographs of thin tissue sections. J. Biophys. Biochem. Cytol. 8: 575-589.

Robertson, J. D. 1981 Membrane structure. J. Cell Biol. 91: 189s-204s.

Singer, S. J. and G. L. Nicholson 1972. The fluid mosaic model of the structure of cell membranes. Science (USA) 175: 720-731.

Susko-Parish, J. S., P. L. Senger and R. H. Hammerstedt 1985 Binding of 125 I-succinylated concanavalin A to bovine spermatozoa. Biol. Reprod. 32: 129-136.

Wassarman, P. M. 1987 Early events in mammalian fertilization. Ann. Rev. Cell Biol. 3: 109-142.

Williams, R. M., J. E. Graham and R. H. Hammerstedt 1987 In vitro induction of the acrosome reaction and hamster egg penetration in epididymal and ejaculated ram sperm. Society Study Reproduction 36 (Suppl 1) Abstract 68.

Wolf, D. E. 1988 Probing the lateral organization and dynamics of membranes. Pp 193-220 in Spectroscopic Membrane Probes, L. M. Loew, ed. Boca Raton, FL: CRC Press.

Zwall, R. F. A., B. Roelofsen, 1976 Applications of pure phospholipases in membrane studies. in Biochemical Analysis of Membranes. Pp 352-377 ed A. H. Maddy New York: John Wiley and Son

MAMMALIAN OOCYTE MATURATION: MECHANISMS FOR REGULATION AND PROSPECTS FOR PRACTICAL APPLICATION OF IN VITRO TECHNOLOGY

John J. Eppig

In 1935, Pincus and Enzmann observed that germinal vesicle (GV) stage oocytes isolated from rabbit follicles underwent spontaneous germinal vesicle breakdown (GVB) and emitted a polar body when cultured in medium without the gonadotropic hormones that promote follicular maturation and ovulation. They suggested that the ovarian follicle exerts a meiosis-arresting influence on the oocyte until the preovulatory surge of gonadotropins. This observation and hypothesis began an era that has used culture systems for mammalian oocytes to resolve the mechanisms that govern oocyte maturation. In addition to providing clues to the molecules and pathways that regulate oocyte maturation, the culture systems may provide the technology for resolving problems of human infertility and for expanding populations of rare and endangered species. This paper will review some current concepts of the regulatory systems that govern mammalian oocyte development, evaluate the prospects for the practical utilization of this information, and suggest some of the future directions for research.

The Ovarian Follicle as the Support System for Oocyte Development In Vivo

After a period of mitotic proliferation in the fetus, oogonia enter meiosis, but the germ cells now called oocytes progress only as far as the diplotene stage where they are arrested until shortly before ovulation. Diplotene oocytes develop a close association with somatic cells that are probably progenitors of granulosa cells and form a primordial follicle. It is likely that oocytes that do not become associated with these pregranulosa cells degenerate. Until an oocyte is recruited into the pool of growing oocytes, the pregranulosa cells remain in a single, often flattened, layer around the oocyte. But when the oocyte is recruited, by unknown mechanisms, into the pool of growing oocytes, granulosa cell proliferation occurs concurrently with the increase in the size of the oocyte. Both growing and nongrowing oocytes appear to be coupled to their companion somatic cells by membrane specializations called gap junctions, which allow the passage of low molecular weight molecules to pass from one cell to another (Anderson and Albertini, 1976).

Gap junctional communication between granulosa cells and the oocyte is essential for oocyte growth, since they are a conduit for nutritional and regulatory substances into the oocyte (Eppig, 1977, 1979; Brower and Schultz, 1982; Herlands and Schultz, 1984; Buccione, et al., 1987). Although gap junctional communication can be established between oocytes and other somatic cell types in vitro, oocyte growth occurs only when oocytes communicate with granulosa cells (Buccione et al., 1987). Some aspects of oocyte development, nevertheless, are independent of oocyte growth or association with granulosa cells. Normally, as oocytes approach completion of their growth phase, they become competent of undergoing spontaneous maturation when liberated from their follicles and cultured (Szybec, 1972; Sorensen and Wassarman, 1976). The developmental program governing the acquisition of meiotic competence, however, is triggered at the onset of oocyte growth, but proceeds in the absence of continued oocyte growth or granulosa cells (Canipari et al., 1984). Once meiotic competence is achieved, granulosa cells, via gap junctional communication, help to maintain the oocyte in meiotic arrest (Eppig and Downs, 1987).

Because cumulus granulosa cells are coupled by gap junctions to the oocyte and to the mural granulosa cells which make up the follicle wall, the follicle may be a functional syncytium allowing the movement of low molecular weight molecules throughout the follicle by diffusion. In theory, therefore, production or mobilization of low molecular weight regulatory substances in mural granulosa cells in response to an agonist could result in the diffusion of these substances to the oocyte, which may itself be unable to produce or mobilize them. These substances could participate in the maintenance of meiotic arrest, the induction of maturation, or both.

Identification of Molecules and Pathways that Participate in the Maintenance of Meiotic Arrest

Cyclic Adenosine Monophosphate (cAMP). In the ovarian follicles of most studied mammalian species, cAMP probably participates in the maintenance of oocyte meiotic arrest. Support for this conclusion is based on the following observations. Membrane permeable analogs of cAMP, such as dibutyryl cAMP (dbcAMP), and cAMP phosphodiesterase inhibitors, such as 3-isobutyl-1-methylxanthine (IBMX), maintain meiotic arrest in vitro (Cho et al., 1974; Wassaman et al., 1976; Magnusson and Hillensjo, 1977). Microinjection of an inhibitor of cAMP-dependent protein kinase (PKI) induces GVB in oocytes cultured in medium containing dbcAMP or IBMX (Bornslaeger et al., 1986). Conversely, microinjection of the active catalytic subunit of cAMP-dependent protein kinase maintains meiotic arrest (Bornslaeger et al., 1986). Finally, a decrease in the cAMP content of mouse (Schultz et al., 1983); and rat (Racowsky, 1984) oocytes appears to precede GVB, but such a decrease is not observed in the oocytes of hamsters (Racowsky, 1985a; Hubbard, 1986), sheep (Moor and Heslop, 1981), or pigs (Racowsky, 1985b). Oocytes of several mammalian species appear able to produce their own cAMP because they possess an adenylate cyclase system (Urner et al., 1983; Ekholm et al., 1984; Bornslaeger and Schultz, 1985a; Racowsky, 1985ab; Kuyt et al., 1988). It is not clear, however, that the oocytes of all of these species can produce levels of cAMP adequate for the maintenance of meiotic arrest. If not, the cAMP molecule is small enough to traverse gap junctions and, theoretically therefore, should be able to enter the oocyte from the cumulus/mural granulosa cell syncytium. Thus far, however, attempts to demonstrate such movement (Racowsky, 1984, 1985b; Bornsleager and Schultz, 1985b) have resulted in ambiguous conclusions because it is not known whether increases in the concentrations of cAMP in oocytes after elevating levels of cAMP in cumulus cells are the result of the transfer of cAMP from the cumulus cells, or the generation of substances in the cumulus cells that increase production of cAMP, or decrease loss of cAMP, in the oocyte.

Oocyte Maturation Inhibitor (OMI) Preparations of follicular fluid from several species prolong the maintenance of the GV stage in cultured oocytes (Tsafriri et al., 1982). Evidence has been presented that the active meiosis-arresting fraction of porcine follicular fluid is a low molecular weight peptide and has been referred to as OMI (Tsafriri et al., 1982). Confirmation of the existence of OMI awaits purification to homogeneity. Crude preparations of OMI, however, apparently act via cumulus cells, and the meiosis-arresting action is reversed by treatment with luteinizing hormone (LH) (Tsafriri et al., 1982).

Mullerian Inhibiting Substance (MIS) MIS, also known as anti-Mullerian hormone (AMH), has been identified in bovine follicular fluid and evidence from one group suggests that partially purified MIS has transient meiosis-arresting activity in rat, but not mouse oocytes (Takahashi et al., 1986). Recently, however, it has been demonstrated that immunopurified, biologically active MIS does not maintain meiotic arrest in rat oocytes (Tsafriri et al., 1988). It was suggested, therefore, that the meiosis-arresting action of he partially purified preparation of MIS could be that of a contaminant (Tsafriri et al., 1988). These studies emphasize the importance of a complete purification of biological preparations before clear identification and characterization of active factors can be possible.

Purines Hypoxanthine was found in preparations of porcine follicular fluid (Downs et al., 1985). This purine accounted for most of the meiosis-arresting activity of porcine follicular fluid on mouse oocytes. Concentrations of hypoxanthine in biological fluids can be increased substantially by a variety of pathological conditions, including anoxia. Since the porcine follicular fluid used for analysis was not collected under conditions that would prevent the artifactual generation of hypoxanthine, the concentration of purines in mouse follicular fluid was assessed when collection was made under more controlled conditions (Eppig et al., 1985). For example, the conversion of adenosine to hypoxanthine was prevented by treatment with an inhibitor of adenosine deaminase; nonetheless, millimolar concentrations of hypoxanthine were found. In addition, murine follicular fluid was found to contain substantial amounts of adenosine. Although it is still possible that the concentration of hypoxanthine in mouse follicular fluid is not as high as reported because of rapid intrafollicular changes, it should be kept in mind that the ovarian follicle is not vascularized central to the thecal layers and a low oxygen concentration may be a natural condition in situ. In addition, hypoxanthine is the predominant purine found in rapidly frozen intact ovaries (Eppig, unpublished results).

When assessed at concentrations estimated to be present in murine follicular fluid, hypoxanthine maintained most mouse oocytes in meiotic arrest. Adenosine had only a transient meiosis-arresting action, but augmented the stable inhibitory effect of hypoxanthine (Eppig et al., 1985). The meiosis-arresting action of hypoxanthine andadenosine was fully reversible. Moveover, oocytes that were maintained in the GV stage by these purines for 24 hours and then matured in medium without purines could be fertilized and underwent embryonic development (Downs et al., 1986a).

Inhibitors of inosine monophosphate (IMP) dehydrogenase induced GVB in oocytes that had been maintained in meiotic arrest by hypoxanthine in vitro. The IMP-dehydrogenase inhibitors, however, did not induce GVB in oocytes arrested in the GV stage by guanosine, the most potent of the purines tested for meiosis-arresting activity (Downs et al., 1986b). These results indicate that the meiosis-arresting action of hypoxanthine in vitro may be mediated through the production of guanyl compounds. Likewise, IMP dehydrogenase inhibitors induced GVB upon injection into mice primed for 24 hours with PMSG. One of these inhibitors, bredinin, induced the maturation of almost all the meiotically competent ovarian oocytes within 6 hours of injection (Downs and Eppig, 1987). It is not likely that the GVB-inducing action of IMP dehydrogenase inhibitors was mediated by gonadotropins since other manifestations of gonadotropin action such as cumulus expansion, thinning of the apical wall of the follicle, or cessation of DNA synthesis in the granulosa cells were not observed after injection of the inhibitors (Downs and Eppig, 1987). These results support the idea that, in mice, the formation of guanyl compounds via the IMP dehydrogenase pathway is essential for the maintenance of meiotic arrest in vivo.

Because of the evidence indicating the importance of cAMP in maintaining meiotic arrest in mouse oocytes, the hypothesis that purines, particularly hypoxanthine and adenosine, may maintain meiotic arrest by promoting meiosis-blocking levels of cAMP in oocytes was tested. The maintenance of meiotic arrest was correlated with elevated cAMP levels in cumulus cell-enclosed oocytes cultured in medium containing hypoxanthine or hypoxanthine plus adenosine. In addition, microinjection of oocytes with an inhibitor of cAMP-dependent protein kinase induced GVB in oocytes cultured in medium containing hypoxanthine (Downs et al., 1989). These results show that hypoxanthine arrests mouse oocytes in the GV stage, at least in part, by maintaining meiosis-blocking levels of cAMP in the oocytes. Hypoxanthine, and also guanosine and adenosine, inhibited cAMP phosphodiesterase activity in lysates of mouse oocytes and augmented the cAMP-elevating action of follicle-stimulating hormone (FSH) in intact oocyte-cumulus cell complexes. These results suggest that one way that hypoxanthine may maintain meiosis-blocking levels of cAMP is by preventing hydrolysis of cAMP. It is likely, though, that purines promote the elevation of cAMP by other mechanisms as well.

Future research on the maintenance of meiotic arrest must focus on the identification of specific molecules in the cAMP-dependent pathway in oocytes. Moreover, a greater understanding is needed regarding the role of purines, such as hypoxanthine, adenosine, and GMP in meiotic arrest, and how the metabolism of these purines is regulated. It is likely that several different interconnecting pathways participate in the maintenance of meiotic arrest and that the relative contribution of these pathways may vary from species to species. The interactions of various factors such as gonadotropins, steroid hormones, and growth factors in the regulation of these pathways will be difficult to study, but their resolution is essential for a basic understanding of the mechanisms that maintain meiotic arrest.

Induction of Oocyte Maturation

Oocyte maturation in vivo is induced in ovulatory follicles by the preovulatory surge of LH. The mechanism by which LH promotes maturation, however, is not well understood. LH could stimulate the granulosa cells to some action that would deprive the oocyte of maturation-arresting substances. For example, LH could terminate the production of meiosis-arresting substances or promote the catabolism of these substances, or it could disrupt the gap junctional system that delivers the meiosis arresting substances to the oocyte. The concentration of cAMP in the intact follicle does not decrease after stimulation with hCG (Schultz et al., 1983). Nor is there a decrease in the concentration of hypoxanthine or adenosine in murine follicular fluid before GVB (Eppig et al., 1985). It is clear, therefore, that a decrease in the concentration of these substances in the whole follicle is not the mechanism of hCG-induced oocyte maturation. Moreover, the gap junctional communication between the oocyte and its companion cumulus cells does not decrease before GVB in vivo (Moor et al., 1981; Eppig, 1982; Salustri and Siracusa, 1983). It is possible that there are differences between the movement of the molecular markers used to measure gap junctional communication and the movement of meiosis-arresting substances, but if they are analogous, then it can be concluded that a disruption in the communication between cumulus cells and the oocyte is not the mechanism of LH-induced oocyte maturation. It has been suggested that LH induces a reduction in the communication between the mural granulosa cells and cumulus cells and thereby reduces the movement of meiosis-arresting substances to the oocyte-cumulus cell complex (Larsen et al., 1987). If cAMP is the meiosis-arresting substance that moves from the follicle wall to the oocyte-cumulus cell complex to maintain meiotic arrest, then a decrease in the cAMP content of the intact complex should precede GVB, or commitment to undergo GVB. Such a decrease in the cAMP content of the intact oocyte-cumulus cell complex was not detected after injecting mice with human chorionic gonadotropin (hCG) (Shultz et al., 1983; Eppig and Downs, 1988). These results, however, do not eliminate the possibility that the passage of other meiosis-arresting substances from the follicle wall to the oocyte-cumulus cell complex could be reduced.

After mice are injected with hCG, the cAMP content of the intact oocyte-cumulus cell complexes increases (Shultz et al., 1983) even though the gap junctional communication between the oocyte and the cumulus cells, which should allow the transfer of cAMP, is not reduced (Eppig and Downs, 1988). One explanation for this observation is that the oocytes become stimulated to secrete or hydrolyze their cAMP by a signal produced by the granulosa/cumulus cells in response to LH/hCG. Alternatively, the cumulus cells may have been stimulated to selectively reduce the transfer of cAMP, and not markers of gap junctional communication, to the oocytes.

There is evidence for a positive, maturation-inducing signal produced by granulosa cells in response to gonadotropin or epidermal growth factor (EGF) (Eppig and Downs, 1987; Downs et al., 1988). This signal overcomes the effect of meiosis-arresting substances. GVB is induced in cumulus cell-enclosed oocytes by either FSH or EGF even in the continual presence of hypoxanthine, dbcAMP, or IBMX. The frequency at which these agonists induced GVB is greater than the frequency of GVB promoted by simply denuding the oocytes in medium containing the meiosis-arresting substance, thus indicating that the agonists do not function merely by promoting the uncoupling of the cumulus cells from the oocyte. The identification of maturation-inducing signals is a major area for future research. Moreover, the possible physiological role, if any, of growth factors such as EGF in the regulation of oocyte maturation needs to be resolved.

Similarities Between Mechanisms that Govern Meiotic and Mitotic Cell Division

There are several lines of evidence suggesting that mechanisms that govern the induction of oocyte maturation are similar to the mechanisms involved in the induction of mitotic cell division. First, an autocatalytic maturation-promoting factor (MPF) is produced in frog oocytes in response to progesterone (Reynhout and Smith, 1974), in starfish oocytes in response to 1-methyladenine (Kishimoto and Kanatani, 1976), and in maturing mammalian oocytes (Sorenson et.al., 1974). A similar factor is also present in somatic cells as they enter M phase of the cell cycle (Kishimoto et al., 1982). Moreover, it has recently been shown that MPF purified from frog oocytes contains the product of a homolog of a mitosis control gene, $cdc2^{+}$, of yeast (Dunphy et al., 1988; Gautier et al., 1988). Second, mRNA coding for the clam protein cyclin A, a protein whose level rises and falls during both meiotic divisions of the oocyte and during the cell cycle of early embryogenesis, induces meiotic maturation upon injection into frog oocytes (Swenson et al., 1986). Third, an increase in free intracellular calium appears to be an essential part of the mechanisms that promote somatic cell division (Poenie et al., 1985; Hafner and Petzelt, 1987; Twigg et al., 1988). Likewise, calcium ionophores, which increase free intracellular calcium by uptake, induce the maturation of mammalian oocytes maintained in meiotic arrest with dbcAMP or IBMX (Powers and Paleos, 1982; Racowsky, 1986). Moreover, inhibitors of calmodulin prevent the spontaneous maturation of murine oocytes, but do not prevent a decline in oocyte cAMP (Bornslaeger et al., 1984). This observation supports the idea that intracellular calcium plays an important role in the initiation of GVB and functions in a pathway distinct from, but perhaps parallel to, the pathway activated by a decline in cAMP levels. Increases in the concentration of free calcium in oocytes could result from either the mobilization of calcium from intracellular stores or the uptake of external calcium. One could speculate that the maturation-inducing signal communicated from granulosa cells to the oocyte could be a factor(s) that promotes an increase in free calcium in the oocyte, by internal mobilization or uptake, or the signal could be calcium itself. It will be important to investigate these parallels between the control mechanism of mitotic and meiotic cell division further.

Murine Oocyte Growth and Development in Culture

Murine oocytes grow and acquire competence to undergo GVB in culture (Eppig, 1977). Oocyte growth is dependent upon metabolic coupling to granulosa cells, but the acquisition of competence to undergo GVB is not (Canipari et al., 1984). Therefore, the developmental program governing the acquisition of GVB competence is independent of oocyte growth. Moreover, the acquisition of GVB competence does not appear to require gonadotropins or steroid hormones (Eppig, 1977). In the rat, however, FSH and estradiol increase the frequency at which GVB competence is acquired (Bar-Ami et al., 1983).

For many years it was thought that oocytes that had matured in vitro were defective because they could not be fertilized and undergo embryonic development. The main reason for failure was thought to be the inability of ova matured in vitro to promote the development of the male pronucleus, but the failure was probably due to inadequate culture systems because it was later demonstrated that cumulus cell-enclosed murine oocytes that mature spontaneously in vitro are competent of fertilization and embryonic development at frequencies equivalent to oocytes that mature in response to superovulatory regimen in vivo (Schroeder and Eppig, 1984). Subsequently, normal development after maturation in vitro has also been reported for oocytes form some other species such as the sheep (Staigmiller and Moor, 1984). Much remains to be done, however, to resolve the optimal conditions for oocyte maturation for each species that will promote maximal developmental potential.

Murine oocytes indicate that oocytes sequentially develop competence for maturation, then fertilization and cleavage to the two-cell stage, and then development to blastocyst (Eppig and Schroeder, 1989). The competence for each of these developmental capacities is acquired in vitro and does not require the presence of gonadotropins. Oocytes isolated in mid-growth phase from 12 day old mice, have been grown, matured, fertilized, and cultured to the 2-cell stage in vitro, and developed to term after transfer to the oviducts of pseudopregnant foster mothers. Continued progress in the development of the technology for oogenesis in vitro is dependent upon basic studies on oocyte nutrition and metabolism, and on the action of hormones, growth factors, and other factors that control the differentiation and function of the functionally integrated oocyte-granulosa cell complex in vivo.

Prospects for the Practical Utilization of Oocyte Culture Systems

Expansion of Populations of Rare or Agriculturally Important Mammals

Very few primary oocytes that are present in the ovary are ovulated and made available for fertilization in mammals. For example, it has been estimated that there are more than 150,000 primordial follicles in the bovine ovary at birth, but not more than 300 of these will ever be ovulated normally (Erickson, 1966). Techniques of superovulation could increase the reproductive capacity of a cow only about tenfold even if it could be superovulated yearly during its reproductive lifespan (Seidel, 1981). This general rule probably holds true for most mammals. There is, therefore, a tremendous pool of oocytes having the potential for development and, therefore, expanding populations if the oocytes could be rescued from their normal fate of degeneration. One approach to such a rescue is to isolate the oocytes from their natural environment and grow them in vitro and, as described above, significant progress has been made to achieving this using mice as a model system. This technology would be applicable to the expansion of populations of agriculturally important animals such as those that have been genetically engineered for desirable characteristics, of important genetically engineered experimental animals, or of endangered species.

Our culture system for murine oocyte maturation in vitro rescues degenerating oocytes from deceased mice. Significant necrotic changes are readily apparent in the oocytes of mice dead for six hours at ambient temperatures but these changes are reversed during oocyte maturation in vitro. The eggs derived from the rescued oocytes can be fertilized and undergo normal development. Thus, it may be possible to salvage the oocytes of rare or agriculturally important animals even after unforeseen death. Eggs derived from murine ova that mature in vitro can also be frozen and show almost normal capacity for fertilization and development after thawing. Thus valuable ova can be preserved after development in vitro if appropriate sperm are not immediately available.

Genetic Engineering. Jenkins and Copeland (1985) have reported that the germ line of SWR/J-RF/J hybrid mice become infected with ecotropic retroviruses during oogenesis. Moreover, Panthier et. al (1988) have produced transgenic mice by inoculation of newborn SWR/J females with an ecotropic murine leukemia virus. These findings, along with the development of oocyte culture systems, could present a unique opportunity for the introduction of genes into the germline. Protocols could be developed to infect oocytes with retroviruses carrying specific genes that would become integrated into the germ line, introducing a significant advancement in the technology for producing transgenic mice. Preliminary studies have produced one mouse with integrated retrovirus after infecting oocytes during maturation in vitro. Perhaps greater success will be achieved in experiments wherein oocytes are infected during more prolonged development or maintenance in culture.

Clinical Applications In vitro fertilization (IVF) has become a realistic clinical resolution to a variety of infertility problems in both women and men. Nevertheless, the success rate could be increased by a radical change in this protocol to utilize oocyte maturation in vitro rather than the protocols of hormonal stimulation currently in use. The stimulatory protocols which promote follicular development and maturation of oocytes in multiple follicles in humans probably produces eggs with a reduced developmental capacity. Moreover, these protocols probably do not promote optimal development of the uterine epithelium for the implantation of the embryo. An alternative to the hyperstimulatory techniques is oocyte maturation in vitro. Our studies with mice indicate that ova with full developmental capacity can be recovered even from degenerating follicles. Thus, it may be possible to recover immature oocytes form several antral follicles, excluding the dominant preovulatory follicle, and mature them in culture while the preovulatory follicle prepares the uterus normally for implantation. Current ultrasonic techniques for follicular visualization and oocyte aspiration do not allow sufficient resolution of the follicles for this procedure, but this is an engineering problem with the ultrasound equipment and could be resolved with appropriate incentive.

Current protocols for human IVF and embryo culture are not based on exhaustive experimentation to define optimal culture conditions for human ova or embryos. Ethical problems have hindered progress in this area. One of these problems involves the source of ova for the studies. However, if immature oocytes could be recovered from ovaries removed for various clinical reasons and matured in culture with full developmental potential, the ethical problems associated with the source of ova for embryo research would be reduced.

Clinical realization of maturation in vitro as an antecedent to IVF will require basic studies to resolve likely differences in conditions for oocyte maturation that impart full developmental capacity between human and mouse oocytes. It would probably be advantageous and appropriate to carry out initial studies using non-human primate oocytes. Moreover, continued basic studies on oocyte metabolism and its regulation by companion somatic cells, hormones, meiotic regulatory substances and growth factors using rodent systems is essential. Mouse ova were used in the pioneering studies on IVF that have produced the clinical successes we now see every day. Certainly, experiments with mouse oocytes will continue to be indicators of what is possible with other species.

Acknowledgments

Research in my laboratory is supported by grants HD20575, HD21970 and HD23839 from the National Institutes of Health. The contents of this paper do not necessarily represent the official views of that agency. My most sincere thanks to Roberto Buccione, Allan Schroeder and Barbara Vanderhyden for their helpful suggestions for this paper.

References

Anderson, A., D.F. Albertini. 1976 Gap Junctions between the oocyte and companion follicle cells in the mammalian ovary. J. Cell Biol. 71:680-686.

Bar-Ami, S.A. Nimrod, A.M.H. Brodie, and A Tsafriri. 1983. Role of FSH and oestradiol-17B in the development of meiotic competence in rat oocytes. J. Steroid. Biochem. 19:965-971

Bornslaeger, E.A., M.W. Wilde, and R.M. Schultz. 1984. Regulation of mouse oocyte maturation: involvement of cyclic AMP phosphodiesterase and calmodulin. Dev. Biol. 105:488-499.

Bornslaeger, E.A., and R.M. Schultz. 1985a. Adenylate cyclase activity in zona-free mouse oocytes. Exp. Cell Res. 156:277-281.

_______. 1985B. Regulation of mouse oocyte maturation: effect of elevating cumulus cell cAMP levels. Biol. Reprod. 33:689-704.

Bornslaeger, E.A., P. Mattie, and R.M. Schultz. 1986. Involvement of cAMP-dependent protein kinase and protein phosphorylation in regulation of mouse oocyte maturation. Dev. Biol. 114:453-462.

Brower, P.T., and R.M. Schultz. 1982. Intercellular communication between granulosa cells and mouse oocytes: existence and possible nutritional role during oocyte growth. Dev. Biol. 90:144-153.

Buccione, R., S. Cecconi, C. Tatone, F. Mangia, and R. Colonna. 1987. Follicle cell regulation of mammmalian oocyte groth. J. Exp. Zool. 242: 351-354.

Canipari, R., Fl Palombi, M. Riminucci, and F. Mangia. 1984. Early programming of maturation competence in mouse oogenesis. Dev. Biol. 102: 519-524.

Cho, W.K., S. Stern, and J.D. Biggers. 1974. Inhibitory effect of dibutyryl cAMP on mouse oocyte maturation in vitro. J. Exp.Zool. 187: 383-386.

Downs, S.M., D.L. Coleman, P.F. Ward-Bailey, and J.J. Eppig. 1985. Hypoxanthine is the principal inhibitor of murine oocyte maturation in a low molecular weight fraction of porcine follicular fluid. Proc. Nat. Acad. Sci. USA 82: 454-458.

Downs, S.M., A.C. Schroeder, and J.J. Eppig. 1986a. The developmental capacity of mouse oocytes following maintenance of meiotic arrest in vitro. Gamete Res. 15: 305-316.

Downs, S.M., D.L. Coleman, and J.J. Eppig. 1986b. Maintenance of murine meiotic arrest: uptake and metabolism of hypoxanthine and adenosine by cumulus cell-enclosed and denuded oocytes. Dev. Biol. 117:174-183.

Downs, S.M., and J.J. Eppig. 1987. Induction of mouse oocyte maturation in vivo by perturbants of purine metabolism. Biol. Reprod. 36:431-437.

Downs, S.M., S.A.J. Daniel, and J.J. Eppig. 1988. Induction of maturation in cumulus cell-enclosed mouse oocytes by follicle-stimulating hormone and epidermal growth factor: evidence for a positive stimulus of somatic cell origin. J. Exp. Zool. 245:86-96.

Downs, S.M., Daniel, S.A.J., Bornslaeger, E.A., Hoppe, P.C., and J.J. Eppig. 1989. Maintenance of meiotic arrest in mouse oocytes by purines: modulation of cAMP levels and cAMP phosphodiesterase activity. (Gamete Res. (in press).

Dunphy, W.G., L. Brizuela, D. Beach, and J. Newport. 1988. The Xenopus cdc2 protein is a component of MPF, a cytoplasmic regulator of mitosis. Cell 54:423-431.

Ekholm, C., T. Hillensjo, C. Magnusson, and S. Rosberg. 1984. Stimulation and inhibition of rat oocyte meiosis by forskolin. Biol. Reprod. 30:537-543.

Eppig, J.J., 1977. Mouse oocyte development in vitro with various culture systems. Dev. Biol. 60:371-388

______. 1979. A comparison between oocyte growth in coculture with granulosa cells and oocytes with granulosa cell-oocyte junctional contact maintained in vitro. J. Exp. Zool. 209:345-353.

______. 1982. The relationship between cumulus cell-oocyte coupling, oocyte meiotic maturation, and cumulus expansion. Dev. Biol. 89:268-272.

Eppig, J.J., Ward-Bailey, and D.L. Coleman. 1985. Hypoxanthine and adenosine in murine ovarian follicular fluid: concentrations and activity in maintaining oocyte meiotic arrest. Biol. Reprod. 33:1041-1049.

Eppig, J.J., and S.M. Downs. 1987. The effect of hypoxanthine on mouse oocyte growth and development in vitro: maintenence of meiotic arrest and gonadotropin-induced oocyte maturation. Dev. Biol. 119:313-321.

______. 1988. Gonadotropin-induced murine oocyte maturation in vivo is not associated with decreased cyclic adenosine monophosphate in the oocyte-cumulus cell complex. Gamete Res. 20:125-131.

Eppig, J.J., and A.C. Schroeder. 1989. Capacity of mouse oocytes from preantral follicles to undergo embryogenesis and development to live young after growth, maturation and fertilization in vitro. Biol. Reprod. (in press).

Erickson, B.H. 1966. Development and senescence of the postnatal bovine ovary. J. An. Sci. 25: 800-805.

Gautier, J., C. Norbury, M. Lohka, P. Nurse, and J. Maller. 1988. Purified maturation-promoting factor contains the product of a Xenopus homolog of the fission yeast cell cycle control gene cdc2+. Cell 54"433-439.

Hafner, M., and C. Petzelt. 1987. Inhibition of mitosis by an antibody to the mitotic calcium transport system. Nature 330:264-266.

Herlands, R.L., and R.M. Schultz. 1984. Regulation of mouse oocyte growth: probable nutritional role for intercellular communication between follicle cells and oocytes in oocyte growth. J. Exp. Zool. 229:317-325.

Hubbard, C.J. 1986. Cyclic AMP changes in the component cells of Graafian follicles: possible influences on maturation in the follicle-enclosed oocytes of hamsters. Dev. Biol. 118:343-351.

Jenkins, N.A., and N.G. Copeland. 1985. High frequency germline acquisition of ecotropic MuLV proviruses in SWR/J-RF/J hybrid mice. Cell. 43:811-819.

Kishimoto, T., and H. Kanatani. 1976. Cytoplasmic factor responsible for germinal vesicle breakdown and meiotic maturation in starfish oocyte. Nature 260:321-322.

Kishimoto, T., R. Kuriyama, H. Kondo, and H. Kanatani. 1982. Generality of the action of various maturation-promoting factors. Exp. Cell Res. 137: 121-126.

Kuyt, J.R.M., T.A.M. Kruip, and M. DeJong-Brink. 1988. Cytochemical localization of adenylate cyclase in bovine cumulus-oocyte complexes. Exp. Cell Res. 174:139-145.

Larsen, W.J., S.E. Wert, and G.D. Brunner. 1987. Differential modulation of rat follicle gap junction populations at ovulation. Dev. Biol. 112: 61-71.

Magnusson, C., and T. Hillensjo. 1977. Inhibition of maturation and metabolism of rat oocytes by cyclic AMP. J. Exp. Zool. 201:138-147.

Moor, R.M., and J.P. Heslop. 1981. Cyclic AMP in mammalian follicle cells and oocytes during maturation. J. Exp. Zool. 216:205-209.

Moor, R.M., J.C. Osborn, D.G. Cran, and D.E. Walters. 1981. Selective effect of gonadotrophins on cell coupling, nuclear maturation and protein synthesis in mammalian oocytes. J. Embryol. Exp. Morphol. 61:347-365.

Panthier, J.J., Condamine, H., and Jacob, F. 1988. Inoculation of newborn SWR/J females with an ecotropic murine leukemia virus can produce transgenic mice. Proc Nat Acad Sci (USA) 85:1156-1160.

Pincus, G., and E.V. Enzmann. 1935. The comparative behavior of mammalian eggs in vivo and in vitro. I. The activation of ovarian eggs. J. Exp. Med. 62:655-675.

Poenie, M.J. Alderton, R. Steinhardt, and R. Tsien. 1986. Calcium rises abruptly and briefly throughout the cell at the onset anaphase. Science 233:996-889.

Powers, R.D., and G.A. Paleos. 1982. Combined effects of calcium and dibutyryl cyclic AMP on germinal vesicle breakdown in the mouse oocyte. J. Reprod. Fert. 66:1-8.

Racowsky, C. 1984. Effect of forskolin on the spontaneous maturation and cyclic AMP content of rat oocyte-cumulus complex. J. Reprod. Fert. 72: 107-116.

_______. 1985a. Effect of forskolin on the spontaneous maturation and cyclic AMP content of hamster oocyte-cumulus complexes. J. Exp. Zool. 234: 87-96.

_______. 1985b. Effect of forskolin on maintenance of meiotic arrest and stimulation of cumulus expansion, progesterone and cyclic AMP production by pig oocyte-cumulus complexes. J. Reprod. Fert. 74:9-21.

_______. 1986. The releasing action of calicum upon cyclic AMP-dependent meiotic arrest in hamster oocytes. J. Exp. Zool. 239:263-275.

Reynhout, J.K., and L.D. Smith. 1974. Studies on the appearance and nature of a maturation-inducing factor in the cytoplasm of amphibian oocytes exposed to progesterone. Dev. Biol. 38:394-400.

Salustri, A., and G. Siracusa. 1983. Metabolic coupling, cumulus expansion and meiotic resumption in mouse cumuli oophori cultured in vitro in the presence of FSH or dbcAMP, or stimulated in vivo by hCG. J. Reprod. Fert. 68:335-341.

Schroeder, A.C., and J.J. Eppig. 1984. The developmental capacity of mouse oocytes that matured spontaneously in vitro is normal. Dev. Biol. 102:493-497.

Schultz, R.M., R. Montgomery, and J. Belanoff. 1983. Regulation of mouse oocyte maturation: implication of a decrease in oocyte cAMP and protein dephosphorylation in commitment to resume meiosis. Dev. Biol. 97:264-273.

Seidel, G.E., 1981. Superovulation and embryo transfer in cattle. Science 211:351-358.

Sorensen, R.A., M.S. Cyert., and R.A. Pedersen. 1985. Active maturation-promoting factor is present in mature mouse oocytes. J. Cell Biol. 100:1637-1604.

Sorensen, R.A., and P.M. Wasserman. 1976. Relationship between growth oocytes meiotic maturtion of the mouse oocyte. Dev. Biol. 50:531-536.

Staigmiller, R.B., and R.M. Moor. 1984. Effect of follicle cells on the maturation and development competence of ovine oocytes matured outside the follicle. Gamete Res. 9:221-229.

Swenson, K.I., K.M. Farrell, and J.V. Ruderman. 1986. The clam embryo protein cyclin A induces entry into M phase and the resumption of meiosis in Xenopus oocytes. Cell 47:861-870.

Szybek, K. 1972. In vitro maturation of oocytes from sexually immature mice. J. Endocrinol. 54:527-528.

Takahashi, M., S.S. Koide, and P.K. Donahoe. 1986. Mullerian inhibiting substance as oocyte meiosis inhibitor. Mol. Cell. Endocrin. 47:225-234.

Tsafriri, A., N. Dekel, and S. Bar-Ami. 1982. The role of oocyte maturation inhibitor in follicular regulation of oocyte maturation. J. Reprod. Fert. 64:541-551.

Tsafriri, A., J. Picard, and N. Josso. 1988. Immunopurified anti-Mullerian hormone does not inhibit spontaneous resumption of meiosis in vitro of rat oocytes. Biol. Reprod. 38:481-485.

Twigg, J., B. Patel, and M. Whitaker. 1988. Translational control of InsP3-induced chromatin condensation during the early cell cycles of sea urchin embryos. Nature 332:366-369.

Urner, F., W.L. Herrmann, E.E. Baulieu, and S. Schorderet-Slatkine. 1983. Inhibition of denuded mouse oocyte meiotic maturation by forskolin, an activator of adenylate cyclase. Endocrinology 113:1170-1172.

Wassarman, P.M., W.J. Josefowicz, and G.E. Letourneau. 1976. Meiotic maturation of mouse oocytes in vitro: inhibition of maturation at specific stages of nuclear progression. J. Cell Sci. 22:531-545.

INTRAGONADAL CONTROL OF TESTIS FUNCTION

William W. Wright

Introduction

There is growing evidence that the establishment and maintenance of fertility in the male requires a series of precise interactions between the cells in the testis (1- 4). Specific interactions have been identified between the major somatic cell types in the testis, the Sertoli cells, the Leydig cells and the peritubular myoid cells. Cellular interactions are not restricted to the somatic cells of this organ as there is substantial morphological and physiological evidence for the interaction of germ cells with one another and with the Sertoli cells. However, we know much less about the interactions involving germ cells than we know about the interactions restricted to the somatic cell types.

This review will examine cellular interactions in the testis with emphasis on interactions involving cell division and differentiation. In this review, we will first describe some of the basic events of the life history of a cell. We will discuss how growth factors and hormones can affect cellular replication and differentiation. Then we will list the growth factors and hormones which have been identified in the testis. Some of these factors will be shown to potentially have pronounced affects on the replication of Leydig and Sertoli cells. Our emphasis will then shift to an examination of the regulation of spermatogenesis. Little is known about the mechanisms which regulate spermatogenesis. However, we will see that the male gamete undergoes a precisely timed set of developmental changes and that this occurs in a highly organized tissue, the seminiferous epithelium. This appreciation will lead to the conclusion that the regulation of spermatogenesis is complex and requires a number of interlocking controlling mechanisms. The last part of this review will describe research ongoing in our laboratory which is aimed at determining how Sertoli cells regulate a specific step in germ cell development.

Regulation of the cell cycle and cell differentiation. The fate of a cell is regulated, in part, by its position in the cell cycle. The cell cycle is divided into four parts: [1] the mitotic phase, when cells divide, [2] G_1, a period of time between mitosis and the next round of DNA replication, [3] S phase, when DNA is synthesized and [4] G_2, the period between DNA replication and mitosis (5) (Fig. 1). Cells can be diverted from the cell cycle during the G_1 phase via two processes. They can enter G_0, in which the cells stop dividing, but remain at their current state of differentiation. Alternatively, at a specific point in G_1, labeled G_D on figure 1, they can enter a pathway which leads to cellular differentiation (6).

Growth factors and hormones regulate the cell cycle at or near the G_1 phase (Fig. 1). A cell enters the cell cycle from the G_o state in response to specific growth factors called competence factors, two of which are platlet derived growth factor (PDGF) and fibroblast growth factor (FGF) (7, 8). Once the cell enters the cell cycle, somatomedin C and epidermal growth factor are often required for the cells to progress through the first half of G_1. Somatomedin C is sufficient to carry the cells through the second half of G_1(5,9). Once a cell has entered S phase, it usually will progress through S, G_2 and mitosis without any requirements for external stimulation (5). Upon completion of mitosis the cells can go through a second cell cycle or exit the cell cycle by entering G_O state. For some cells, entry into G_O is inhibited by interleukin 1 (10). Cells can also exit the cell cycle by following a pathway leading to cellular differentiation. This process has been shown to be stimulated by hormones, some paracrine factors, especially SmC, and may be inhibited by the same factors, the competence factors, which promote entry of cells into the cell cycle.

A recently described example of a competence factor inhibiting the differentiation of a cell is the effect of platlet derived growth factor on progenitors of oligodendrocyte in the developing optic nerve. Nobel and colleagues (11,12) demonstrated that when oligodendrocyte progenitors were placed in culture in the absence of platlet derived growth factor (a competence factor), these cells quickly ceased cell division and differentiated into oligodendrocytes. In contrast, when oligodendrocyte progenitors were cultured with platlet derived growth factor, these cells continued to divide for the same period of time that was observed <u>in vivo</u>. After that they differentiated into oligodendrocytes. The authors concluded that PDGF stimulated the progenitor cells to progress through the maximum possible number of cell cycles for these cells. In the testis, the proliferation and differentiation of spermatogonia, may be regulated in a similar manner (see below).

<u>Growth factors and hormones in the testis</u>. Having seen which factors generally regulate the cell cycle, it is appropriate to ask whether factors with the ability to regulate cell proliferation and differentiation are made in the testis. Table 1 demonstrates that a large number of growth factors and hormones are synthesized in the testis. It is noteworthy that included in this list are somatomedin C, an EGF-like molecule, fibroblast growth factor and interleukin 1, all of which have demonstrated effects on the cell cycle. Indeed, it is possible that the sequential changes in the synthesis of these factors may be responsible for the maturational changes in Sertoli cell and Leydig cell number and function.

Potential Role of Growth Factors and Hormones in Replication and Differentiation of Leydig and Sertoli Cells. The potential significance of a number of the growth factors and hormones to establishment of the mature number of functional Leydig and Sertoli cells may be seen when we consider the time course of the replication of these cells during pubertal maturation of the rat testis (Fig 2). Sertoli cells are actively replicating in the 22 day old embryo (38). Subsequent to this, the labeling index of Sertoli cells with 3H-thymidine decreases linearly, indicating either that the percent of the Sertoli cells in the cell cycle decreases, or that the duration of the cell cycle increases. What ever the case, by 20 days of age, DNA synthesis by Sertoli cells has ceased.

The replication of Leydig cells appears to follow a completely different time sequence than is observed with Sertoli cells. Zirkin and Ewing (39), demonstrated that Leydig cells numbers per testis did not begin to increase until 2 to 3 days after birth and then increased to only 20% of their adult numbers by 21 days. Subsequent to this, however, there was a rapid rise in Leydig cell numbers, and thus, possibly the rate of replication of these cells. By 35 days of age, Leydig cell numbers had reached 83% of adult numbers. Thus, the replication of Sertoli cells and Leydig cells appear to exhibit opposite patterns, with numbers of Leydig cells increasing rapidly only after the end of Sertoli cell proliferation.

An examination of the source and biological activities of some of the growth factors and hormones in the testis lead to the hypothesis that the replication of Leydig cells and Sertoli cells are related processes. A defense of this hypothesis follows.

Fibroblast growth factor, which occurs in high concentration in the testis (32) stimulates Sertoli cell replication and stimulates the synthesis of FSH receptors (40). The stimulation of FSH receptor numbers might be important since FSH has been shown to be mitogenic to immature Sertoli cells in culture (41). Thus, we propose that the conditions in the late fetal testis may be optimal for stimulating Sertoli cell proliferation. Since Sertoli cells synthesize Somatomedin C (13,14), the increased numbers of Sertoli cells might be expected to lead to an increase in the intratesticular SmC concentration. This event would stimulate the synthesis of LH receptors on the Leydig cells (42) which should cause an increase in Leydig cell numbers since LH is a mitogenic to immature Leydig cells (43). This increased numbers of Leydig cells would partially explain the increase in beta endorphin concentration during pubertal maturation of the testis (30). Since Sertoli cells have receptors for beta endorphins (44) and as immunoneutralization of beta endorphin in the immature testis causes an increase in DNA synthesis by the Sertoli cells (31), it follows that the Leydig cells must suppress Sertoli cell proliferation. Thus, growth in Sertoli cell and Leydig cells numbers during prepubertal maturation of the testis may be regulated by the effect of paracrine factors originating from both cell types. If this hypothesis is true, it must represent a very precise regulatory mechanism, for at the end of this process, the numbers of Sertoli cells and Leydig cells per testis are almost identical (45).

A General Review of Cell-Cell Interactions in the Seminiferous Epithelium. While the location of the Sertoli cells and Leydig cells constrains them to interact via secreted factors, all basic forms of cellular interaction are apparent in the seminiferous epithelium. These interactions are: secretion of paracrine factors (ie, hormones and growth factors) which bind to receptors on a second cell, cell adhesion, stromal-epithelial interactions and gap junction communication. Table 1 lists the growth factors and hormones which have been identified in the testis and the source of these factors. There are only a few reports of growth factor receptors within the seminiferous epithelium. One of these reports is the observation of somatomedin C receptors on human spermatocytes and spermatids (46). It should be noted, though, that the presence of these receptors on rat germ cells has been both affirmed (47) and denied (48). However, if germ cells do contain somatomedin C receptors and since somatomedin C has been shown to be both a progression factor in the cell cycle and to stimulate differentiated function in specific cells (please refer to figure 1), somatomedin C may play a biologically significant role in germ cell proliferation and development.

Cell adhesion has been found to be a powerful regulator of cell survival and differentiation during organogenesis. In the development of the cerebellum, binding of neurons to one other via the use of nerve cell adhesion molecules, promotes the survival of these cells while cells which do not adhere to other neurons die (49). The surviving neurons go on to form the functional cerebellum. Cell adhesion may also promote survival and differentiation of spermatogenic cells. A specific cell adhesion molecule has been identified on pachytene spermatocytes and shown to mediate adhesion of these cells to Sertoli cells in vitro (50). There is also evidence for the presence of similar molecules on round spermatids (51). The biological importance of cell adhesion to spermatogenesis is suggested by experiments of Parvinen et al (52) in their studies of the development of germ cells in vitro when those cells were cultured within intact segments of seminiferous tubules. Those investigators reported that late meiotic cells would complete meiosis and early spermiogenesis when these cells remained bound to the Sertoli cells within the tubule. They also noted that this binding required the maintenance of the normal morphology of the Sertoli cells.

The Sertoli cell is a highly polarized epithelial cells with numerous cytoplasmic process, which, in vivo surround and make contact with the surrounding germ cells. The polarity of this cell is promoted by the basement membrane (53, 54). This basement membrane has been shown to be synthesized in vitro by the Sertoli cells and the peritubular myoid cells when these two cell types are in direct contact with one another (55).

Finally, gap junctions have been described between Sertoli cells and shown to electrically couple these cells (56, 57). In general, gap junctions allow for the direct passage of ions and cyclic nucleotides between cells. This can allow for the direct stimulation of the function of one cell by another (58).

Functional Interaction of Germ Cells and Sertoli Cells. The above review indicates that Sertoli cells have mechanisms which allow them to interact with surrounding germ cells. It is therefore pertinent to ask whether such interactions have an impact on the function of either the Sertoli or the germ cells. That Sertoli cells can stimulate germ cell function is apparent from data from Rivarola et al (59, see Fig. 3). In this study, pools of germ cells (containing predominantly pachytene spermatocytes and round spermatids) were isolated from immature rats. These germ cells were cultured alone for one day and then transferred to one of four culture conditions: Condition 1 was the culture of germ cells alone without FSH, condition 2 was the culture of germ cells alone with FSH, condition 3 was the coculture of germ cells with Sertoli cells in the absence of FSH and condition 4 was the coculture of germ cells with Sertoli cells in the presence of FSH. After an additional 24 hours of culture, or coculture, ^{3}H-thymidine was added to some cells to measure the rate of DNA synthesis by the germ cells while ^{3}H-uridine was added to the other cultures to measure the rat of RNA synthesis by the germ cells. After two hours of incubation with the radionucleotides, the germ cells were collected from the dishes and RNA and DNA synthesis determined. This analysis demonstrated that germ cells cocultured with Sertoli cells exhibited substantially more RNA and DNA synthesis than did germ cells cultured alone. Additionally, this experiment demonstrated that Sertoli cells cultured with FSH had a greater effect on RNA and DNA synthesis of germ cells than did Sertoli cells cultured without FSH.

As Sertoli cells can influence germ cell function, so to, germ cells can modulate Sertoli cell function. Magueresse and Jegou (51) demonstrated that the addition of a pool of germ cells to cultured Sertoli cells (obtained from 45 day old rats) stimulated androgen binding protein secretion by the Sertoli cells (Fig. 4). They also noted that both pachytene spermatocytes and round spermatids were stimulatory to this function of Sertoli cells. In contrast, they noted that germ cells suppressed the conversion of testosterone to estradiol by Sertoli cells. Thus, it was concluded that germ cells have different affects on different Sertoli cell functions.

The process of spermatogenesis and the organization of the seminiferous epithelium. While it is established that germ cells and Sertoli cells can interact, we do not know how these interactions regulate the diverse cellular events required for spermatogenesis. Neither do we know how the high degree of organization of the cells within the seminiferous epithelium is achieved. An appreciation of both aspects of testis cell biology are necessary before we can appreciate how complex and precise must be the intragonadal regulation of spermatogenesis. Thus, it is pertinent to review the basic aspects of spermatogenesis and the organization of the spermatogenic cells in the seminiferous epithelium.

There are three basic processes in spermatogenesis. Once the stem spermatogonia have differentiated into A1 spermatogonia, these spermatogonia undergo 6 mitotic cell cycles. In the rat, these cycles are completed every 42 hours and yield 64 preleptotene spermatocytes, assuming there is no death of cells during this process (60). These preleptotene spermatocytes then enter the meiotic cell cycle. At the end of 19.4 days, the meiotic cells complete the first meiotic cell cycle and then rapidly traverse the second meiotic cell cycle, giving rise to haploid, round spermatids (61). For the next 22.1 days. these haploid cells differentiate into immature testicular spermatozoa (61). At the end of this time, they are released into the lumen of the seminiferous tubule.

Spermatogenic cells are not distributed randomly throughout the seminiferous epithelium. Rather they occur in specific cellular associations, called the stages of the cycle. In the rat, there are 14 such stages (61). Each stage contains specific types of spermatogonia, spermatocytes and spermatids at particular phases of development. All of the germ cells within a specific segment of tubule mature synchronously. For example, in the 3.3 days that are required for a tubule to progress from stage VI to stage VIII of the cycle, the B spermatogonia complete the last mitotic cell cycle and give rise to the preleptotene spermatocytes. Additionally, the pachytene spermatocytes found in the stage VI tubule significantly increase in volume. Also, during this time the round spermatids complete the synthesis of the constituents of the acrosome. Finally, the compacted spermatids move towards the tubular lumen and then are released from the epithelium. The cycle of the seminiferous epithelium is completed every 12.9 days in the rat (61). The continuing slow proliferation of the stem spermatogonia and the differentiation of some of these cells into Type A1 spermatogonia insures the continuous production of sperm in the fertile testis.

<u>Strategy for identifying Sertoli cell products which regulate specific events during spermatogenesis</u>. The analysis of the process of spermatogenesis and the stages of the cycle of the seminiferous epithelium suggests that there must be some central organizer within this epithelium. For years, it has been assumed that this organizer was the Sertoli cell. However. it has not been known how the Sertoli cell exerted its control over specific events during spermatogenesis. The purpose of the research in our laboratory has been to identify, purify and characterize secretory products from Sertoli cells which may be involved in regulating specific steps in spermatogenesis. We assumed that Sertoli cells products which regulated specific steps in the development of spermatogenic cells would be secreted at a time when these factors were required by those germ cells. Based on this assumption, it followed that such factors would be secreted at specific stages of the cycle of the seminiferous epithelium. This hypothesis could be tested because Dr. Martti Parvinen had developed a technique for the isolation of seminiferous tubules at discrete stages of the cycle of the seminiferous epithelium (62). Therefore, in collaboration with Dr.

Parvinen, we set out to identify secretory products which were secreted in a stage dependent manner by Sertoli cells within intact seminiferous tubules (62). To do this, tubules at each stage of the cycle were cultured in the presence of ^{35}S-methionine and the proteins in the medium analyzed by two dimensional gel electrophoresis. This analysis revealed that the proteins recovered from the medium changed dramatically with progression of the cycle of the seminiferous epithelium and suggested that there were profound changes in the synthesis of Sertoli cell secretory products with progression of the cycle (63). Of particular note was an extremely heterogeneous protein or family of proteins we called Cyclic protein 2 or CP-2 which had a mean pI of 5.5 and mean molecular size of 37,000. CP-2 was the predominant protein resolved at stages Vl and Vll. Subsequent to these stages, however, the amount of CP-2 detected by 2D gel analysis decreased dramatically and the protein was undetectable by stage Xll (Fig. 5). Thus, Cyclic protein-2 fit the criteria we had set for a molecule which might regulate specific events in spermatogenesis, for its appearance changed dramatically with progression of the cycle of the seminiferous epithelium. Therefore, CP-2 deserved further analysis and we set about to isolate the protein and generate a specific antiserum to it.

Isolation Cyclic Protein-2 and Generation of an Antiserum. A four step chromatographic procedure was developed for the isolation of CP.2 from rat seminiferous tubule fluid (64). In order to obtain sufficient protein for the generation of an antiserum, 225 ml of this fluid were committed to the purification procedure. Analysis of the purified protein by SDS gel electrophoresis demonstrated that a single protein, CP-2 had been isolated from the pool of proteins present in seminiferous tubule fluid (65, Fig. 6). The purified protein was then used to generate an antiserum in one rabbit. Using this antiserum, we demonstrated that CP-2 was a single, but heterogeneous glycoprotein and that the only cell within the seminiferous epithelium that synthesized CP-2 was the Sertoli cell.

Stage-specific synthesis of CP-2. The first question we posed using the antiserum was, are the changes in the accumulation of CP-2 in medium surrounding cultured tubules the result of a change in the rate of synthesis of the protein? This question was important for it was just as likely that the rate of synthesis of CP-2 was constant but that the rate of degradation of the protein in the tubules varied in a stage-specific manner. To answer this question, we measured both the in vitro rates of synthesis and secretion of CP-2 by cultured seminiferous tubules at specific stages of the cycle (see table 2). This analysis demonstrated that there were dramatic stage-specific changes in both the rates of synthesis and secretion of CP-2. However, it was apparent that the rate of synthesis, as measured, was in excess of the apparent rate of secretion (65). One possible explanation for this result was that all of CP-2 was secreted into the lumen of the tubule and that export of the protein out the lumen and into the surrounding culture medium required a considerable period of time.

To address this possibility, we examined the rate of export of CP-2 out of cultured tubule. Tubules at stage VI and VII of the cycle were cultured for one hour with ^{35}S-methionine, the tubules removed from the radioactive amino acid and the amount of radiolabeled CP-2 in the medium and in the tubules measured at varying times thereafter (65). This analysis demonstrated that all CP-2 was secreted but that at least 17 hours were required for the export of CP-2 from the tubules. This observation could be explained in part by the fact that fluid flow through the tubules is very slow; the entire volume of the fluid in segment of tubule is replaced only in 9 hours (65, 66).

The conclusion that there is a very slow fluid flow through the tubules is important to an understanding of the physiology of paracrine factors within the testis. Once a molecule is secreted into the lumen of the tubule, it will not be flushed out the testis, but rather will have the opportunity to diffuse up and down the length of the tubule, and thus come in contact with cells in tubules at many different stages of the cycle. How then could such factors exert their effect only on germ cells at specific stages of development? First, it is possible that the factors, once secreted, are degraded by proteases. In this regard, preliminary results using Western blotting techniques demonstrate that there is considerable degradation of CP-2 in vivo in the seminiferous tubule (data not shown). An alternative means by which paracrine factors may have specific effects on particular types of germ cells is the expression of the receptors for these factors on specific types of those cells. Finally, it is possible that many paracrine factors are secreted into microenvironments within the seminiferous tubule and that these factors never reach the lumen of the seminiferous tubule.

Summary and Conclusions

This review has stressed the general mechanisms of cell-cell interactions and has examined how these mechanisms are manifest in the testis. We have seen that changes in the secretion or concentration of specific growth factors may regulate the growth of the Leydig cell and Sertoli cell population. We have seen that all forms of cell-cell communication are active in the seminiferous epithelium and have argued that the control of spermatogenesis must be a precise but complex process. Finally, we have reviewed work in this laboratory to identify Sertoli cell products which have a specific effect on germ cell development. A protein which we have discovered, isolated and to which an antiserum has been generated is Cyclic Progein-2. The synthesis and secretion of this protein has been shown to exhibit profound stage-specific changes. We have shown that it is retained within the tubule, in part because of the slow fluid through these structures. As in the case with many other testis products, the function of CP-2 in the regulation of spermatogenesis has not been documented. However, sequencing the purified protein and a cDNA for the mRNA which encodes this protein will allow us to determine its primary structure and thus search for sequence homology between CP-2 and proteins of known function. Use of the antiserum as an immunohistochemical reagent should begin to delineate potential cellular targets for this protein.

REFERENCES

1. Parvinen, N., K.K. Vihko and J. Toppari, 1986. Cell interactions during the seminiferous epithelial cycle. Int. Rev. Cytol. 104:115-151.

2. Russell, L.D., 1980. Sertoli-Germ Cell Interrelations: A review. Gamete Research 3:179-202.

3. de Kretser, D.M., 1987. Local Regulation of Testicular Function. Int. Rev. Cytol. 109:89-112.

4. Saez, J.M., M.H. Perrard-Sapori, P.G. Chatelain, E. Tabone and M.A.Rivarola, 1987. J. steroid Biochem. 27:317-329.

5. Pledger, W.J., 1985. Regulation of Cell Proliferation: Serum Growth factors control an ordered series of Gl events. In C.M. Veneziale, ed. Control of Cell Growth and Proliferation, Van Norstrand Reinhold Co., New York, pp.108-131.

6. Scott, R.E., J.J. Wille, Jr., M.L. Wier, B.J. Hoerl and P.B. Maercklein, 1985. Carcinogenesis and defects in the integrated control of cell proliferation and differentiation. In C.M. Veneziale, ed. Control of Cell Growth and Proliferation, Van Norstrand Reinhold Co., New York, pp.132-146.

7. Stiles, C.D., W.J. Pledger, R.W. Tucker, R.G. Martin and C.D. Scher, 1980. Regulation of the Balb/c-3T3 cell cycle-effects of growth factors..J. Supremolecular Structure 13:489-499.

8. Gospodarowicz, D. N. Ferrara, L. Schweigerer and G. Neufeld, 1987. Structural characterization and biological functions of fibroblast growth factor. Endocrine Rev. 8:95-114.

9. Russell, W.E., J.J. van Wyk, and W.J. Pledger, 1984. Inhibition of the mitogenic effects of plasma by a monoclonal antibody to somatomedin C. Proc. Natl. Acad. Sci. 81:2389-2392.

10. Singh, J.P., L.D. Adams and P.D. Bonin, 1988. Mode of fibroblast growth enhancememt by human interleukin-1. J. Cell Biology 106:813-819.

11. Nobel, M., K. Murray, P. Stroobant, M.D. Waterfield and P. Riddle, 1988. Platlet-derived growth factor promotes division and motility and inhibits premature differentiation of the oligodendrocyte/type 2-astrocyte progenitor cell. Nature 333:560-562.

12. Raff, M.C., L.E. Lillien, W.D. Richardson, J.F. Burne and M.D. Nobel, 1988. Platlet-derived growth factor from astrocytes drives the clock that times oligodendrocyte development in culture. Nature 333:562-565.

13. Chatelain, P.G., D. Naville and J.M. Saez, 1987. Somatomedin-C/Insulin-like growth factor I-like material secreted by porcine Sertoli cells in vitro: characterization and regulation. Biochem. Biophys Res. Commun. 146:1009-1017.

14. Smith, E.P., M.E. Svoboda, J.J. van Wyk, A.L. Kierszenbaum and L.L. Tres, 1987. Partial characterization of a somatomedin-like peptide from the medium of cultured rat Sertoli cells. Endocrinology 120:186-193.

15. Holmes, S.D., G. Spotts and R.G. Smith, 1986. Rat Sertoli cells secrete a growth factor that blocks epidermal growth factor (EGF) binding to its receptor. J. Biol. Chem. 261:4076-4080.

16. Buch, J.P., D.J. Lamb, L.I. Lipshultz and R.G. Smith, 1988. Partial characterization of a unique growth factor secreted by human Sertoli cells. Fertility and Sterility 49: 658-665.

17. Rivier, C., S. Cajander, J. Vaughan, A.J.W. Hsueh and W. Vale, 1988. Age-dependent changes in physiological action, content and immostaining of inhibin in male rats. Endocrinology 123: 120-126.

18. Meunier, H.C. Rivier, R.M. Evans and W. Vale, 1988. Gonadal and extragonadal expression inhibin alpha, beta A and beta B subunits in various tissues predicts diverse function. Proc. Natl. Acad. Sci. 85:247-251.

19. Feig, L.A., A.R. Bellve, N.H. Erickson and M. Klagsbrun, 1980. Sertoli cells contain a mitogenic polypeptide. Proc. Natl. Acad. Sci. 77:4774-4778.

20. Feig, L.A., M. Klagsbrun and A.R. Bellve, 1983. Mitogenic polypeptide of the mammalian seminiferous epithelium: biochemical characterization and partial purification. J. Cell Biol. 97: 1435-1443.

21. Bellve, A.R., and L.A. Feig. 1984. Cell proliferation in the mammalian testis: biology of the seminiferous growth factor (SGF). Recent Prog. Hormone Res. 40:531-560.

22. Kahn, S.A., O. Soder, V. Syed, K. Gustaffsson, M. Lindh and E.M. Ritzen, 1987. The rat testis produces large amounts of an interleukin-1-like factor. Int. J. of Andrology 10:495-503.

23. Skinner, M.K. and I.B. Fritz, 1985. Testicular peritubular cells secrete a protein under androgen control that modulates Sertoli cell function. Proc. Natl. Acad. Sci. 82:114-118.

24. Skinner. M.K. and P.M. Fetterolf and C.T. Anthony, 1988. Purification of a paracrine factor, F.Mod-S, Produced by testicular peritubular cells that modulates Sertoli cell function. J. Biol. Chem. 263:2884-2890.

25. Ayer-leLievre C., L. Olson, T. Ebendal, F. Hallbook and H. Persson, 1988. Nerve growth factor mRNA and protein in the testis and epididymis of mouse and rat. Proc. Natl. Acad. Sci. 85:2628-2632.

27. de Rooij, D.G., D. Lok and D. Weenk, 1985. Feedback regulation of the proliferation of the undifferentiated spermatogonia in the chinese hamster by the differentiation spermatogonia. Cell. Tissue Kinet. 18:71-81.

28. Clermont, Y. and A. Mauger, 1984. Existence of a spermatogonial chalone in adult irradiated testes. Cell Tissue Kinet. 12:81-89.

29. Klinefelter, G.R. and L.L. Ewing, 1988. Optimizing testosterone production by purified adult rat Leydig cells in vitro. In Vitro Cellular and Developmental Biology.24:545-549.

30. Bardin, C.W., C-L.C. Chen, P.L. Morris, I.Gerendai, C. Boitani, A.S. Liotta, A. Margioris and D.T. Krieger, 1987. Proopiomelanocortin-derived peptides in testis, ovary and tissues of reproduction. Rec. Prog. Hormone Res. 43:1-28.

31. Orth, J.M., 1986. FSH-induced Sertoli cell proliferation in the developing rat is modified by beta endorphin produced in the testis.Endocrinology 119:1876-1878.

32. Ueno, N., A. Baird, F.Esch, N. Ling and R. Guillemin, 1987. Isolation and partial characterization of basic fibroblast growth factor from bovine testis. Mol. Cell. Endocr., 49:189-194.

33. Berry, S.A., and O.H. Pescovitz, 1988. Identification of a rat GHRH-like substance and its messenger mRNA in rat testis. Endocrinology 123: 661-663.

34. Yoon, D.J., C. Sklar, and R. David, 1988. Presence of immunoreactive corticotropin-releasing factor in rat testis. Endocrinology 122:757-761.

35. Pekary, A.E., T. Yameda, B. Sharp, S. Basin, R.S. Swerdloff, and J. Hershman, 1984. Somatostatin-14 and -28 in the male reproductive tract. Life Sci 34:939-945.

36. Morley, J.M., N. Meyer, A.E. Pekery, S. Melmed, H.E. Carlson, J.E. Briggs and J.M. Hershman, 1980. A prolactin inhibitory factor with immunocharacteristics similar to thyrotropin releasing factor (TRH) is present in rat pituitary tumors (GH3 and W5), testicular tissue and a plant material, alfafa. Biochem. Biophys. Res. Commun. 96:47-53.

37. Bhasin, S., D. Herber, M. Peterson and R. Swerdloff, 1983. Partial isolation and characterization of testicular Gn-RH like factors. Endocrinology 112: 1144-1146.

38. Orth, J.M., 1982. Proliferation of Sertoli cells in fetal and post natal rats: A quantitative autoradiographic study. Anat. Rec. 203:485-492.

39. Zirkin, B.R. and L.L. Ewing, 1987. Leydig cell differentiation during maturation of the rat testis: a stereological study of cell number and ultrastructure. Anatomical Record 219:157-163.

40. Jaillard, C., P.G. Chatelain and J.M Saez, 1987. In vitro regulation of pig Sertoli cell growth and function: Effects of fibroblast growth factor and somatomedin C. Biol. Reprod. 37:665-674.

41. Rich, K.A., C.W. Bardin, G.L. Gunsalus, and J.P. Mather, 1983. Age dependent secretion of androgen binding protein by cultured Sertoli cells. Endocrinology 113:2284-2293.

42. Perrard-Sapori, M-H., P.G. Chatelain, C. Jaillard and J.M. Saez, 1987.Characterization and regulation of somatomedin-C/insulin-like growth factor I(Sm-C/IGF-1) receptors on cultured pig Leydig cells. Eur. J. Biochem. 165:209-214.

43. Orth, J.M., 1984. The role of FSH in controlling Sertoli cell proliferation in testes of fetal rats. Endocrinology 115:1248-1255.

44. Fabbri, A., C.H. Tsai Morris, S. Luna, F. Fraioli and M.L Dufau, 1985. Opiate receptors are present in the rat testis. Identification and localization in Sertoli cells. Endocrinology 117:2544-2546

45. Wing, T.Y., anc A.K. Christensen, 1982. Morphometirc studies on rat seminiferous tubules. Amer. J. Anat. 165:13-25.

46. Vannelli, B.G., T. Barni, C. Orlando, A. Natali, M. Serio and G.C. Balboni, 1988. Insulin-like growth facotor-1 (IGF-1) and IGF-1 receptor in human testis: An immunohistochemical study. Fertility and Sterility 49:666-669.

47. Tres, L.L., E.P. Smith, J.J. van Wyk and A.L. Kierszenbaum, 1986. Immunoreactive sites and accumulation of somatomedin-C in rat Sertoli-spermatogenic cell co-cultures. Exp. Cell Res. 162:33-50.

48. Oonk, R.B. and J.A. Grootegoed, 1988. Insulin-like growth factor I (IGF-I) receptors on Sertoli cells from immature rats and age-dependent testicular binding of IGF-I and insulin. Mol. Cell Endocrinol. 55:33-43.

49. Rutishauser, A. B. Acheson, A.K. Hall, D.M. Mann and J. Sunshine, 1988. The neural cell adhesion molecule (NCAM) as a regulator of cell-cell interactions. Science 240:53-57.

50. D'Agostino, and M. Stefanini, 1987. A rat spermatocyte surface protein is involved in adhesion of pachytene spermatocytes to Sertoli cells in vitro. Mol. Cell Biol. 7:1250-1255.

51. le Magueresse, B and B. Jegou, 1988. In vitro effects of germ cells on the secretory activity of Sertoli cells recovered from rats of different ages. Endocrinology 22:1672-1680.

52. Parvinen, M., W.W. Wright, D.M. Phillips, J.P. Mather, N.A. Musto and C.W. Bardin, 1983. Spermatogenesis in vitro: completion of meiosis and early spermiogenesis. Endocrinology 112:1150-1152.

53. Tung, P.S. and I.B. Fritz, 1984. Extracellular matrix promotes rat Sertoli cell histotypic expression in vitro. Biol. Reprod. 23:207-217.

54. Hadley, M.A., S.W. Byers, C.A. Suarez-Quian, H.K. Kleinman and M. Dym, 1985. Extracellular matrix regulates Sertoli cell differentiation, testicular cord formation, and germ cell development in vitro. J. Cell. Biology 101:1511-1512.

55. Tung, P.S. and I.B. Fritz, 1988. Morphogenetic restructuring and formation of basement membranes by Sertoli cells and testis peritubular cells in co-culture: Inhibition of the morphogenetic cascade by cyclic AMP derivatives and by blocking direct cell contact. Developmental Biology 120:139-153.

56. Eusebi, F. E. Ziparo, G. Fratamico, M.A. Russo and M. Stefanini, 1983. Intercellular communication in rat seminiferous tubules. Developmental Biology 100:249-255.

57. Gilula, N.B., D.W. Fawcett and A. Aoki, 1976. The Sertoli cell occluding junctions and gap junctions in mature and developing mammalian testis. Developmemtal Biology 50:142-168.

58. Lawrence, T.S., W.H.Beers and N.B. Gilula, 1978. Transmission of hormonal stimulation by cell-to-cell communication. Nature 272:501-506.

59. Rivarola, M.A., P. Sanchez, and J.M. Saez, 1985. Stimulation of ribonucleic acid and deoxyribonucleic acid synthesis in spermatogenic cells by their coculture with Sertoli cells. Endocrinology 117:1796-1802.

60. Huckins, C., 1971. The spermatogonial stem cell population in adult rats. Their morphology, proliferation and maturation. Anat. Rec. 169:533-558.

61. Clermont, Y., 1972. Kinetics of spermatogenesis in mammals: seminiferous epithelium cycle and spermatogonial renewal. Physiol Rev. 52:198.110.

62. Parvinen. M., and A. Runkonen, 1982. Endogenous steroids in rat seminiferous tubule. Comparison of different spermatogenic stages isolated by transillumination-assisted microdissection. J. Androl. 3:211.220.

63. Wright, W.W., M. Parvinen, N.A. Musto, G.L. Gunsalus, D.M. Phillips, J.P. Mather and C.W. Bardin. 1983. Identification of stage.specific proteins synthesized by rat seminiferous tubules. Biol. Reprod. 29:257.270.

64. Wright, W.W. and M.L. Luzarraga, 1986. Isolation of cyclic protein-2 from rat seminiferous tubule fluid and Sertoli cell culture medium. Biol. Reprod. 35:761-772.

65. Wright, W.W., 1988. Germ Cell-Sertoli Cell Interactions: Analysis of the biosynthesis and secretion of cyclic protein-2. Developmental Biology, 130:45-56.

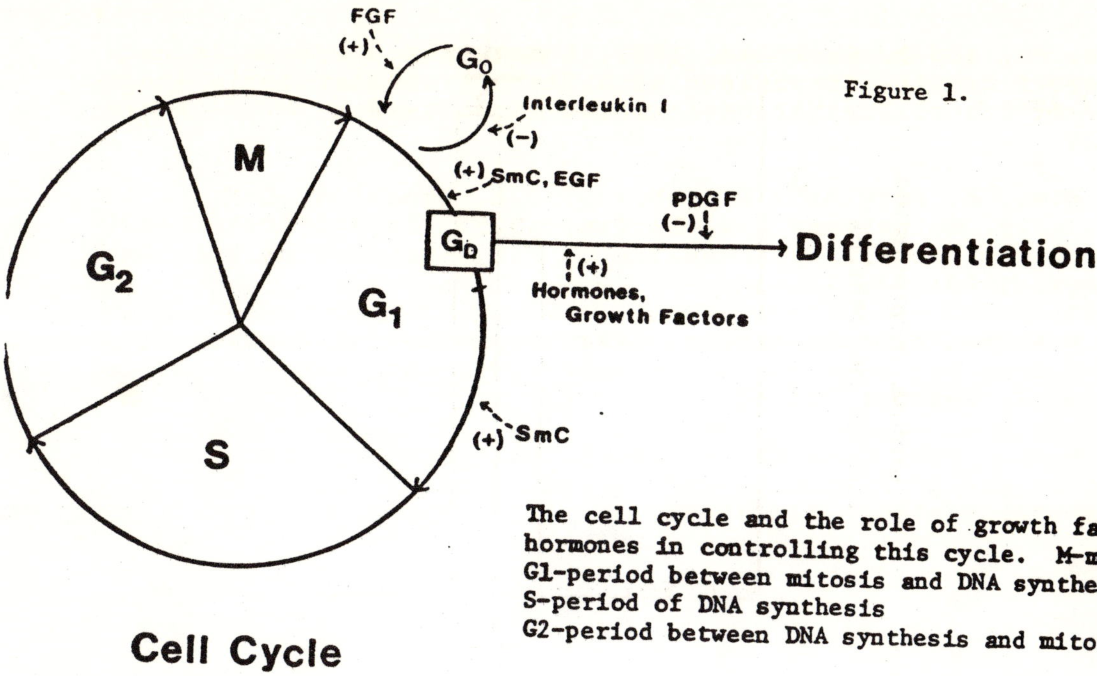

Figure 1.

The cell cycle and the role of growth factors and hormones in controlling this cycle. M-mitosis; G1-period between mitosis and DNA synthesis; S-period of DNA synthesis G2-period between DNA synthesis and mitosis.

Figure 2.

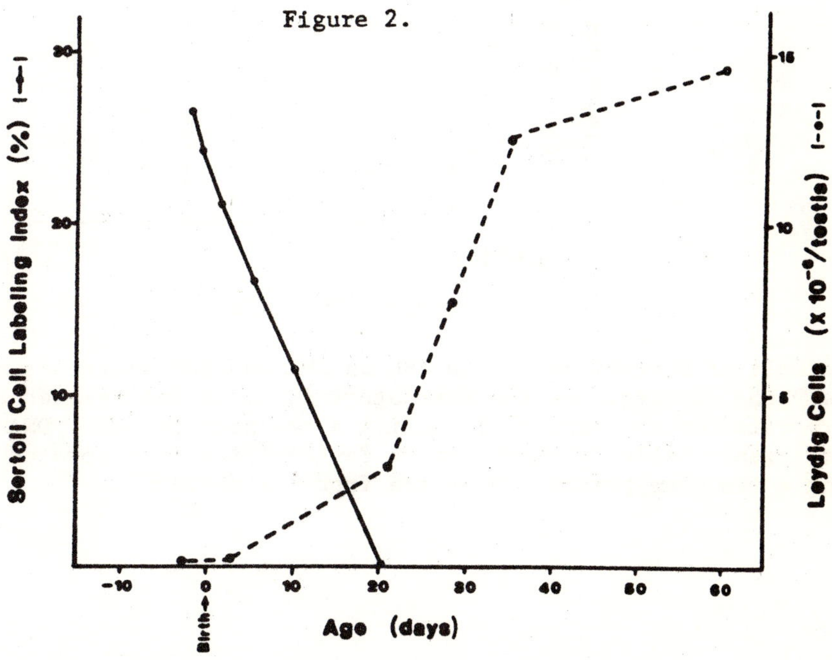

Replication of Sertoli cells and Leydig cells during prepubertal development of the rat testis. The replication of the Sertoli cells is expressed as the labeling index of these cells when they are exposed In Vivo to 3H-thymidine. Leydig cell numbers are obtained from stereological analysis of different age testes. Data are from references 38 and 39.

Figure 3

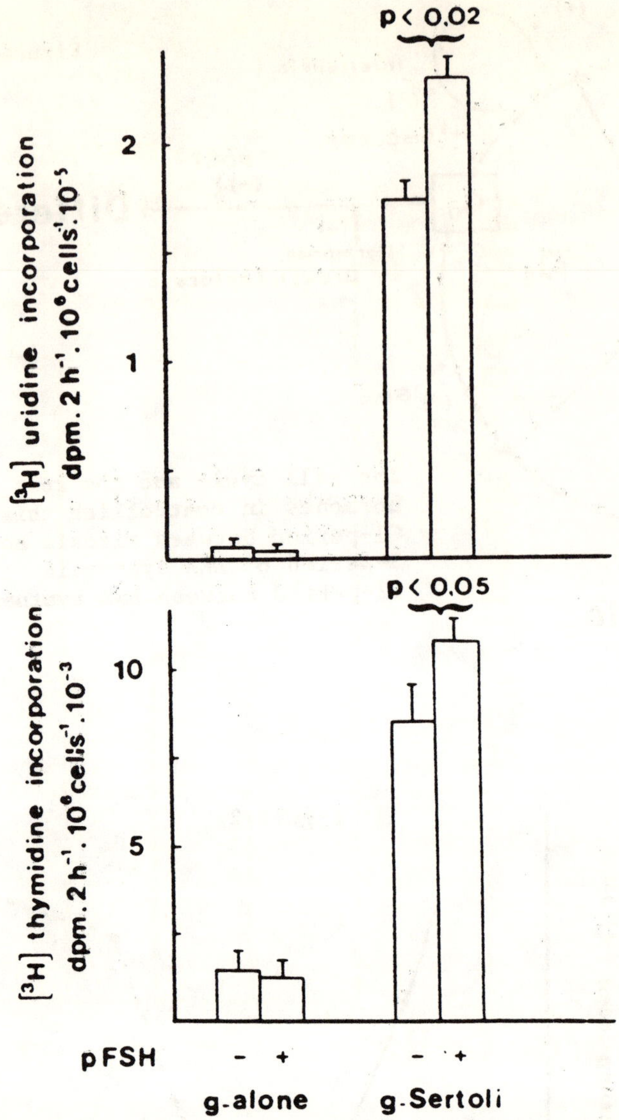

The effect of Sertoli cells and FSH on DNA and RNA synthesis by rat spermatogenic cells. RNA synthesis by the spermatogenic cells isexpressed as 3H-uridine incorporated by the germ cells during a two hour period. DNA synthesis by the spermatogenic cells is expressed as 3H-thymidine incorporated by the germ cells during a two hour period. Data are from reference 59.

Figure 4

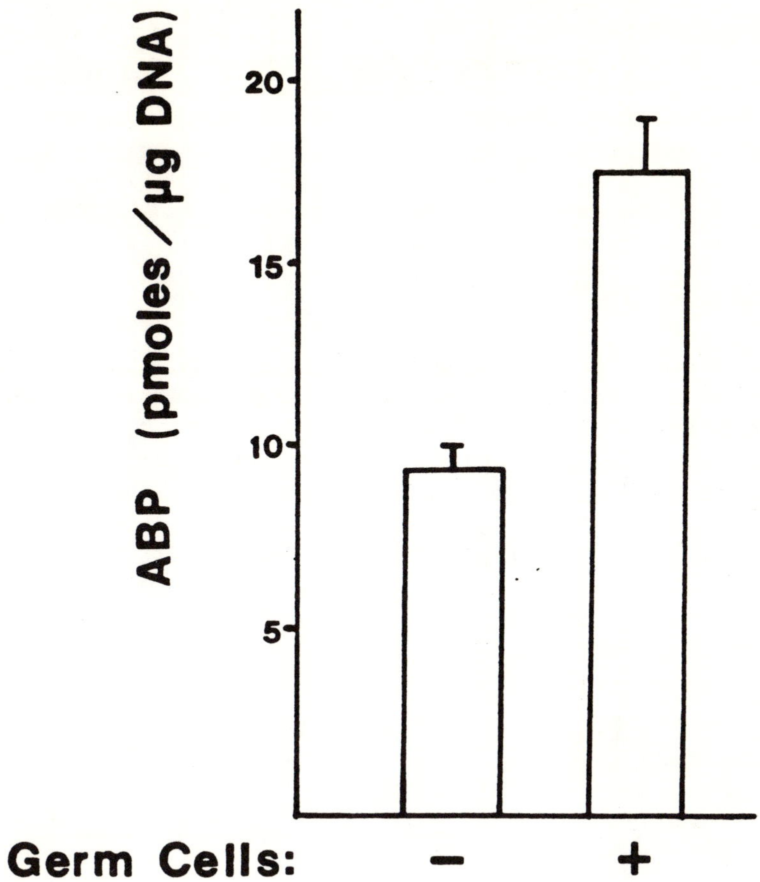

The effect of rat spermatogenic cells on ABP secretion by Sertoli cells. Data are expressed as the amount of ABP secreted in 24 hours per microgram of Sertoli cell DNA. Data are from reference 51.

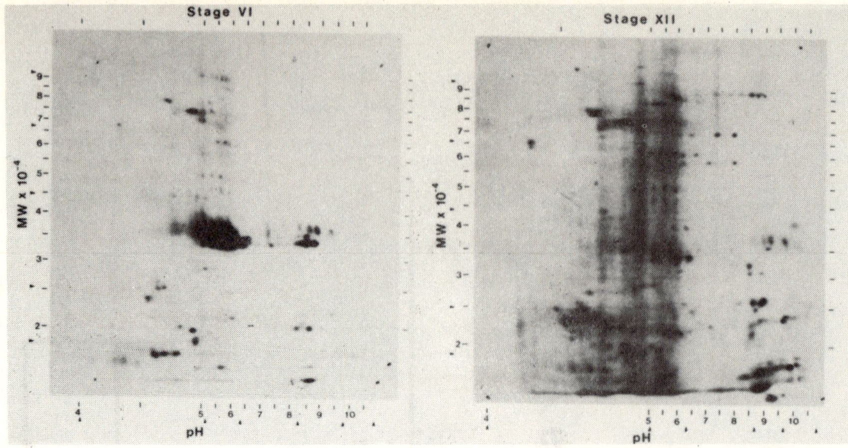

Figure 5. Analysis by two dimensional gel electrophoresis of radiolabeled proteins released into culture medium from stage VI and XII tubules when these tubules are cultured in 35S-methionine for 17 hours. Cyclic Protein-2 is the major protein recovered from the Stage VI tubules. It has a mean pI of 5.5 and a mean molecular weight of 37,000. Note that CP-2 is not detected in the medium from stage XII tubules. Data are from reference 63.

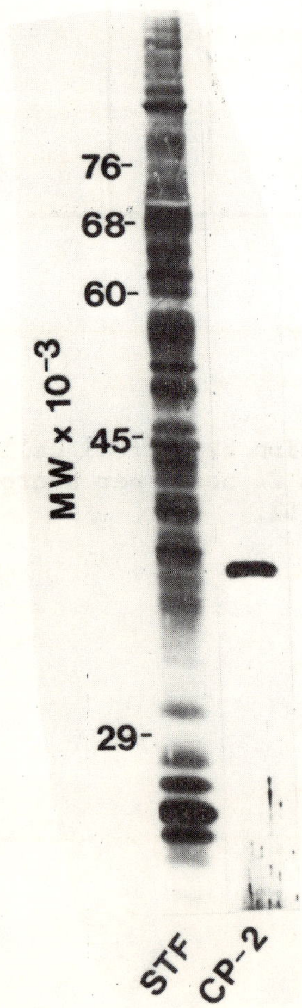

Figure 6. Purification of CP-2. CP-2 was purified from seminiferous tubule fluid by a four step chromatographic procedure. This figure shows SDS-gel analysis of proteins in STF and in the purified preparation of CP-2. Proteins were detected by staining with silver.

TABLE 1: GROWTH FACTORS AND HORMONES IN THE TESTIS

Cell Source	Factor	Attributes	Reference
Sertoli cell	Somatomedin C	Progression factor	13,14
	EGF-like factor	Progression factor	15,16
	Inhibin	Related to hormones which induce differentiation	17,18
	Seminiferous growth factor	Unique to testis	19-21
	Interleukin I	Suppresses entry of cells into G_0	22
Peritubular myoid cells	P-Mod S	Stimulatory to Sertoli cells	23,24
	Somatomedin C	Progression factor	14
Germ cells	Nerve Growth Factor	Unknown in testis	25
	Chalone	Suppresses replication of stem spermatogonia	27,28
Leydig Cells	Testosterone	Required for spermatogenesis	29
	POMC-related peptides	Suppress replication of immature Sertoli cells	30,31
Unknown	Fibroblast growth factor	A competence factor	32
	GHRH	Hormone	33
	Corticotropin-releasing hormone	Hormone	34
	Somatostatin-14 and -28	Hormone	35
	TRH	Hormone	36
	LHRH	Hormone	37

TABLE 2: STAGE SPECIFIC SYNTHESIS AND SECRETION OF CP-2 (X+SEM)

STAGE:	II	VI	VIa, b	VIII	XI
Synthesis	1765	15005	14860	2670	119
cpm x 10^{-3}/hr	596	1775	850	387	739
Secretion	624	4015	4143	702	19
cpm x 10^{-3}/hr	444	670	372	131	22

5 cm of seminiferous tubules were cultured for 1 hr (synthesis) or 16 h (secretion) and CP-2 was immunoprecipitated from tubular homogenates (synthesis) or media (secretion). Data are expressed as cpm of ^{35}S-methionine synthesized or secreted per hour.

PARACRINE REGULATION IN THE OVARY

James M. Hammond

Introduction

The growth and development of the ovarian follicle entails a coordinated process of cell replication and differentiation during which the number of granulosa cells lining the follicle may increase by several logs and daughter cells emerge with a different complement of hormone receptors and steroidogenic enzymes (reviewed in Richards, 1980). During this process maturational events in the several follicular components - theca, granulosa, and oocyte - must be synchronized. During the reproductive life of mammals these events are clearly entrained to the pituitary signals of the reproductive cycle. However, few follicles are able to respond to these signals and reach an ovulable state. In addition, a considerable body of data suggests that initial stages of follicular growth begin before the cyclical hypothalamic-pituitary feedback mechanisms are established and that the initiation of follicular growth continues to some extent during periods in which cyclical gonadotropin secretion is interrupted (eg. pregnancy) (Richards, 1980).

These physiological facts required ovarian physiologists to invoke intraovarian control systems now termed autocrine or paracrine mechanisms, long before these terms were developed and popularized. The definition of these mechanisms has emerged as one of the dominant topics in ovarian research in the last decade. In recent years, refined _in vitro_ systems and molecular approaches to identify and characterize such mechanisms have allowed an unprecedented rate of data accumulation. It is now generally agreed that paracrine mechanisms are critical to follicle development and selection, but definitive proof of this hypothesis remains elusive.

To come to grips with this amount of information it has been necessary to limit the scope of this review. Issues central to our own research area - paracrine and autocrine control of somatic cells of the ovarian follicle by estrogens and growth factors - have been emphasized, and our own data have been used as examples for many concepts presented. Discussion of oocyte maturation will be left to individuals at this symposium who have been more directly involved in defining these mechanisms. For a survey of paracrine mechanisms in the corpus luteum, the reader is referred to another recent review (Khan-Dawood and Dawood, 1986). In reviewing paracrine mechanisms in follicular cell function several classes of regulators have been identified. These are distinguished in part by their biochemical nature and in part by the strategies employed in their elucidation.

Intraovarian Steroids

It was obvious in the 1940's that ovarian steroids were strong candidates for ovarian regulatory principles (Pencharz, 1940; Simpson et al, 1941). This hypothesis was confirmed and refined by a number of critical in vivo experiments performed in the 1970's (Goldenberg et al, 1972a,b and 1973; Louvet et al 1975; Rao et al, 1978). These studies, using various combinations of gonadotropins, androgens, estrogens and steroid antagonists in immature and/or hypophysectomized rats demonstrated a requirement for estrogens in follicular development and gonadotropin action; in general, androgens were found to be inhibitory or atretogenic. The effects of estrogens on granulosa cell replication in vivo were particularly impressive and these observations led to the evolution of the DES-treated rat model which has now provided biological material for countless studies of granulosa cell function in vitro. In addition, estrogens were found to interact synergistically with gonadotropins in promoting follicular cell differentiation, with LH receptor induction serving as a convenient and widely studied end point (See Richards, 1980).

More or less simultaneously, the obligatory cooperation of LH-dependent thecal androgen biosynthesis in providing substrate for FSH-dependent granulosa estrogen synthesis was defined (Reviewed by Dorrington 1977; Ryan 1979). Collectively these observations provided the framework for current thinking on the integration of steroid biosynthesis and action in the ovarian follicle.

More recent observations have provided a much more detailed picture of the action of steroids on follicle cell function. Estrogen action, in particular, has been investigated in considerable depth. Data on estrogen receptors and their regulation in rat ovarian cells is particularly abundant and has led to a persuasive case for these binding sites as an important regulatory locus for hormone interaction (Richards, 1975). However, a recent conscientious effort to delineate estrogen receptors in monkey ovarian cells failed to disclose such structures with immunochemical techniques (Hild-Petito et al, 1988). These studies, along with the failure to demonstrate the mitogenic action of estrogens in vivo in several species other than the rat (Kim and Greenwald, 1987) may indicate significant differences in the nature and importance of estrogen effects on the ovary of various mammals.

The studies performed in my laboratory and elsewhere by Johannes Veldhuis are typical of experiments conducted by several groups which have pin-pointed the action of estrogens on the cAMP-dependent steroidogenic pathway in ovarian cells. These studies, in cultured porcine granulosa cells, revealed a self-limited inhibitory action of estradiol (E_2) (Veldhuis and Hammond, 1980) which appears to be mediated at the 3--hydroxysteroid dehydrogenase step (Veldhuis, et al 1986,a) and which is followed by a brisk stimulatory action of estrogens on progestin biosynthesis (Veldhuis and Hammond, 1980; Veldhuis, et al, 1981). The latter effects are time-dependent, inhibited by anti-estrogens, and involve biosynthesis of steroidogenic enzymes, cholesterol side chain cleavage components in particular (Veldhuis and Hammond, 1980, Veldhuis et al 1981, 1982a,b; Toaff et al 1983; 1985). In addition, a substantial increase in substrate flow to the side chain cleavage site is generated by alteration in patterns of lipoprotein utilization (Veldhuis et al, 1982b, 1984). Effects of estrogens in this system are particulary apparent in the presence of gonadotropins (Veldhuis et al 1981; 1982a;b). In the porcine system this interaction appears to occur predominantly, if not exclusively at a locus distal to the generation of cAMP: it has been impossible for us to show an effect of estrogens on basal or FSH-dependent cAMP levels (Spicer and Hammond, 1988,b) and E_2 significantly enhances the effects of cAMP analogues (Veldhuis et al, 1981). In most respects, our studies in the porcine granulosa cell system agree well with data concerning stimulatory effects of E_2 on steroidogenesis in rat granulosa cells which were conducted at the same time in other laboratories (Fangul et al, 1984). In addition, our results complement the studies of Richards and others who have demonstrated important stimulatory interactions of gonadotropins and estrogens at the level of cAMP-binding proteins and of functional protein kinase activity (Ratoosh and Richards, 1985). These latter results suggest a plausible intracellular mechanism by which the steroidogenic actions of estrogens might be mediated. While the initial gene product leading to these estrogen effects has not been determined, the more distal effects are now understood in considerable detail.

More recent studies in our laboratory have focused on the catecholestrogens, an intriguing class of compounds which may function to modulate and/or mediate some ovarian actions of estrogens. The activity of estrogen-2-hydroxylase (E-2-H), the enzyme responsible for converting E_2 to catecholestrogens, is as high or higher in the preovulatory porcine follicle than in other extrahepatic tissues; importantly, the activity of E-2-H is dramatically enhanced in the walls of large well-differentiated ovarian follicles, in the membrana granulosa in particular, with levels 10-100 fold lower measured in immature follicles and corpora lutea (Hammond et al, 1986). A physiological function for the brief surge of catecholestrogen biosynthesis in the preovulatory follicle has been suggested by our studies of the actions of catecholestrogens on cultured porcine granulosa cells. These studies have shown that the catecholestrogens, in the presence of FSH, have potency comparable to that of E_2 in stimulating progesterone secretion; however, their mode of action is quite distinct - the two classes of steroids are significantly additive at maximally effective concentrations and the actions of catecholestrogens are not inhibited by anti estrogens, while those of E_2 are (Spicer and Hammond, 1987). Further the catecholestrogens have stimulatory effects on cAMP levels and adrenergic receptors which cannot be mimicked by the parents estrogens (Spicer and Hammond, 1988a,b). Thus the stage is set for an additional autocrine/paracrine system in the preovulatory follicle by which locally secreted catecholestrogens could function, along with estrogens to amplify the stimulatory actions of gonadotropins on a local level.

Although space does not permit a detailed review, both androgens and progestins also have clearly demonstrated receptors and actions on ovarian cells (reviewed in Hsueh et al, 1984).

In summary, steroids of several classes can be regarded as established paracrine and/or autocrine regulators of ovarian function by the criteria which can be currently applied in this field. Both the biosynthesis and action of these compounds are understood in considerable detail. In general, the control of these processes can be integrated into a coherent and plausible construct of ovarian physiology during the reproductive cycle. However, important conceptual issues remain to be resolved concerning the bipotential actions of androgens and estrogens and their relative importance in vivo. On one hand, the steroids serve as a model for the level of understanding which needs to be achieved concerning many of the ovarian peptides discussed in subsequent sections. On the other hand, the areas of persistent uncertainty point out stubborn difficulties in extrapolating findings in vitro to the complex milieu of the ovary in vivo.

The Role of Ovarian Growth Factors

As discussed above, gonadotropins and gonadal studies have unambiguous stimulatory effects on follicular growth and granulosa cell replication in vivo. However, at least in our hands, these hormones lack significant mitogenic actions when added to cultured granulosa cells in vitro (Hammond and English, 1987). In contrast, peptide growth factors have been known to exert potent mitogenic effects on cultured granulosa cells for at least a decade. Thus, our own interest, and that of others, in the ovarian growth factors was stimulated in part by the speculation that such factors might provide the mechanism by which pituitary and gonadal hormones regulated ovarian cell replication. In the years which have followed, important actions of growth factors on cytodifferentiation of ovarian cells have also become apparent. In the paragraphs which follow, the evidence concerning paracrine effects of several growth factor families -- the insulin-like growth factors (IGFs), the epidermal growth factor (EGF)/transforming growth factor (TGF)-alpha family, the TGF-beta family, and the fibroblast growth factors (FGFs) are reviewed. For each of these families the minimal criteria of action on follicular cells and presence in the ovarian follicle have been met. However, it has not yet been possible to prove that any of these peptides are obligatory mediators of follicular growth. Because of our own research interests and because of the amount of evidence available, the IGFs are considered in greatest detail.

Insulin-like Growth Factors. The IGFs (sulphation factors, somatomedins) were originally discovered and named based on their insulin-like effects and on their presumed role as the circulating mediators of growth hormone action on cartilage (reviewed in Daughaday, 1984). However, it has also become clear that IGFs are synthesized widely in peripheral tissues, and recent interest in the field has emphasized their role as local mediators of tissue growth and differentiation (Underwood et al., 1986). Among the organs for which an autocrine or paracrine function of the IGFs has been postulated, the data concerning their effects in the ovary may be the most comprehensive and persuasive.

Since our original description of the effects of rat IGF-II (MSA) on the level of ornithine decarboxylase in porcine granulosa cells (Veldhuis and Hammond, 1979), more than 50 papers have been published concerning the action of the IGFs on the gonad. Accordingly, an encyclopedic review is impossible in the current space. Briefly stated, it seems that these peptides have a stimulatory action on every endpoint which has been examined. These segregate readily into the general areas of growth-related actions and effects on differentiated function. The first area includes stimulatory effects of the IGFs on ornithine decarboxylase (Veldhuis and Hammond, 1979), cell number in culture (Savion et al., 1981; Hammond et al., 1983), thymidine incorporation into DNA (Baranao and Hammond, 1984; Hammond and English, 1987), glucose oxidation (Weber and LaBarbera, 1988), and stimulation of mucopolysaccharide secretion (Adashi et al., 1986c).

In addition to these growth-related phenomena, insulin and the IGFs have been found to enhance differentiated function of both theca cells (Barbieri et al., 1986; Cara and Rosenfield, 1988) and granulosa cells. With granulosa cells, the IGFs or insulin seem to be essential for optimal effects of FSH on LH receptor induction (Adashi et al., 1985c), generation of cyclic AMP (Adashi et al., 1986b) and for FSH action on steroidogenesis (Baranao and Hammond, 1984,1985b; Adashi et al., 1985b). Several specific points in the steroidogenic pathway are enhanced by IGFs including internalization and processing of exogenous cholesterol (Veldhuis et al., 1987), the side-chain cleavage enzyme (Veldhuis et al., 1986b), and aromatase (Adashi et al., 1985a; Davoren et al., 1985; Veldhuis et al., 1985). Based on the magnitude of the effects described, it would appear that the differentiative influences of the IGFs on follicular cells are at least as important as their mitogenic actions. In addition, the relative importance of the growth-related and cytodifferentiative effects of the IGFs depend on the level of granulosa differentiation and the influence of other hormones and growth factors (Hammond et al., 1988b): in the presence of FSH and in the milieu of the preovulatory follicle, IGFs promote steroidogenesis; in the presence of other growth factors mitogenesis predominates.

For these actions of the IGFs to be mediated in an autocrine or paracrine fashion, it was necessary for them to be present and/or secreted in the ovarian follicle. This topic has been a particular interest of our laboratory, and our studies have recently been confirmed and extended by other groups. Regarding ovarian IGF levels <u>in vivo</u>, our studies have indicated that 1) IGFs and IGF binding proteins were easily measurable in porcine follicular fluid (Hammond et al., 1982; 1983), 2) IGF-II-like molecules and IGF-I-like immunoactivity were both present (Hammond et al., 1985, Mondschein et al 1988c), 3) the concentrations of the IGFs were sufficient to mimic the actions of these peptides in follicular culture systems (Hammond et al., 1983,1985), 4) the concentration of IGFs in follicular fluid increased during follicular growth (Hammond et al., 1985, Mondschein et al 1988c), and 5) the concentration of IGF-I in follicular fluid was increased by gonadotropins (Hammond et al., 1988a), and growth hormone (Bryan et al., 1988). Related data on ovarian IGF-I levels in the rat and their stimulation by growth hormone have been developed through organ extractions by Davoren and Hsueh

(1986). In addition, high concentrations of IGF-II have been found in human follicular fluid, (Ramasharma et al., 1986).

Evidence concerning the secretion of IGFs by ovarian cells and its hormonal regulation has been provided more directly in cultured granulosa cells of porcine and human species. Our data with porcine cells suggest that IGF secretion is sustained under serum-free conditions for at least 10 days in culture (Hammond et al., 1985), and that immunoreactive IGF-I predominates over IGF-II (Mondschein et al, 1988c). The secretion of immunoreactive IGF-I could be enhanced by FSH, estradiol, LH, growth hormone (Hsu and Hammond, 1987a,b), and EGF (Mondschein and Hammond, 1988). In contrast, transforming growth factor-beta appears to inhibit IGF secretion <u>in vitro</u> except at very low concentrations (Mondschein and Hammond, 1988c; Mondschein et al, 1988a). Our data with follicular fluid, recently complemented by analysis of ovarian mRNA levels, suggest that IGF-I and II are both synthesized in the porcine ovary (Mondschein et al 1988; Hammond et al unpublished). This situation may differ from that in the human where assays for the peptides and their mRNAs suggest that the dominant ovarian IGF is IGF-II (Ramasharma and Li, 1987; Voutilainen and Miller, 1987). In contrast, hybridization data from ovarian extracts (Hernandez et al., 1987a; Murphy et al., 1987) suggest that IGF-I is the dominant, if not the exclusive, ovarian IGF in the rat. The diversity of expression of these peptides in the ovary among species is intriguing and somewhat unexpected.

The evidence reviewed above concerning simultaneous secretion and action of IGFs in the follicle strongly suggests a role for IGFs in follicular growth and development. However, it remains challenging to prove that they are obligatory mediators in these processes. Studies with monoclonal antibodies to the IGFs have provided an additional approach. In our porcine granulosa cell culture system (Hsu and Hammond 1987b), growth hormone, E_2 and gonadotropins interact synergistically to stimulate IGF-I and progesterone secretion. Recent studies have shown that the effects of these hormones on steroidogenesis can be substantially diminished by a monoclonal antibody to IGF-I (Mondschein et al , 1988b). These data constitute a direct demonstration of an autocrine action of IGF <u>in vitro</u>. Related studies, indicating an inhibitory effect of the IGF-I antibody on the steroidogenic actions of follicular fluid have also shown the importance of IGFs in the stimulatory milieu of the preovulatory follicle (Mondschein et al, 1988b). These data are consistent with a role for IGF in follicular development <u>in vivo</u> and constitute the most direct evidence available for a critical role for IGFs as local amplifiers of the effects of pituitary and gonadal hormones. To our knowledge, such issues have not been addressed for the other growth factors.

<u>EGF and EGF-Like Peptides</u>. As reviewed by others (Carpenter and Zendegui, 1986; Cohen, 1987), the primary structure of EGF, as well as that of its receptor, have been known for some time. While many cells in the body respond to this peptide, the nature and source of the endogenous ligand for these actions is unclear. Detectable levels of EGF precursor mRNA were found in submaxillary gland, kidney, mammary gland, pancreas, small intestine,

pituitary, lung, spleen, brain, and ovary in order of decreasing abundance (Rall et al., 1985). In addition to peptides similar to salivary EGF, related peptides have been recognized through the use of radioreceptor assays for EGF. The most widely studied of these is TGF alpha. This factor has been found mainly in transformed cell types, but peptides which appear to be similar have been found in platelets (Assoian et al., 1984), pituitary (Samsoondar et al., 1986), and mammary epithelium (Valverius et al., 1987). As reviewed below, TGF alpha appears to be the EGF-like peptide in the ovary.

The mitogenic actions of EGF on ovarian cells, first demonstrated by Gospodarowicz et al. (1977a) in bovine granulosa cells, have now been shown in granulosa cells of other species and in a variety of culture conditions, (Gospodarowicz and Bialecki, 1979, Franchimont et al., 1986; Hammond and English, 1987). In contrast to the clear stimulatory actions of EGF on cell replication, the actions of EGF on differentiated function are more complex and generally inhibitory. Several groups have demonstrated a negative effect on estrogen secretion by granulosa cells (Hsueh et al., 1981; Knecht and Catt, 1983a) and on FSH-induced LH receptors (Mondschein and Schomberg, 1981; Knecht and Catt, 1983b). As anticipated, comparable results were achieved with TGF-alpha (Adashi and Resnick, 1986) which binds to the same receptor. The multiple inhibitory effects of EGF on FSH action suggested the possibility of a global antagonism of FSH effects. However, EGF can enhance FSH binding to pig granulosa cells (May et al., 1987), and the data on EGF effects and EGF/FSH interactions on the progestin biosynthetic pathway have been quite variable. The latter results are at least partially accounted for by concomitant stimulation of progesterone metabolism to 20-hydroxyprogesterone and stimulation of overall progestin biosynthesis at the side-chain cleavage reaction and the 3-hydroxysteroid dehydrogenase step (Jones et al., 1982).

The data reviewed above made EGF-like peptides attractive candidates as intraovarian regulatory molecules. This possibility has been further supported by recent studies in several laboratories indicating EGF- or TGF-alpha like activity in the ovary. Using a radioreceptor assay which recognizes both EGF- and TGF alpha, we detected physiologically significant levels of EGF-like activity in porcine follicular fluid (Hsu et al., 1987). The concentrations in follicular fluid were substantially higher than those in porcine plasma and higher in fluid from small than from preovulatory follicles. More recently, these studies have been extended by other groups who have shown that TGF aplha is secreted by cultured theca cells (Skinner et al 1988) and localized to theca cells immunohistochemically (Kudlow et al, 1988). Collectively, these data make it extremely likely that EGF-like peptides will be assigned an autocrine or paracrine function in ovarian physiology. We postulate that their dominant effects will be exerted in small follicles where they should act to promote cell replication and restrain differentiation. However, an understanding of the true significance of these peptides will require more detailed studies of their nature and regulation.

Transforming Growth Factor- TGF-beta was initially defined as a peptide which, in conjunction with EGF or TGF-alpha, allowed reversible transformation of normal rat kidney cells to anchorage-independent growth (Sporn et al., 1986). Structural analyses have shown a substantial degree of sequence homology between TGF-beta, inhibins, activins and Mullerian-inhibiting substance(Massague, 1987). TGF-beta is now known to be present in, or secreted by, a number of normal as well as neoplastic cell types and is present in abundance in platelets (Assoian et al., 1984). The action of this factor is bipotential, facilitating growth in some cell types and culture conditions, but inhibiting growth in other circumstances (Moses et al., 1984; Roberts et al., 1985).

As purified preparations of TGF-beta became available, its effects on rat granulosa cells were described more or less simultaneously by several laboratories (Adashi and Resnick, 1986; Knecht et al., 1986; Ying et al., 1986; Dodson and Schomberg, 1987). In the aggregate, these studies suggested a stimulatory action of TGF-beta on FSH action, reflected in an increase in estrogen secretion (Adashi and Resnick, 1986; Ying et al., 1986), LH receptor activity (Knecht, et al., 1986; Dodson and Schomberg, 1987), and progesterone biosynthesis (Dodson and Schomberg, 1987). More recently, a stimulatory interaction of TGF-beta and FSH on DNA synthesis by rat granulosa cells has also been demonstrated (Dorrington et al, 1988). While the stimulatory actions of TGF-beta on rat granulosa cells seem clear, the work of Knecht et al. (1987) has shown that these effects depend critically on the concentration of the growth factor and of FSH. At high FSH concentrations a dose-dependent inhibitory action of TGF-beta was demonstrated. In our porcine granulosa cell culture system, in which replication, thymidine incorporation, IGF secretion, and gonadotropin-dependent progesterone secretion were studied, an inhibitory effect of TGF-beta was found on all the parameters examined (Mondschein and Hammond, 1988; Mondschein et al., 1988a). The effects of TGF-beta in this system were particularly obvious when the peptide was combined with stimulatory concentrations of EGF (with regard to thymidine incorporation and IGF-I secretion) or with FSH (with regard to progesterone secretion). Our results are in agreement with the growth inhibitory effects of TGF-beta on bovine granulosa cells (Skinner et al., 1987) as well as with the inhibitory effects of TGF-beta on growth and differentiation of other normal and neoplastic cell types (reviewed in Moses et al., 1984) including adrenocortical cells (Hotta and Baird, 1986). In summary, TGF-beta has been found to have potent but variable effects on both replication and differentiated function of ovarian granulosa cells. It seems that the action of this peptide will depend critically on the species and/or the presence or absence of other modulators of granulosa cell function. This is reflective, on the one hand, of the well described bipotential nature of TGF-beta's action on numerous cell types and, on the other, of the complex interaction of multiple factors which are involved in ovarian cellular function.

The importance of these actions of TGF-beta is underscored by recent data from several sources indicating probable secretion of TGF-beta in the ovary. Skinner et al. (1987) have demonstrated the secretion of immuno- and receptor-active TGF-beta by cultured bovine and rat theca cells. In addition, the mRNA for TGF-beta has been found in the ovary by Derynck and Rhee (1987) and Hernandez et al. (1987b). Although the details of the secretion of TGF-beta by ovarian cells remains to be worked out, this growth factor can probably be regarded as an important autocrine/paracrine influence in the ovary on the basis of present data.

Fibroblast Growth Factor. The existence of FGF was deduced from the mitogenic activity of pituitary extracts and partially purified pituitary hormone preparations. Since these original studies the peptide has been purified to homogeneity from a number of tissues, including the ovary, and its primary structure analyzed (reviewed by Baird et al., 1986; Gospodarowicz et al., 1986b). The FGF peptides have potent growth-promoting activities on many cells of mesenchymal origin, and the mitogenic action of the FGFs on endothelial cells, implying angiogenic activity, has attracted particular attention. This action may be of importance in the development of the ovarian follicle as well as the corpus luteum.

Shortly after purified preparations of FGF became available, the peptide was found to be mitogenic for granulosa cells from bovine (Gospodarowicz et al., 1977a), rabbit, porcine, human, and guinea pig follicles (Gospodarowicz and Bialecki, 1979). Cells from corpora lutea also retained sensitivity to FGF but not EGF (Gospodarowicz et al., 1977b). In addition, FGF supports the growth of luteal endothelial cells (Gospodarowicz et al., 1986a). Effects of FGF on differentiated function and steroidogenesis by ovarian cells have not been studied as systematically, but the available data suggests that the net effects are generally inhibitory (Savion and Gospodarowicz, 1980, Baird et al., 1986, Adashi et al, 1988).

The first clear demonstration of FGF in the ovary was by Gospodarowicz et al. (1985) who purified an angiogenic factor from bovine corpus luteum that had biochemical, biological, and sequence homology to pituitary FGF. More recently, Neufeld et al. (1987) have also shown that bovine granulosa cells produce a basic FGF similar to that from pituitary in immunochemical, electrophoretic and receptor binding activity. Related data have been presented by Baird et al. (1986) who extracted an FGF-like peptide from non-luteinized rat ovary. The relationship of these well characterized molecular species to other peptides which possess one or more of the properties of FGF remains a subject of active speculation. A heparin-binding theca cell mitogen/angiogenic factor, demonstrated by Makris et al. (1984), apparently has some similarities to FGF in chemical composition (Makris, personal communication). In addition, other factors from granulosa cell-conditioned medium have angiogenic activity (Koos, 1986). With the improvement in specific receptor and immunoassays for the FGFs, a more direct comparison of these factors to FGF should be forthcoming.

In summary, ovarian growth factors have been intensely investigated in recent years. The data available indicates that several types of growth factors are synthesized, active, and probably important in the ovarian follicle. Data concerning the IGFs is most complete, and our level of understanding of the role these peptides rivals that achieved for steroids. With regard to the IGFs, the emerging picture suggests an ovarian amplification system interfaced with pituitary and gonadal hormones at several levels which could enhance the development of follicles selected for ovulation. Data regarding the secretion and action of other growth factors is less abundant, but it is accumulating rapidly. A more coherent picture of the physiological importance of these peptides should be possible shortly.

Other Regulatory Peptides in the Ovarian Follicle

When I first reviewed this area for a symposium in 1980 (Hammond, 1981), thinking in the field was dominated by a number of factors, identified in follicular fluid or ovarian extracts and defined in terms of in vitro bioassays which were assumed to reflect paracrine relationships in the ovary - "oocyte maturation inhibitor", "luteinization inhibitor", "luteinization stimulator", gonadotropin binding inhibitors, and GnRH-like substances (see Channing and Segal, 1982). At the time, these factors engendered considerable excitement based on the assumption that they were unique ovarian peptides which, when purified, would allow safe and specific approaches to fertility regulation. Particularly from this historical perspective, progress in the definition of these factors has been disappointing. Although the phenomenological and conceptual framework of these observations has persisted (see reviews by Tonetta and deZeriga, Tsafriri and Pomerantz, and Khan-Dawood and Dawood, in Franchimont, 1986) none of the peptides themselves has been purified sufficiently to permit structural analysis or definition of mechanism of action. In the absence of such data, it seems possible that such effects may be mediated by other, more recently defined paracrine regulators, acting individually or in concert. In particular, we found that most of the stimulatory activity, for granulosa cell steroidogenesis in follicular fluid from preovulatory follicles, previously termed "luteinization stimulator" (Ledwitz-Rigby and Rigby, 1983), could be neutralized by a monoclonal antibody to IGF-I. Actions of several of the other growth factors described above overlap the inhibitory effects ascribed to various follicular factors on somatic cells of the follicles, and the TGFs and EGF also have well described effects on oocyte maturation (Dekel and Sherizly, 1985, Feng et al, 1988). Of the follicular factors, recent data is most abundant concerning a GnRH-like peptide identified and partially purified by Behrman and colleagues (Aten et al, 1986), and a broad spectrum inhibitor of steroidogenesis named Follicular Regulatory Peptide by deZeriga and colleagues (reviewed in Tonetta and diZerega, 1986). Both of these factors appear to represent discrete entities; however, their biochemical nature and similarity or dissimilarity to other proteins previously identified in the gonad or elsewhere remains to be determined. The cumbersome bioassays for apparently unique ovarian peptides may continue to hamper purification efforts. However, it would be unfortunate if such efforts were totally suspended, since they still offer the best chance to isolate peptides unique to the gonad and useful for regulation of gonadal function. As proved to be the case for inhibin, these difficult purifications should be technically feasible when classical bioassay procedures are reinforced with modern fractionation and molecular techniques.

In more recent years, progress on the definition of peptide regulators in the gonad has proceeded with premises and techniques quite different than those which underlay these initial efforts. The tacit assumption of most current studies is that peptide regulators identified and purified from other organs are present and important in the ovary. The utilization of immunochemical and molecular probes for such previously identified peptides has made possible accumulation of information on gonadal peptides at a rate previously impossible. Variants of this strategy led to identification of many of the ovarian growth factors described previously. Conversely, surveys of other tissues for gonadal gene products, e.g. inhibin, indicated the potential for a much broader physiological role than initially expected (Meunier et al, 1988). Finally, such techniques have made possible a more rapid understanding of the relation of gonadal peptides to related gene families of considerable interest to ovarian physiologists. As again exemplified by the inhibin/activin field, some of these were unknown at the initiation of these studies (Ying, 1988). Other gonadal peptides identified by these strategies include transferrin, fibronectin, angiotensin II, neuropeptide Y, opioid peptides, vasopressin, growth hormone releasing hormone and tumor necrosis factor. It is impossible to summarize the state-of-the-art regarding each of these regulatory substances here. Some of these topics have been addressed in other recent reviews that may be helpful in covering areas inadequately developed here (Hsueh et al., 1984; Hsueh, 1986; Tonetta and deZeriga, 1986; Tsafriri and Pomerantz, 1986; Smith and Funders, 1988, Symonds, 1988).

Summary of Current Efforts and Future Perspectives

It will be clear from the foregoing review that the autocrine/paracrine influences which impinge on ovarian cells are considerably more numerous and almost certainly as important as the classic endocrine feedback loops. On the other hand, the endocrine, autocrine, and paracrine relationships involved in ovarian development have reached a level of complexity that currently defies comprehensive analysis, and the true situation is undoubtedly substantially more complex. While continuing the descriptive phase, concerned principally with identification and characterization of new regulatory mechanisms in the ovary, we also need to integrate this information into physiological constructs, to establish criteria for judging which of the putative regulatory systems are most important, and to begin to develop new investigative strategies which may ultimately lead to pharmacological control of such mechanisms.

By general agreement the minimal criteria for accepting paracrine/autocrine regulatory systems is local secretion and local action of putative regulatory principles. These criteria have been met for the principles considered in this review with the exception of some of the more recently discovered peptides mentioned in the previous paragraph. However, considerably more detailed information about the mechanisms involved in the synthesis and action of effect or molecules is needed for most of the paracrine systems. Of many topics deserving of mention, two are particularly intriguing to me: 1) the multiplicity of mRNHA species and/or large peptide precursors for several ovarian factors is consistent with biosynthetic and/or post-translational regulation which could be tissue specific and hormone regulated; 2) the biochemical mechanisms which govern the interplay between replication and differentiation of ovarian cells represents a critical topic for further study. Activation of protooncogenes may be one fruitful approach in this area. Next, it seems essential to understand the manner in which the secretion and action of these local regulators is interfaced with the hormonal signals of the reproductive cycle and coordinated with developmental changes in the several ovarian compartments. These issues have been best addressed with regard to steroids and the IGFs. Increasingly refined cell culture techniques will continue to contribute to our understanding of the secretion and action of paracrine factors and the control of these processes by the classic ovarian regulators -- pituitary hormones and gonadal steroids. Ultimately, it seems necessary to integrate the information from these *in vitro* studies, in which progress has been much more rapid, with the situation *in vivo*. Analysis of follicular fluid concentrations of the several steroid and peptide regulators during follicular growth has become a classic strategy. However, this strategy has lead to an emphasis on the role of such factors in the terminal stages of follicular development and differentiation. To assess the likely importance of such influences in folliculogenesis and the initiation of follicular growth, other techniques will be required, the most obvious being immunohistochemical analysis for relevant peptides and hybridization histochemistry for their mRNA. It is acknowledged that such studies are intrinsically descriptive and correlative and unable to clearly delineate cause and effect. Nonetheless, a good deal more of this information will be required to piece together a coherent physiological picture of the role and importance of these factors.

Ultimately it will be necessary to develop techniques to block or amplify the effects of such factors. In the first instance, such strategies are needed to clearly determine which of the many putative regulators are essential to ovarian function. Subsequently, these pilot studies may point the way to a more systematic application of such techniques to regulate fertility. One approach is illustrated by our use of monoclonal antibodies to IGF *in vitro* and *ex vivo* (with follicular fluid). With more widespread availability, antibodies to the ligands and/or their receptors should provide useful tools for pharmacological ablation of these systems *in vitro*, and, under some circumstances, *in vivo*. Ultimately, immunochemical techniques may be supplanted by molecular methods such as binding of functional RNAs by anti-sense probes and site-directed mutagenesis of ovarian peptides or their receptors; such techniques are in relatively widespread practice in the study of growth factor effects in transformed cell lines.

However, even with the availability of appropriate blocking agents, systematic manipulation of these paracrine relationships *in vivo* seems remote. Several of the limitations on these strategies are intrinsic to the very nature of the paracrine/autocrine hypothesis and may be extremely difficult to overcome. A cardinal feature of these relationships is their restriction in time and space to allow their impact to be specific. At present, we lack the technological capability to deliver either the regulators or blocking agents in a comparably specific fashion. In addition, all of the ovarian paracrine regulators discovered to date exist in or impact on, many if not most body tissues. The specificity in their ovarian actions derive from their interactions with pituitary hormones or gonadal steroids which either enhance their secretion or modify their effects. As reviewed above, the actions of many steroid and peptide regulators on ovarian cells are quite variable, with the effects encountered critically dependent on the concomitant presence of circulating hormones, other local regulators and/or the level of differentiation of the ovarian cells examined. These complex relationships seem well designed for fine control of the developing ovarian follicle, but they may prove frustrating to ovarian physiologists who seek to manipulate them.

ACKNOWLEDGEMENTS

I acknowledge the helpful comments of Dr. J.S. Mondschein and devoted service of Mrs. Wendy Merkey, Mrs. Sandra Christian, Mrs. Sandra Canning and Ms. Marlene Thompson in the preparation of this manuscript.

Literature Cited

Adashi, E.Y. and C.E. Resnick. 1986. Antagonistic interactions of transforming growth factors in the regulation of granulosa cell differentiation. Endocrinology 119:1879.

Adashi, E.Y., C.E. Resnick, C.S. Croft, J.V. May, Gospodarowicz. 1988. Basic fibroblast growth factor as a regulator of ovarian granulosa cell differentian: a novel non-mitogenic role. Mol. Cell. Endocrinology. 55:7.

Adashi, E.Y., C.E. Resnick, A.M.H. Brodie, M.E. Svoboda, and J.J. Van Wyk. 1985a. Somatomedin-C-mediated potentiation of follicle-stimulating hormone-induced aromatase activity of cultured rat granulosa cells. Endocrinology 117:2313.

Adashi, E.Y., C.E. Resnick, M.E. Svoboda, and J.J. Van Wyk. 1985b. Somatomedin-C synergizes with follicle-stimulating hormone in the acquisition of progestin biosynthetic capacity by cultured rat granulosa cells. Endocrinology 116:2135.

Adashi, E.Y., C.E. Resnick, M.E. Svoboda, and J.J. Van Wyk.1985c. Somatomedin-C enhances induction of luteinizing hormone receptors by follicle-stimulating hormone in cultured rat granulosa cells. Endocrinology 116:2369.

Aten, R.F., A.T. Williams, H.R. Behrman. 1986. Ovarian gonadotropin-releasing hormone-like protein(s): demonstration and characterization. Endocrinology 118:1148.

Adashi, E.Y., C.E. Resnick, M.E. Svoboda and J.J. Van Wyk. 1986b.Somatomedin-C as an amplifier of follicle-stimulating hormone action: Enhanced accumulation of adenosine 3',5'-monophosphate. Endocrinology 118:149.

Adashi, E.Y., C.E. Resnick, M.E. Svoboda, J.J. Van Wyk, V.C. Hascall, and M. Yanagishita. 1986c. Independent and synergistic actions of somatomedin-C in the stimulation of proteoglycan biosynthesis by cultured rat granulosa cells. Endocrinology 118:456.

Assoian, R.K., A.B. Roberts, L.M. Wakefield, M.A. Anzano, and M.B. Sporn.1984. Transforming growth factors in nonneoplastic tissues and their role in controlling cell growth. In: J. Feramisco, B. Ozanne, and C. Stiles (Eds.) Cancer Cells 3/Growth Factors and Transformation, pp. 59-64. Cold Spring Harbor Laboratories, Cold Spring Harbor, New York.

Baird, A., F. Esch, P. Mormede, N. Ueno, N. Ling, P. Bohlen, S.-Y. Ying, W.B. Wehrenberg, and R. Guillemin. 1986. Molecular characterization of fibroblast growth factor: Distribution and biological activities in various tissues. Rec. Prog. Horm. Res. 42:143.

Baranao, J.L.S. and J.M. Hammond. 1984. Comparative effects of insulin and insulin-like growth factors on DNA synthesis and differentiation of porcine granulosa cells. Biochem. Biophys. Res. Comm. 124:484.

Baranao, J.L.S. and J.M. Hammond. 1985b. Multihormone regulation of steroidogenesis in cultured porcine granulosa cells: Studies in serum-free medium. Endocrinology 116:2143.

Bryan, K.A., J.M. Hammond, S.Canning, J.S, Mondschein, E. Carbaugh, A.M.Clark, and D.R. Hagen. 1989. Reproductive and growth responses of gilts to exogenous porcine pituitary growth hormone. J. Anim. Sci. 67:196.

Cara, J.F., and R.L. Rosenfield. 1988. Insulin-Like Growth Factor I and Insulin Potentiate Luteinizing Hormone-Induced Androgen Synthesis by Rat Ovarian Thecal-Interstitial Cells. Endocrinology 123:733.

Carpenter, G., and J.G. Zendegui. 1986. Epidermal growth factor, its receptor, and related proteins. Exp. Cell. Res. 164:1.

Channing and Segal (Eds.). 1982. Intraovarian Control Mechanisms. Plenum Publishing Corporation, New York.

Cohen, S. 1987. Epidermal growth factor. In Vitro Cell. Develop. Biol. 23:239.

Daughaday, W.H. (ed.) 1984. Tissue Growth Factors. Clinics in Endocrinology and Metabolism, vol. 13, no. 1, W.B. Saunders Company, Philadelphia.

Davoren, J.B., and A.J.W. Hsueh. 1986. Growth hormone increases ovarian levels of immunoreactive somatomedin C/insulin-like growth factor I in vivo. Endocrinology 118:888.

Davoren, J.B., A.J.W. Hsueh, and C.H. Li. 1985. Somatomedin C augments FSH induced differentiation of cultured rat granulosa cells. Am. J. Physiol. 249:E26.

Dekel, N., and I. Sherizly. 1985. Epidermal growth factor induces maturation of rat follicle-enclosed oocytes. Endocrinology 116:406.

Derynck, R., and L. Rhee. 1987. Sequence of the porcine transforming growth factor-beta precursor. Nucleic Acids Res. 15:3187.

Dodson, W.C., and D.W. Schomberg. 1987. The effect of transforming growth factor-beta on follicle-stimulating hormone-induced differentiation of cultured rat granulosa cells. Endocrinology 120:512.

Dorrington, J., A.V. Chuma, and J.J. Bendell. 1988. Transforming Growth Factor and Follicle-Stimulating Hormone Promote Rat Granulosa Cell Proliferation. Endocrinology 123:353.

Dorrington, J.H. 1977. Steroidogenesis in vitro. In: Zuckerman L, Weir BJ, (eds) The Ovary, Vol 3, Academic Press, New York, p 359.

Fanjul, L.F., C.M. Ruiz De Galarreta, A.J.W. Hsueh. 1984. Estrogen Regulation of Progestin Biosynthetic Enzymes in Cultured Rat Granulosa Cells. Biology of Reproduction 30:903.

Feng, P., K.J. Catt, M. Knecht. 1988. Transforming Growth Factor-beta Stimulates Meiotic Maturation of the Rat Oocyte. Endocrinology 122:181.

Franchimont, P., M.T. Hazee-Hagelstein, C. Charlet-Renard, and J.M. Jaspar. 1986. Effect of mouse epidermal growth factor on DNA and protein synthesis, progesterone and inhibin production by bovine granulosa cells in culture. Acta Endocrinol. 111:122.

Goldenberg, R.L., E.O. Reiter, and G.T. Ross. 1973. Follicle response to exongenous gonadotropins: an estrogen mediated phenomenon. Fertil. Steril. 24:121.

Goldenberg, R.L., E.O. Reiter, J.L. Vaitukaitis, and G.T. Ross. 1972a. Interaction of FSH and hCG on follicle development in the ovarian augmentation reaction. Endocrinology 91:533.

Goldenberg, R.L., J.L. Vaitukaitis, and G.T.Ross. 1972b. Estrogen and follicle-stimulating hormone interactions on follicle growth in rats. Endocrinology 90:1492.

Gopodarowicz, D., and H. Bialecki. 1979. Fibroblast and epidermal growth factors are mitogenic agents for cultured granulosa cells of rodent, porcine, and human origin. Endocrinology 104:757.

Gospodarowicz, D., J. Cheng, G.M. Lui, A. Baird, F. Esch, and P. Bohlen. 1985. Corpus luteum angiogenic factor is related to fibroblast growth factor. Endocrinology 117:2383.

Gospodarowicz, D., C.R. III, and C.R. Birdwell. 1977a. Effects of fibroblast and epidermal growth factors on ovarian cell proliferation <u>in vitro</u>. I. Characterization of the response of granulosa cells to FGF and EGF. Endocrinology 100:1108.

Gospodarowicz, D., C.R. Ill, and C.R. Birdwell. 1977b. Effects of fibroblast and epidermal growth factors on ovarian cell proliferation <u>in vitro</u>. II. Proliferative response of luteal cells to FGF but not EGF. Endocrinology 100:1121.

Gospodarowicz, D., S. Massoglia, J. Cheng, and D.K. Fujii. 1986a. Effect of fibroblast growth factor and lipoproteins on the proliferation of endothelial cells derived from bovine adrenal cortex, brain cortex, and corpus luteum capillaries. J. Cell. Physiol. 127:121.

Gospodarowicz, D., G. Neufeld, and L. Schweigerer. 1986b. Fibroblast growth factor. Mol. Cell. Endocrinol. 46:187.

Hammond, J.M., J.S. Mondschein, and S.F. Canning. 1988. Insulin-like Growth Factors (IGFs) as Autocrine/Paracrine Regulators in the Porcine Ovarian Follicle. In Paracrine Communication in the Ovary, Proceedings of the VII Ovarian Workshop, A.N. Hirschfield ed., Serono Foundation Press. (In Press).

Hammond, J.M. 1981. Peptide regulators in the ovarian follicle. Australian J Biol Sci 34:491.

Hammond, J.M., J.L.S. Baranao, D. Skaleris, A.B. Knight, J.A. Romanus, and M.M. Rechler. 1985. Production of insulin-like growth factors by ovarian granulosa cells. Endocrinology 117:2553.

Hammond, J.M., and H.F. English. 1987. Regulation of deoxyribonucleic acid synthesis in cultured porcine granulosa cells by growth factors and hormones. Endocrinology 120:1039.

Hammond, J.M., R. Hershey, M. Walega, and J. Weisz. 1986. Catecholestrogen production by porcine ovarian cells. Endocrinology 118:2292.

Hammond, J.M., C-J. Hsu, J. Klindt, B.K. Tsang, and B.R. Downey. 1988. Gonadotropins increase concentrations of immunoreactive insulin-like growth factor-I in porcine follicular fluid _in vivo_. Biol. Reprod. 38:304.

Hammond, J.M., J.D. Veldhuis, T.W. Seale, and M.M. Rechler. 1982. Intraovarian regulation of granulosa-cell replication. In: Channing and Segal (Eds.) Intraovarian Control Mechanisms, pp. 341-356. Plenum Publishing Corporation, New York.

Hammond, J.M., K. Yoshida, J.D. Veldhuis, M.M. Rechler, and A.B. Knight. 1983. Intrafollicular role of somatomedins: Comparison with effects of insulin. In: G.S. Greenwald and P.F. Terranova (Eds.) Factors Regulating Ovarian Function, pp. 197-201. Raven Press, New York.

Hernandez, E.R., E. Hoyt, J.J. Van Wyk, and E.Y. Adashi. 1987a. The somatomedin-C/insulin-like growth factor-I (Sm-C/IGF-I) gene is expressed in the rat ovary. Ann. Mtg. Endocr. Soc., Abstr. 821.

Hild-Petito, S., R.L. Stouffer, N.B. West, R.M. Brenner. 1988. Localization of Progesterone, but not Estradiol, Receptors in Follicular and Luteal Tissues During the Menstrual Cycle. Ann. Mtg. Endocr. Soc., Abstr. 1121.

Hernandez, E.R., D.R. Twardzik, A. Purchio, and E.Y. Adashi. 1987b. Gonadotropin-dependent ovarian transforming growth factor-beta gene expression. Biol. Reprod. 36 (Suppl. 1): 58.

Hotta, M., and A. Baird. 1986. Differential effects of transforming growth factor type beta on the growth and function of adrenocortical cells _in vitro_. Proc. Natl. Acad. Sci. 83:7795.

Hsu, C.-J., and J.M. Hammond. 1987a. Gonadotropins and estradiol stimulate immunoreactive insulin-like growth factor-I production by porcine granulosa cells in vitro. Endocrinology 120:198.

Hsu, C.-J., and J.M. Hammond. 1987b. Concomitant effects of growth hormone on secretion of insulin-like growth factor-I and progesterone by cultured porcine granulosa cells. Endocrinology 121:1343.

Hsu, C.-J., S.D. Holmes, and J.M. Hammond. 1987. Ovarian epidermal growth factor-like activity. Concentrations in porcine follicular fluid during follicular enlargement. Biochem. Biophys. Res. Comm. 147:242.

Hsueh, A.J.W. 1986. Paracrine mechanisms involved in granulosa cell differentiation. In: P. Franchimont (Ed.) Clinics in Endocrinology and Metabolism. pp. 117-134. W.B. Saunders, Philadelphia.

Hsueh, A.J.W., E.Y. Adashi, P.B.C. Jones, and T.H. Welsh, Jr. 1984. Hormonal regulation of the differentiation of cultured ovarian granulosa cells. Endocr. Rev. 5:76.

Hsueh, A.J.W., T.H. Welsh, and P.B.C. Jones. 1981. Inhibition of ovarian and testicular steroidogenesis by epidermal growth factor. Endocrinology 108:2002.

Josso, N. 1986. Anti-mullerian hormone. Clin. Endocrinol. 25:331.

Jones, P.B.C., T.H. Welsh, Jr., and A.J.W. Hsueh. 1982. Regulation of ovarian progestin production by epidermal growth factor in cultured rat granulosa cells. J. Biol. Chem. 257:11268.

Khan-Dawood, F.S., Dawood, M.Y. 1986. Paracrine Regulation of Luteal Function. Clinics in Endocrinology and Metabolism 15:171.

Kim, I., G.S. Greenwald. 1987. Effect of estrogens on follicular development and ovarian and uterine estrogen receptors in the immature rabbit, guinea pig and mouse. Endocrinol Jpn 34:871.

Knecht, M., and K.J. Catt. 1983a. Modulation of cAMP-mediated differentiation in ovarian granulosa cells by epidermal growth factor and platelet-derived growth factor. J. Biol. Chem. 258:2789.

Knecht, M., and K.J. Catt. 1983b. Epidermal growth factor andgonadotropin releasing hormone inhibit cyclic AMP-dependent luteinizing hormone receptor formation in ovarian granulosa cells. J. Cell. Biochem. 21:209.

Knecht, M., P. Feng, and K.J. Catt. 1986. Transforming growth factor-beta regulates the expression of luteinizing hormone receptors in ovarian granulosa cells. Biochem. Biophys. Res. Comm. 139:800.

Knecht, M., P. Feng, and K.J. Catt. 1987. Bifunctional role of transforming growth factor-beta during granulosa cell development. Endocrinology 120:1243.

Koos, R.D. 1986.Stimulation of endothelial cell proliferation by rat granulosa cell-conditioned medium. Endocrinology 119:481.

Kudlow, J.E., M.S. Kobrin, A.F. Purchio, D.R. Twardzik, E.R. Hernandez. S.L. Asa, E.Y. Adashi. 1987. Ovarian Transforming Growth Factor-alpha Gene Expression: Immunohistochemical Localization to the Theca-Interstitial Cells. Endocrinology 121:1577.

Ledwitz-Rigby, F., and B.W. Rigby. 1983. The actions of follicular fluid factors on steroidogenesis by cultured ovarian granulosa cells. J. Steroid Biochem. 19:127.

Louvet, J.P., S.M. Harman, J.R. Schreiber, G.T. Ross. 1975. Evidence for a role of androgens in follicular maturation. Endocrinology 97:366.

Makris, A., K.J. Ryan, T. Yasumizu, C.L. Hill, and B.R. Zetter. 1984. The nonluteal porcine ovary as a source of angiogenic activity. Endocrinology 115:1672.

Massague, J. 1987. The TGF-beta family of growth and differentiation factors. Cell 49:437.

May, J.V., P.A. Buck, and D.W. Schomberg. 1987. Epidermal growth factor enhances [^{125}I]iodo-follicle-stimulating hormone binding by cultured porcine granulosa cells. Endocrinology 120:2413.

Meunier, H., C. Rivier, R.M. Evans, W. Vale. 1988. Gonadal and extragonadal expression of inhibin alpha beta-A and beta-B subunits in various tissues predicts diverse functions. Proc Nat Acad Sci, USA 85:247.

Mondschein, J.S., S.F. Canning, and J.M. Hammond. 1988.Effects of Transforming Growth Factor Beta on the Production of Immunoreactive Insulin-Like Growth Factor-I and Progesterone and 3H-Thymidine Incorporation in Porcine Granulosa Cell Cultures. Endocrinology 123:1970.

Mondschein, J.S., S.F. Canning, and J.M. Hammond.1988. The role of insulin-like growth factors in granulosa cell steroidogenesis: Studies with neutralizing antibody to IGF-I. VII The Ovarian Workshop Serono Symposia. In Press.

Mondschein, J.S., and J.M. Hammond. 1988. Growth Factors Regulate Immunoreactive Insulin-Like Growth Factor-I Production by Cultured Porcine Granulosa Cells. Endocrinology 123:463.

Mondschein, J.S., J.M. Hammond, S.F. Canning, C.H. Li, and K. Ramasharma. 1988. Profiles of immunoreactive insulin-like growth factors-I and -II in procine follicular fluid and granulosa cell-conditioned medium. Biol. Reprod. Suppl. 1. Abst 429.

Mondschein, J.S., and D.W. Schomberg. 1981. Growth factors modulate gonadotropin receptor induction in granulosa cell cultures. Science 211:1179.

Moses, H.L., R.F. Tucker, E.B. Leof, R.J. Coffey, Jr., J. Halper, and G.D. Shipley. 1984. Type-beta transforming growth factor is a growth stimulator and a growth inhibitor. In: J. Feramisco, B. Ozanne, and C. Stiles (Eds.) Growth Factors and Transformation, pp. 65-71. Cold Spring Harbor Laboratory, Cold Spring Harbor, New York.

Murphy, L.J., G.I. Bell, and H.G. Friesen. 1987. Tissue distribution of insulin-like growth factor I and II messenger ribonucleic acid in the adult rat. Endocrinology 120:1279.

Neufeld, G., N. Ferrara, L. Schweigerer, R. Mitchell, and D. Gospodarowicz. 1987. Bovine granulosa cells produce basic fibroblast growth factor. Endocrinology 121:597.

Pencharz, R.I. 1940.Effect of estrogens and androgens alone and in combination with chorionic gonadotropin on the ovary of the hypophysectomized rat. Science 91:554.

Rall, L.B., J. Scott, G.I. Bell, R.J. Crawford, J.D. Penschow, H.D. Niall, and J.P. Coghlan. 1985. Mouse prepro-epidermal growth factor synthesis by the kidney and other tissues. Nature 313:228.

Ramasharma, K., C.M. Cabrera, and C.H. Li. 1986.Identification of insulin-like growth factor-II in human seminal and follicular fluids. Biochem. Biophys. Res. Comm. 140:536.

Ramasharma, K., and C.H. Li. 1987. Human pituitary and placental hormones control human insulin-like growth factor II secretion in human granulosa cells. Proc. Natl. Acad. Sci. 84:2643.

Rao, M.C., J.S. Richards, and A.R. Midgley, Jr. 1978. Hormonal regulation of cell proliferation in the ovary. Cell 14:71.

Ratoosh, S.L., and J.S. Richards. 1985. Regulation of the content and phosphorylation of R_{II} by adenosine 3', 5'-monophosphate, follicle-stimulating hormone, and estradiol in cultured granulosa cells. Endocrinology 117:917.

Richards, J.S. 1975. Content of nuclear estradiol receptor complex in rat granulosa cells during follicular development: modification by estradiol and gonadotropins. Endocrinology 97:1174.

Richards, J.S. 1980. Maturation of Ovarian Follicles: Actions and interactions of pituitary and ovarian hormones on follicular cell differentiation. Physiological Reviews 60:51.

Roberts, A.B., M.A. Anzano, L.M. Wakefield, N.S. Roche, D.F. Stern, and M.B. Sporn. 1985. Type transforming growth factor: A bifunctional regulator of cellular growth. Proc. Natl. Acad. Sci. 82:119.

Ryan, K.J. 1979. Granulosa-thecal cell interaction in ovarian steroidogenesis. J. Steroid Biochem. 11:799.

Samsoondar, J., M.S. Kobrin, and J.E. Kudlow. 1986. Transforming growth factor-beta secreted by untransformed bovine anterior pituitary cells in culture. J. Biol. Chem. 261:14408.

Savion, N., and D. Gospodarowicz. 1980. Factors controlling the proliferation and phenotypic expression of cultured bovine granulosa cells. In: V.B. Mahesh, T.G. Muldoon, B.B. Saxena and W. Sadler (Eds.), Functional Correlates of Hormone Receptors in Reproduction, pp. 437-461. Elsevier North Holland, Inc.

Savion, N., G-M. Lui, R. Laherty, and D. Gospodarowicz. 1981. Factors controlling proliferation and progesterone production by bovine granulosa cells in serum-free medium. Endocrinology 109:409.

Simpson, M.E., H.M. Evans, H.L.Fraenkelconrat, and C.H. Li. 1941. Synergism of estrogens with pituitary gonadotropins in hypophysectomized rats. Endocrinology 28:37.

Skinner, M.K., J. Keski-Oja, K.G. Osteen, and H.L. Moses. 1987. Ovarian thecal cells produce transforming growth factor-beta which can regulate granulosa cell growth. Endocrinology 121:786.

Skinner, M.K., D. Lobb, J.H. Dorrington. 1987. Ovarian Thecal/Interstitial Cells Produce an Epidermal Growth Factor-Like Substance. Endocrinology 121:1892.

Smith, A.I., and J.W. Funder. Proopiomelanocortin Processing in the Pituitary, Central Nervous System, and Peripheral Tissues. Endocrine Reviews 9:159.

Spicer, L.J., and J.M. Hammond. 1987. Catechol estrogens stimulate progestin secretion by cultured porcine granulosa cells. Mol. Cell. Endocrinol. 50:139.

Spicer, L.J., and J.M. Hammond. 1988a. Effects of 2-hydroxyestradiol on the number of porcine granulosa cell beta -adrenergic binding sites. Endocrinology 122:2710.

Spicer, L.J., and J.M. Hammond. 1988b. Mechanism of action of 2-hydroxyestradiol on steroidogenesis in ovarian granulosa cells: Interactions with catecholamines and gonadotropins involve cyclic AMP. Biol. Reprod. In Press.

Sporn, M.B., A.B. Roberts, L.M. Wakefiel, and R.K. Assoian. 1986. Transforming growth factor-beta: Biological function and chemical structure. Science 233:532.

Symonds, E.M. 1988. Renin and Reproduction. Am. J. Obstet. Gynecol. 158:754.

Toaff, M.E. J.F. Strauss III, and J.M. Hammond. 1983. Regulation of cytochrome $P\text{-}450_{scc}$ in immature porcine granulosa cells by FSH and estradiol. Endocrinology 112:1156.

Toaff, M.E., J.F. Strauss, III., P.F. Hall, J.L. Barano, J.M. Hammond. Regulation of cytochrome P-450$_{scc}$ synthesis in porcine granulosa cells. In: Lipoprotein and Cholesterol Metabolism in Steroidogenic Tissues, J.F. Strauss and K.M. Mennon (eds.), George F. Stickley Co., Philadelphia, 1985, pp. 39-41.

Tonetta, S.A., and G.S. diZerega. 1986. Paracrine regulation of follicular maturation in primates. In: P. Franchimont (Ed.) Clinics in Endocrinology and Metabolism. pp. 135-156. W.B. Saunders, Philadelphia.

Tsafriri, A., and S.H. Pomerantz. 1986. Oocyte maturation inhibitor. In: P. Franchimont (Ed.) Clinics in Endocrinology and Metabolism. pp. 157-170. W.B. Saunders, Philadelphia.

Underwood, L.E., A.J. D'Ercole, D.R. Clemmons, and J.J. Van Wyk. 1986. Paracrine functions of somatomedins. In: P. Franchimont (Ed.) Clinics in Endocrinology and Metabolism. pp. 59-78. W.B. Saunders, Philadelphia.

Valverius, E., S.E. Bates, D.S. Salomon, M. Stampfer, R. Clark, F. McCormick, M.E. Lippman, and R.B. Dickson. 1987. Production of transforming growth factor alpha by human mammary epithelial cells and sublines. Ann. Mtg. Endocr. Soc., Abstr. 306.

Veldhuis, J.D., P. Azimi, D. Juchter, and J. Garmey. 1986a. Mechanisms Subserving the Bipotential Actions of Estrogen on Ovarian Cells: Studies With a Selective Anti-Estrogen, LY156758, and the Sparingly Metabolizable Estrogen Agonist, Moxestrol. J. Steroid Biochem. 24:977.

Veldhuis, J.D., L.M. Demers, J. Garmey, D. Juchter, and P. Azimi. 1985. A role for somatomedin C as a differentiating hormone and amplifier of hormone action on ovarian cells: Studies with synthetically pure human somatomedin C and swine granulosa cells. Biochem. Biophys. Res. Comm. 130:234.

Veldhuis, J.D., J.T. Gwynne, J.F. Strauss III, L.M. Demers, and P.A. Klase. 1984. Role of Estradiol as a Biological Amplifier of Gonadotropin Action in the Ovary: In Vitro Studies Using Swine Granulosa Cells and Homologous Lipoproteins. Endocrinology 114:2312.

Veldhuis, J.D., and J.M. Hammond. 1979. Multiplication stimulating activity regulates ornithine decarboxylase in isolated porcine granulosa cells in vitro. Endocr. Res. Comm. 6:299.

Veldhuis, J.D., and J.M. Hammond. 1980. Estrogens regulate divergent effects of Prolactin in the ovary. Nature 284:262.

Veldhuis, J.D., P.A. Klase, and J.M. Hammond. 1981. Direct actions of 17-estradiol on progesterone production by highly differentiated procine granulosa cells in vitro: II. Regulatory interactions of estradiol with luteinizing hormone and cyclic nucleotides. Endocrinology 109:433.

Veldhuis,J.D., P.A. Klase, J.F. Strauss III, and J.M. Hammond. 1982a. The role of estradiol as a biological amplifier of the actions of follicle-stimulating hormone. In vitro studies in swine granulosa cells. Endocrinology 111:144.

Veldhuis, J.D., P.A. Klase, J.F. Strauss III, and J.M. Hammond. 1982b. Facilitative interactions between estradiol and luteinizing hormone in the regulation of progesterone production by cultured swine granulosa cells: Relation to cellular cholesterol metabolism. Endocrinology 111:441.

Veldhuis, J.D., J.E. Nestler, J.F. Strauss, III, P. Azimi, J. Garmey, and D. Juchter. 1987. The insulin-like growth factor, somatomedin-C, modulates low density lipoprotein metabolism by swine granulosa cells. Endocrinology 121:340.

Veldhuis, J.D.,R.J. Rodgers, A. Dee, and E.R. Simpson. 1986b. The insulin-like growth factor, somatomedin C, induces the synthesis of cholesterol side-chain cleavage cytochrome P-450 and adrenodoxin in ovarian cells. J. Biol. Chem. 261:2499.

Voutilainen, R.,and W.L. Miller. 1987. Coordinate tropic hormone regulation of mRNAs for insulin-like growth factor II and the cholesterol side-chain-cleavage enzyme P450ssc, in human steroidogenic tissues. Proc. Natl. Acad. Sci. 84:1590.

Weber, S.L., and A.R. LaBarbera. 1988. Insulin, Somatomedin-C, Human Chorionic Gonadotropin, and Forskolin Enhance Glucose Oxidation by Granulosa Cells. Biol Reprod 38:812.

Ying, S.-Y., A. Becker, N. Ling, N. Ueno, and R. Guillemin. 1986. Inhibin and beta type transforming growth factor (TGF) have opposite modulating effects on the follicle stimulating hormone (FSH)-induced aromatase activity of cultured rat granulosa cells. Biochem. Biophys. Res. Comm. 136:969.

Ying, S.-Y. 1987. Inhibins and Activins: Chemical Properties and Biological Activity (42611A). Proc Soc Exp Biol Med 186:253.

EXPERIMENTAL APPROACHES TO THE STUDY OF EARLY DEVELOPMENTAL FAILURE IN HUMAN REPRODUCTION

JONATHAN VAN BLERKOM

EARLY DEVELOPMENTAL FAILURE IN THE HUMAN

One of the most unfortunate facts of human reproduction is that the majority of normally fertilized eggs will not develop into a viable, continuing pregnancy (Edwards, 1986). This conclusion is strengthened by the relatively low rate at which an ongoing pregnancy is achieved after in vitro fertilization and multiple embryo transfer at expert IVF programs (between 1# and 25%: Liu et al, 1988; Gibbons, 1988). These values are derived from a typical population of infertile couples that present with a variety of etiologies (male factor, female factor, or both).

Not infrequently, the failure of fertilization to occur or of development to persist can be attributed to abnormal immunological or endocrine patterns, or to morphological perturbations in the structure of the uterine epithelium. Likewise, compromised sperm, which include such factors as low count, aberrant or reduced motility, and abnormal morphology that results from spermatogenic or maturational dysfunction, can influence both the frequency of fertilization and the prognosis for postfertilization embryogenesis ((Baccetti, 1983; Zamboni, 1987). While certain classes of reproductive dysfunction can often be corrected by relatively simple means (e.g., intrauterine insemination, sperm antibody suppression), more complex surgical or endocrinological, or both, intervention is often necessary. However, surgical and endocrinological approaches are frequently precluded because of severe oligospermia, or as a result of blocked, damaged or absent fallopian tubes. In this case in vitro fertilization is indicated.

Why the vast majority of embryos created by in vitro fertilization fail to establish a pregnancy after transfer to the uterus is one of the fundamental unanswered questions in the application of this technology to human infertility. This is especially relevant when multiple early cleavage stage embryos that appear normal at the level of the light microscope are transferred without the detectable occurrence of implantation. On some occasions, endocrinological, immunological and uterine morphological factors can be identified as contributing to the failure of implantation. For most embryos transfers, however, it is very difficult to ascribe a specific reason for the failure.

One approach to an investigation of early developmental failure in human reproduction is to ask whether the inability of an oocyte to be fertilized, or of an embryo to continue development or to implant is an inherent rather than an external inadequacy or dysfunction, that is to say that the developmental capacity of an oocyte or embryo is limited from the outset. This question is of particular relevance when, in the presence of spermatozoa of known fertility, penetration of some or all of the normal appearing, meiotically mature oocytes obtained from hyperstimulated human ovaries fails to occur. If fertilization or normal development is inherently limited in the human species, then to what degree might this occur during both in vivo and in vitro attempts to achieve a pregnancy?

The study of human oocytes and early embryos requires clear and focused experimental analyses that address questions which will be ultimately of clinical relevance in the treatment or understanding of the causes of infertility. The types of analyses discussed in this paper focus on two such aspects of early human development. The first type of analysis addresses the question of the degree of chromosomal abnormality in the preovulatory oocyte. Some investigators have indicated that more than half of all preovulatory oocytes produced by ovarian hyperstimulation are chromosomally aberrant (see Table 1). If this is indeed the prevailing situation, then attempts to achieve a pregnancy, and more importantly a normal pregnancy, may be compromised even before sperm and oocyte meet. The following two questions also arise from the analysis of the chromosomal status of the preovulatory oocyte: (1) is the frequency of chromosomal aberration higher in women whose oocytes persistently fail to fertilize in vitro, and (2) is the occurrence of chromosomally abnormal oocytes more patient specific than a ubiquitous and obligate consequence of exogenous ovarian stimulation? With some noted exceptions, the findings related to the chromosomal status of human oocytes are derived from the examination of oocytes that fail to fertilize after insemination in vitro (see Table 1).

The second type of analysis focuses on the failure of newly fertilized eggs to progress in development. Early studies of embryonic development in vivo by Hertig, et al (1954) noted that in humans an unexpectedly high proportion of embryos arrested development prior to the blastocyst stage. More recent studies by Buster et al (1985) described a very high frequency of embryonic demise during the preimplantation stages of in vivo development in women undergoing ovarian hyperstimulation and intrauterine insemination for subsequent embryo transfer. Early developmental failure is clearly evident during in vitro culture, where approximately 10-15% of the normally fertilized human eggs actually develop to the blastocyst stage (Fishel et al, 1985: Lindenberg and Hytell, 1988). Even for blastocyst-stage embryos, however, the normality of development is often questionable owing to suspected quantitative (inappropriate cell number) and qualitative disorders (e.g. multinucleate cells) in the inner cell mass and trophectoderm (Buster et al, 1985; Edwards, 1986; Lindenberg and

Hytell, 1988). That cleavage arrest occurs with similar frequencies in IVF cycles that involved different ovarian stimulation protocols and culture media suggests that developmental failure is embryo-related rather than a consequence of how oocytes are generated or the nature of the environment in which embryos are fertilized and grown.

Collectively, a considerable body of evidence indicates that even when a normal fertilization has occurred, whether in vivo or in vitro, development to the stage of implantation is by no means assured. When, why, and to what extent early developmental arrest is associated with the human species is one of the fundamental issues in understanding human reproduction. Because of the moral, ethical and potential legal ramifications that attend the analytical or experimental use of presumably normal zygotes and preimplantation embryos, study of early human development usually involves embryos that (1) are derived from abnormally fertilized oocytes (polyspermic), (2) clearly develop in an aberrant pattern, and (3) have unambiguously ceased to progress in vitro. In spite of the restricted nature of "analyzable specimens," a considerable amount of information relevant to normal and abnormal developmental processes has been obtained. As discussed below, some of these findings not only indicate why development fails but also provide insight into the more subtle origins of infertility.

THE FREQUENCY OF CHROMOSOMAL ANOMALIES IN HUMAN OOCYTES AFTER OVARIAN HYPERSTIMULATION

Table 1 presents current estimates of the frequency of aneuploidy (hypohaploidy and hyperhaploidy) in meiotically mature human oocytes that were karyotyped after failed fertilization. It is apparent that these values vary widely from a low of 11% to a high of 65%. In comparison to other studies, the extraordinarily high rate of aneuploidy reported by Wramsby and Fredga (1987) appears to be an overestimate that may have resulted from the comparatively low number of oocytes examined, from the patient population included in the study and, for the hypohaploid assignments, perhaps from artifactual loss during preparation of the oocytes for chromosome quantitation (Plachot et al, 1988; Van Blerkom and Henry, 1988). The reports of Bongos et al (1988), Pellestor and Sele (1988), Plachot et al. (1988), Van Blerkom and Henry (1988) and Van Blerkom (1989a) provide a frequency of aneuploidy that is fairly consistent and is derived from the analysis of nearly 1000 oocytes obtained from hyperstimulated ovaries. At present, it would appear that between 20 and 25 percent of the meiotically mature oocytes that are derived from stimulated ovaries and which fail to fertilize in vitro are chromosomally aberrant. This value appears to be a constant that is independent of the protocol of ovarian stimulation (Plachot et al, 1988; Van Blerkom, 1989a).

A significant limitation in the interpretation of the above finding is that the frequency of aneuploidy is derived almost entirely from the examination of oocytes that failed to fertilize. It may be reasonably argued that such oocytes are already compromised, and in such a selected population of oocytes chromosomal anomalies may be only one manifestation of a developmentally perturbed state. In addition, Van Blerkom (1989a) noted that the question of whether chromosomal perturbations are an unavoidable consequence of the stimulation and development of multiple follicles is difficult to answer at present because the baseline frequency of oocyte aneuploidy that occurs in natural (unstimulated) cycles has not been determined for a large and clinically varied population. However, one approach to obtain this background value is to determine the frequency of chromosomal abnormalities in oocytes that matured in vitro from the GV to MII-stages. The most extensive analysis of in vitro matured human oocytes was reported by Jagiello et al. (1976). These investigators detected only 6 chromosomal anomalies (1.5%) in 411 MII-stage oocytes that had resumed meiosis spontaneously in vitro after collection at the GV stage. This finding tends to suggest that the current protocols for the recruitment and developmental stimulation of multiple follicles/oocytes are associated with an elevated frequency of aneuploidy.

Another approach to the question of the background frequency of aneuploidy is to determine the chromosomal status of meiotically mature oocytes obtained from the ovary and prepared for karyotyping immediately preceding the anticipated time of ovulation. Wramsby et al. (1987) noted that approximately 50% of the meiotically mature oocytes obtained in this fashion were aneuploid. By contrast, Van Blerkom and Henry (1988) reported that 11% of normal-appearing preovulatory (uninseminated) oocytes were aneupoloid. As discussed previously, the extent of aneuploidy indicated in the studies of Wramsby and his collaborators appears to be an overestimate. However, while the frequency of aneuploidy reported by Van Blerkom and Henry (1988) is approximately half of the frequency described for failed fertilizations (20-25%), it is nevertheless significantly higher than the 1.5% value reported by Jagiello et al (1976) for human oocytes obtained from unstimulated ovaries, and which spontaneously resumed and completed meiotic maturation entirely in vitro.

While it seems likely that ovarian hyperstimulation is associated with an elevated frequency of aneuploidy, a rather basic question that has not been fully addressed is whether such an increased genetic risk is an unavoidable consequence of attempts to generate multiple oocytes, or whether chromosomally aberrant oocytes are more likely to be patient-specific. The notion that a rather small number of women account for a disproportionately large share of chromosomally aberrant oocytes was addressed on a limited basis (163 women) in the study of Van Blerkom and Henry (1988). These investigators reported that nearly 27% of the hypohaploid oocytes and 33% of the oocytes that exhibited chromosomes not associated with the metaphase II spindle (i.e., potential aneuploids) were derived from two patients during single attempts at IVF. Thus, the frequency with which chromosomally abnormal oocytes may occur, or may be anticipated after ovarian hyperstimulation, may be difficult to estimate with respect to a particular individual.

The current ambiguity associated with the risk of aneuploidy after ovarian stimulation could be resolved through a concerted, multicenter analysis of oocytes, both in the living state and by karyotyping after destructive fixation. Oocytes from both failed fertilizations and under circumstances where a truly excessive number are available for fertilization, and where embryo cryopreservation is not an option, could be used for this purpose. The report of Plachot et al (1988) discusses the results of such a multicenter effort. In this respect, Van Blerkom and Henry (1988) described a method for oocyte cytogenetic evaluation that could be performed on living oocytes which failed to fertilize or where fertilization was not undertaken. This method entails the use of fluorescent, chromosomal DNA stains. Not only were aneuploid oocytes detected by this procedure, but also oocytes that exhibited major disorders in chromosome structure and association with the metaphase spindle were observed. Such oocytes probably contribute significantly to the overall frequency of aneuploidy. A focused effort to examine the cytogenetics of oocytes that would otherwise be discarded, first in the living state and subsequently by conventional karyotyping methods, would go a long way to definitively establishing (1) the frequency of aneuploidy in stimulated cycles, (2) whether it was patient and/or age related, and (3) whether specific protocols of stimulation would be more appropriate in order to reduce the risk of generating a genetically abnormal oocyte(s). The production of an aneuploid oocyte is not a trivial concern because it cannot be assumed, _a priori_, that such an oocyte is either unfertilizable or, if fertilized, is incapable of implantation. Indeed, even triploid embryos that result from dispermic penetration are capable of apparently normal preimplantation development (Van Blerkom et al, 1984) and can develop to term on rare occasions (Werteleki et al, 1976). In this regard, Van Blerkom and Henry (1988) examined the chromosomal state of both uninseminated and unfertilized oocytes obtained from women whose oocytes, on multiple attempts at IVF, persistently failed to fertilize in the presence of spermatozoa of known fertility (both husband and donor). The frequency of structural aberration and aneuploidy was no different from that observed in IVF cycles (for different patients) when only some of the meiotically mature oocytes failed to fertilize. This finding suggests that failed fertilization may not have a direct chromosomal association. By contrast, structurally aberrant oocytes, such as those in which one or more chromosomes have become detached from the metaphase spindle, should be fertilizable, especially if all other aspects of preovulatory maturation have been completed successfully. In this regard, Angell et al (1983) reported that a rather high percent of cleavage-stage embryos produced by in vitro fertilization are chromosomally abnormal. One interpretation of this result is that some of these genetic anomalies pre-exist in an oocyte prior to fertilization. Clearly, the extent to which an increased frequency of genetic perturbation that can compromise pre- or early postimplantation development is associated with ovarian hyperstimulation needs to be determined with oocytes that are routinely available and, for the described purposes, currently discarded.

DEVELOPMENTAL FAILURE DURING THE PERIFERTILIZATION PERIOD

While fertilization in vitro of meiotically mature human oocytes obtained from hyperstimulated ovaries occurs at a relatively high frequency (typically between 60 and 80%), most of these zygotes will arrest development during the preimplantation stages (Edwards, 1986; Van Blerkom, 1988b). The perifertilization period is a particularly critical stage of development because a series of cellular and molecular changes must occur in an time-dependent, progressive pattern in order for the zygote to initiate and complete the first mitotic or cleavage division. For example, major cellular changes involve the cortical granule reaction, incorporation into the ooplasm of the fertilizing spermatozoon, abstriction of the second polar body, and formation and translocation of pronuclei. Some of the major molecular changes associated with this stage of development are decondensation of the sperm chromatin, replication of DNA, and differential modification of male and female genomes at the pronuclei stage (genomic imprinting; Surani, 1987). Developmental failure at the 1-cell stage is of particular interest owing to the importance of the successful completion of each of these processes for subsequent embryogenesis. Consequently, much of our analytic efforts have been directed to human developmental failure at the 1-cell stage (for details, see Van Blerkon, 1988).

After closer inspection by high-resolution, differential interference contrast microscopy, approximately 5-10% of the oocytes that were thought to be unfertilized did indeed have one or more sperm in the perivitelline space (Van Blerkom, 1988b). At the electron microscopic level, a feature typical of these oocytes is the absence of binding between the sperm and oocyte microvilli- -the first step in the actual process of fertilization. For such oocytes, three fundamental aspects of sperm-oocyte interaction may be abnormal. First, microvilli-associated cell surface glycoproteins that recognize sperm-associated surface proteins (or vice versa) may either be absent, altered, or distributed abnormally. Therefore, although the juxtaposition of the gametes appropriate for attachment can occur, the molecular interactions necessary for incorporation of the fertilizing spermatozoon may be absent or inadequate. This particular hypothesis is speculative at present because information concerning the biochemical and spatial nature of the cell surface molecules involved in the fertilization process is lacking for the human oocyte. Clearly, an understanding of the temporal and spatial aspects of the first molecular interactions between human gametes is a necessary prerequisite to determining whether, for some oocytes, subtle biochemical perturbations are responsible for developmental failure.

A second potential cause of the failure of the male gamete to penetrate the oocyte may be associated with the organization of cortical cytoskeletal elements, such as microfilaments. Actin filaments not only provide the structural support for the elaboration of microvilli, but also appear to be closely involved in the motility of the cortical cytoplasm during the process of sperm penetration (for review, see Van Blerkom, 1988). Consequently, an anomaly in either structure, organization or distribution of these elements may preclude the penetration of the fertilizing spermatozoon after a presumably normal attachment to the oocyte surface. Although detailed information concerning the cytostructural organization of the mature human oocyte is lacking, a few studies have shown the importance of the spatial distribution of actin filaments for human oocyte development and fertilizability. Sathananthan et al, 1985 described the presence of a cortical band of microfilaments that, in the GV-to-GVB (germinal vesicle-to-germinal vesicle breakdown)-stage human oocyte, appeared to provide a physical barrier that prevented the premature deposition of cortical granules in the subplasmalemmal cytoplasm. These investigators suggested that the development of such a barrier precludes a premature (preovulatory) cortical reaction that would make an oocyte refractory to sperm penetration through the zona pellucida (zona reaction) after ovulation.

Van Blerkom and Henry (1988) and Van Blerkom (1988) described an anomaly in the structural organization of the cytocortix that was associated with the failure of fertilization in grossly normal-appearing, meiotically mature human oocytes. In this situation, nearly the entire cell surface was devoid of microvilli, and, in the majority of the subplasmalemmal cytoplasm, cortical granules, microtubules and actin filaments were absent. By contrast, all of these elements were present at high density in narrow region that surrounded the portion of the oocyte from which the first polar body had been abstricted. The occurrence of this phenotype in meiotically mature preovulatory oocytes demonstrated that it did not necessarily develop as a consequence of culture or in vitro exposure to spermatozoa. Nearly all of the oocytes that displayed this pattern of cytocortical polarization were observed in women that had a history of idiopathic infertility, and whose oocytes failed to fertilize after multiple attempts by in vitro methods.

While the precise origin of this phenotype is not known, indirect evidence suggests that it arises during the stage at which the first polar body is abstricted (Van Blerkom and Henry, 1988). The polarization of the cellular and molecular components involved in the fertilization process could be a significant factor associated with persistent fertilization failure in this patient population. A clear indication of when during oogenesis this abnormality may develop could be derived from the analysis of fully-grown but meiotically immature oocytes, obtained for diagnostic purposes from women in which this condition has either been previously identified or is suspected. Under these circumstances, oocytes could be examined at the germinal vesicle stage and at timed intervals during meiotic maturation in vitro (Van Blercom, 1989b). The detailed examination of oocytes from women with persistent fertilization failure, and especially where a male factor has been precluded (as far as possible), should indicate the frequency with which this type of unfertilizable oocyte occurs both in unstimulated (in vitro matured) and stimulated cycles (with and without the hormonal induction of ovulation). The reorganization and polarization of the cortical cytoplasm and plasma membrane is also viewed as a fundamental developmental and cell biological phenomenon. An understanding of the molecular and cellular mechanisms by which this perturbation occurs could provide fundamental insights into how developmental changes in the structure and organization of the cortical cytoplasm and plasma membrane lead to the attainment of the fertilizable state in the human species.

Two other cellular conditions associated with peri-fertilization failure in the human that are often observed after insemination in vitro are (1) absent or incomplete decondensation of spermatozoal DNA, and (2) incomplete formation, migration or juxtaposition of pronuclei that is required for syngamy (for details, see Van Blerkom, 1988). The failure of sperm to decondense could be attributed to a fundamental defect in the packaging or organization of DNA during spermatogenesis. Our studies have attempted to reduce this possibility by focusing the analysis on those developmentally arrested 1-cell embryos that are derived from grossly normal-appearing, meiotically mature oocytes inseminated with sperm with normal morphology, motility and of established fertility. In many of the cases that have been examined, routine light microscopic inspection suggested that fertilization had not occurred. As a standard protocol, these putative oocytes were stained for chromosomal DNA and examined by fluorescence microscopy (Van Blerkom et al, 1987). For approximately 5% of these oocytes, fertilization was indicated by the presence of a sperm head within the cytoplasm. Electron microscopy revealed sperm DNA in various stages of decondensation (Figs. 1 and 2). However, in none of the eggs examined was the state of decondensation comparable to that typically observed in a normal human egg (Van Blerkom, 1988). For human eggs exhibiting this type of early developmental failure, it is difficult to eliminate completely the possibility that the fertilizing spermatozoa contained subtle defects in DNA packaging. With this caveat noted, work from animal species (for review of pronuclear development, see Longo, 1985) suggests that for normal development to occur, the cytoplasm of most mammalian species acquires the ability to decondense sperm DNA during the latter stages of preovulatory maturation. This ability appears to involve the appearance or activation of cytoplasmic enzymes and/or other proposed "factors" that rapidly promote DNA decondensation (Masui and Clark, 1979; Longo, 1985). The varying degrees of DNA decondensation observed in inseminated human oocytes may be a manifestation of an absent or inadequate biochemical ability of the mature human ooplasm to interact appropriately with sperm DNA. The cellular and molecular biology of this interaction represents another area of early mammalian development where available experimental information is of a limited nature. For example, the precise number and identity of the biochemical factors associated with decondensation have not been established definitively. Likewise, whether changes in the structure of the cytoplasm in the region of the penetrated sperm are required for, or associated with DNA decondensation or pronuclear formation, or both, have not been determined.

A comprehensive understanding of the cellular and molecular processes involved in sperm DNA decondensation and pronuclear formation are clearly a prerequisite to beginning to comprehend why some penetrated human oocytes fail to progress in development. It is also apparent that routine light microscopic inspection may not recognize the situation where an oocyte is indeed penetrated but does not form pronuclei. The frequency with which oocytes are penetrated but fail to initiate or promote DNA decondensation/male pronuclear formation is unknown for the human species. The occurrence of this type of early developmental failure may be especially relevant in understanding the subtle origins of infertility and, consequently, its occurrence should be correlated both with the protocol of ovarian hyperstimulation and the prior fertility/infertility (e.g., idiopathic) history of the couple.

If sperm decondensation and male pronuclear formation have occurred, the next major developmental landmark is the migration and juxtaposition of the pronuclei. Pronuclear juxtaposition and the subsequent breakdown of the pronuclear membranes is a necessary prerequisite for the association of the maternal and paternal chromosomes at syngamy. A representative sequence of normal pronuclear movements is shown in Figures 3-6. Based on studies from other mammals, it is assumed that replication of parental DNA occurs in the human during the pronuclear stages that precede juxtaposition. After the completion of syngamy and the formation of the mitotic spindle, the first cleavage division usually occurs within hours.

A relatively small percent of fertilized human eggs (less than 5%) arrest development during the pronuclear stage. While an occasional egg will display pronuclei that have formed but fail to migrate, the majority of eggs with this phenotype arrest development with nuclei juxtaposed. Electron microscopic studies indicate that the changes in pronuclear membrane geometry that typically precede dissolution do not occur in eggs arrested with juxtaposed pronuclei (Van Blerkom et al, 1987; Van Blerkom, 1988).

Clearly, the dissolution of the pronuclear membranes is of fundamental importance in the formation of the zygote, because without this event the maternal and paternal chromosomes remain in separate compartments. While the causes of developmental failure at the pronuclear stage are not readily apparent, the analysis of fertilized eggs that arrest at this stage of development should provide basic insight into the mechanisms and differentiative processes that establish conditions permissive for the actual joining of parental genomes. In the human, to investigate the molecular and cellular factors involved in pronuclear formation during normal development generally requires invasive analyses that would destroy the newly fertilized egg. By contrast, pronuclear- arrested eggs may be suitable candidates for analysis once it is apparent that development has truly ceased. Van Blerkom et al (1987) described a series of morphodynamic events involving time-dependent changes in nuclear geometry and nucleolar distribution that were correlated with and predictive of subsequent development and developmental potential (see also Figs.). With this basis, analysis of arrested embryos can be undertaken with reasonable assurance that a particular egg is indeed nonprogressive.

One relatively simple question that we have asked with respect to pronuclear-arrested human embryos is whether DNA replication had occurred in one or both pronuclei prior to juxtaposition (Van Blerkom, 1988). At present, semiquantitative chromosomal fluorescence has been used for 4 eggs to assess whether or not replication had occurred. While this particular approach is useful for approximating ploidy, it is not of sufficient sensitivity to determine the absolute amount of DNA (e.g., whether a particular chromosome is absent). The demonstration of haploidy in the male and female pronuclei of 3 of the 4 eggs indicated that replication had failed to occur or was incomplete. Because of the low number of pronuclear eggs examined, these results are preliminary. However, they do suggest the possibility that the ability of the pronuclear membrane to undergo progressive changes that lead to fragmentation at pronuclear juxtaposition is associated with the occurrence or completion of DNA replication. Perhaps such a mechanism exists in the human to prevent syngamy when replication has failed to occur in one or both pronuclei.

The ethical questions attendant to experimentation on normally fertilized human eggs preclude the types of analyses and experimental design that may provide a direct answer to whether DNA replication is a prerequisite for pronuclear membrane dissolution. However, this question is one type of question that can be examined in model systems such as the mouse. Aphidicolin is a potent and relatively specific inhibitor of the enzyme DNA polymerase I (Spardi et al, 1982). If this inhibitor is present during the in vitro fertilization and subsequent culture of 1-cell mouse eggs, pronuclear formation, migration and juxtaposition occur on schedule (Howlett, 1986; Van Blerkom, unpublished observations). However, in the absence of DNA replication, dissolution of the pronuclear membranes and therefore syngamy do not occur. The inhibition of replication in one or both pronuclei with ultraviolet activated agents that crosslink DNA is another experimental approach to determine whether an association exists between DNA synthesis and pronuclear membrane breakdown. Recent studies from my laboratory indicate that culture of newly fertilized mouse eggs in the presence of trioxsalen (4,5′, 8-Trimethylpsoralen), a UV-activated DNA crosslinker, develop to the stage of pronuclear juxtaposition, but neither fragmentation nor changes in the membrane geometry that typically precede pronuclear fragmentation occur. For these studies, either male or female, or both, pronuclei were irradiated with a UV microbeam. At this rather gross level of analysis, the aphidicolin and trioxsalen findings suggest that inhibition of pronuclear DNA replication in the mouse produces a developmental arrest that is phenotypically similar to the situation that is observed in newly fertilized human eggs.

Collectively, the above observations suggest that the newly fertilized mouse egg may serve as an appropriate model to examine some of the cause(s) of specific developmental abnormalities common to early human embryogenesis. For example, what are the temporal and spatial aspects of DNA organization and association with the pronuclear membrane in normally developing and replication-inhibited eggs? With such information available from animal systems, comparisons to the situation that occurs in human eggs that arrest at this stage of development can be reasonably attempted.

SUMMARY

In this paper, the discussion of real and potential developmental abnormalities in oogenesis (Van Blercom, 1989b) and early human embryogenesis was intentionally limited to the chromosomal status of the preovulatory oocyte and the possible origins of failure in the 1-cell, presyngamic embryo. These particular topics represent only a few of the chromosomal and cellular perturbations that are encountered in the preimplantation human embryo (for a comprehensive discussion, see Van Blerkom, 1988;1989b). Other areas of human developmental failure that are currently receiving both analytical and experimental attention include (1) the association between the arrest of development at the 4-cell stage and the failure of the embryonic genome to become expressionally activated (Braude et al, 1988; and Braude, this volume), (2) the process by which early human embryos become genetically mosaic, i.e., develop a mixture of diploid and aneuploid mononucleate blastomeres (Angell et al, 1983; Van Blerkom et al, 1984), (3) the mechanism by which normally fertilized eggs, at surprisingly high frequency, develop multinucleated blastomeres during cleavage (Plachot, 1985; Tesarik et al, 1987), and (4) the etiology of abnormal inner cell mass and trophectoderm development at the blastocyst stage (Lindenberg and Hytell, 1988).

The areas of research discussed in this paper represent some of the critical stages of early human development that, because they are subject to frequent and often obvious alteration, preclude the establishment of a normal pregnancy. The recognition and understanding of how specific developmental steps in early human ontogeny become perturbed are necessary for a realistic appreciation of the extent to which clinical protocols that arise from basic science findings (e.g., microinjection of spermatozoa, removal of accessory pronuclei, twining, etc.) can be (or should be) applied in a meaningful and effective fashion to initiate a pregnancy that has a reasonable potential of development to birth.

ACKNOLWEDGEMENT

This work was supported by funds from Reproductive Genetics in Vitro, Denver, Colorado and by grants from the National Institutes of Health (HD 21582) and the National Science Foundation (BBS-8601231).

REFERENCES

Angell, R.R., Aitken, R.J., Van Look, P.F.A., Lumsden, M.A. and Templeton, A.A. (1983). Chromosome abnormalities in human embryos after in vitro fertilization. Nature, 303: 336-337.

Baccetti, B. (1984). The human spermatozoon, In Ultra-structure of Reproduction (J. Van Blerkom and P. Motta, eds). pp. 110-126. Martinus Nijhoff Publishers Boston and The Hague.

Bongso, A., Soon Chye, N., Ratman, S., Sathananthamm, H. and Wong, P.C. (1988). Chromosome abnomalies in human oocytes failing to fertilize after insemination in vitro. Human Repro. 3: 645-649.

Braude, P., Bolton, V. and Moore, S. (1988). Human gene expression between four- and eight-cell stages of preimplantation development. Nature 332: 459-461.

Buster, J.E., Busillo, M., Rodi, I.A., Cohen, S.W., et.al. (1985). Biologic and morphoslogic development of donated human ova recovered by non-surgical uterine lavage. Am. J. Obstet. and Gynecol. 153: 211-217.

Edwards, R. G. (1986). Causes of pregnancy loss. Human repro. 1: 185-198.

Fishel, S.B., Cohen, J., Fehilly, C., Purdy, J.M., et al. (1985). Factors influencing human embryonic development in vitro. NY Academy of Science. 442: 342-356.

Gibbons, J.H. (1988). Infertility: Medical and Social Choices. Office of Technology Assessment, U.S. Gov't Printing Office, Washington, D.C.

Hertig, A.T., Rock, J., Adams, E.C., and Mulligan, W.J. (1954). On the preimplantation stages of the human ovum: A description of four normal and four abnormal specimens ranging from the second to the fifth day of development. Contrib. Embryol. 35: 201-220.

Howlett, S.K. (1986). The effect of inhibiting DNA replication in the one-cell mouse embryo. Roux's Arch Dev. Biol. 195: 499-505.

Jagiello, G., Ducayen, M., Fang, J. and Graffeo, J. (1976). Cytogenetic observations in mammalian oocytes: In Chromosomes Today (P.K. Pearson and K.R. Lewis, eds). vol. 5, pp 43-63, John Wiley and Sons, New York.

Lindenberg, S. and Hyttel, P. (1988). In vitro studies of the peri-implantation phases of human embryos In Ultrastructure of Human Gametogenesis and Embryogenesis (J. Van Blerkom and P. Motta, eds.) Kluwer Academic Publishers, Boston (in press).

Liu, H.-C., Jones, G., Jones, H., and Rosenwaks, Z. (1988). Mechanisms and factors of early pregnancy wastage in in vitro fertilization-embryo transfer patients. Fertil. Steril 50:

Longo, F.J. (1985). Pronuclear events during fertilization In Biology of Fertilization (C.B. Metz and A. Monroy, eds.). vol. 3, pp. 251-298. Academic Press, New York

Masui, Y. and Clark, H. (1979). Oocyte maturation Int. Rev. Cytol. 57: 185-282, Academic Press, New York

Pellestor, F. and Sele, B. (1988). Assessment of aneuploidy in the human female by using cytogenetics of IVF failures. Am. J. Hum. Genet. 42: 274-283.

Plachot, M. (1985). Contribution a l'etude de la fecondation et du development in vitro de l'oeuf humain. These de doctorat d'etat es sciences naturalles. Paris: Universite Paris VI, pp. 1-114.

Plachot, M., Veiga, A., Montagut, J., de Grouchy, J., Calderon, G., and others (1988). Are clinical and biological parameters correlated with chromosomal disorders in early life: a multicentric study. Human Reprod. 3: 627-635.

Sathananthan A.H., Ng, S.C., Chia, C.M., Law, H.Y., Edirisinghe, W.R. and Ratnam, S.S. (1985). The origin and distribution of cortical granules in human oocytes with reference to Golgi, Nucleolar and microfilament activity. Annal NY Acad. Sci. 442: 251-264.

Spielmann, H., Kruger, C., Stauber, M., and Vogel, R. (1985). Abnormal chromosome behavior in human oocytes which remained unfertilized during human in vitro fertilization. IVF 2: 138-142.

Spardi, S., Sala, F., and Pedroli-Noy, G. (1982). Aphidicolin: A specific inhibitor of nuclear DNA replication in eukaryotes. Trends in Biochemical Sciences. 7: 29-31.

Surani, M.A.H. (1987). Evidence and consequence of differences between maternal and paternal genomes during embryogenesis in the mouse. In Experimental Approaches to Mammalian Embryonic Development. (J. Rossant and R. Pederson, eds). pp 410-435. Academic Press, New York.

Tesarik, J., Kopecny, V., Plachot, and Mandelbaum, J. (1987). Ultrastructural and autoradiographic observations on multinucleated blastomeres of human cleaving embryos obtained by in vitro fertilization. Human Reprod. 2: 127-136.

Van Blerkom, J. (1988). Developmental failure in human reproduction associated with preovulatory oogenesis and preimplantation embryogenesis. In Ultrastructure of Human Gametogenesis and Early Embryogenesis (J. Van Blerkom and P. Motta, eds). pp 125-180. Kluwer Academic Publishers, Boston and London.

Van Blerkom, J. (1989a). The origin and detection of chromosomal abnormalities in meiotically mature human oocytes obtained from stimulated follicles and after failed fertilization in vitro. In Developments in Ultrastructure of Reproduction (P. Motta, ed). Alan Liss, New York. In press.

Van Blerkon, J. (1989b). The occurence and developmental consequences of abberant cellular organization in meiorically mature human oocytes after exogenous ovarian hyperstimulation. JEMT (in press).

Van Blerkom, J. and Henry, G. H., (1988). Cytogenetic analysis of living human oocytes: cellular basis and developmental consequences of perturbations in chromosomal organization and complement. Human Reprod. 3: 777-790.

Van Blerkom, J. and Henry, G. H., and Porreco, R.P. (1984). Preimplantation human embryonic development from polypronuclear eggs after in vitro fertilization. Fertil. Steril. 41: 686-696.

Van Blerkom, J., Bell, H., and Henry, G.H. (1987). The occurrence, recognition and developmental fate of pseudo-pronuclear eggs after in vitro fertilization of human oocytes. Human Reprod. 2: 217-225.

Werteleki, W., Graham, J.M., and Sergovich, F.R. (1976). The clinical syndrome of triploidy. Obstet. Gynecol. 47: 69-73.

Wramsby, H, and Fredga, K. (1987). Chromosome analysis of human oocytes failing to cleave after insemination in vitro. Human Reprod. 2: 137-142.

Wramsby , H., Fredga, K. and Liedholm, P. (1987). Chromosome analysis of human oocytes recovered from preovulatory follicles in stimulated cycles. N. Engl. J. Med. 316: 121-124.

Zamboni, L. (1987). The ultrastructural pathology of the spermatozoon as a cause of infertility: the role of electron microscopy in the evaluation of semen quality. Fertil. Steril. 48: 711-734.

TABLE 1

REPORTED FREQUENCIES OF ANEUPLOIDY IN MEIOTICALLY MATURE HUMAN OOCYTES FROM FAILED IN VITRO FERTILIZATION

AUTHORS	% ANEUPLOID*
JAGIELLO ET AL (1976)**	1.5
SPIELMAN ET AL (1985)	11
VAN BLERKOM AND HENRY (1988)	15
PELLESTOR AND SELE (1988)	19
BONGOS ET AL (1988)	21
PLACHOT ET AL (1988)	26
MARTIN ET AL (1986)	34
WRAMSBY ET AL (1987)	50
WRAMSBY AND FREDGA (1987)	65

* percent combines both hypohaploid and hyperhaploid situations

** from analysis of uninseminated oocytes that matured in vitro from the GV to the MII stage. (from Van Blerkom, 1989a)

FIGURE LEGENDS

Figures 1 and 2:

Failed (Fig. 1) and incomplete (Fig. 2) decondensation of sperm DNA (S) after successful penetration of the human ooplasm. M, mitochondria Fig. 1: 28,000X; Fig. 2: 44,000X. from Van Blerkom (1988b).

Figures 3-6:

The normal progression of changes in pronuclear (PN) membrane geometry (arrows) and nucleolar (n) distribution during the perisyngamic stage of human development (14-22 hrs post-insemination) as observed by differential interference contrast microscopy. PB, second polar body. from Van Blerkom et al (1987).

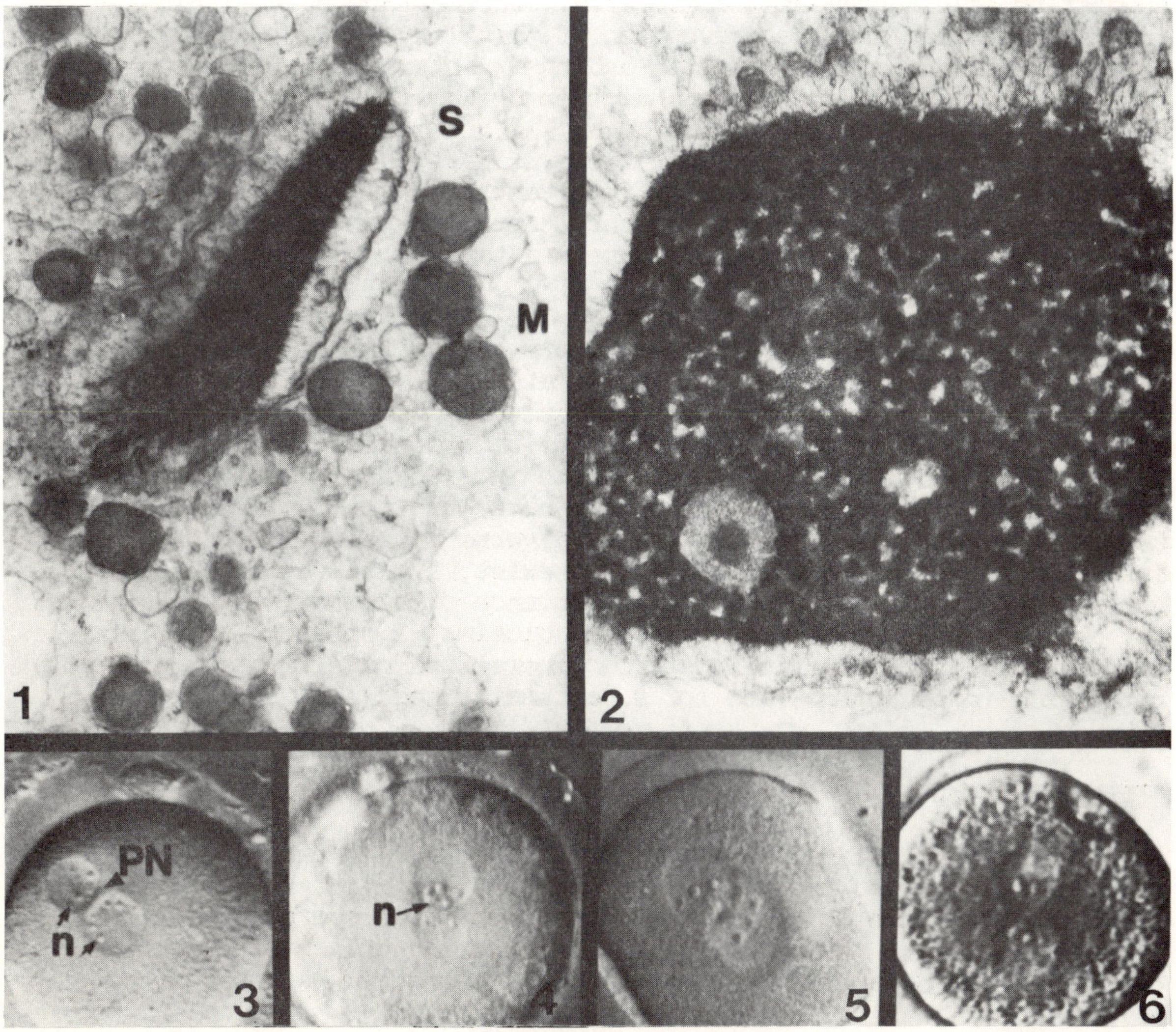

THE FERTILIZING SPERM: STRUCTURE, MATURATION AND FUNCTION

Patricia M. Saling

INTRODUCTION

It is becoming increasingly apparent that highly conserved molecular recognition and extracellular signal transducing mechanisms are used by cells in widely divergent systems. Investigation of mammalian gametes at the cellular and molecular levels has been extensive during the last decade, revealing that the process of fertilization comprises an elegant example of cellular interaction and communication. The considerable information that has emerged makes it possible to begin comparing various aspects of both sperm and eggs from different species and to attempt detection of common patterns and principles. Since sperm-egg interaction constitutes one of the oldest recognition systems known, gametes' use of conserved mechanisms does not seem unlikely.

Insight into cellular and molecular homology across species boundaries is a particularly important topic with regard to human in vitro fertilization. The basic methodology and conditions used for that procedure derive entirely from parallel work in animal model systems, suggesting some fundamental conformity. Further advances clinically, both for the promotion and the prevention of fertility, now demand a more detailed understanding of gametes and their interaction. The same is true, of course, in veterinary and other animal sciences. But for some mammalian species, including the human, it is difficult to conduct novel, fundamental research due to limitations of material as well as ethical considerations. Thus, it becomes important to evaluate the animal model systems under study currently for their appropriateness in understanding the molecular basis of mammalian gamete interaction generally.

Recently, a host of reviews on gamete structure and function have appeared. A list of some of these is provided at the end of this chapter. Rather than attempt to duplicate that work, I will stress throughout this manuscript homologies and common themes that have been recognized across many mammalian species regarding sperm morphology, sperm maturation in the epididymis, and gamete interaction leading to fertilization. Superfical differences have led various investigators to suggest that different species employ diverse mechanisms for gamete interaction. However, sufficient data exist, I believe, to suggest a general scheme for the sub-cellular sequence of gamete interaction that may apply to all mammals. Finally, in closing, I will try to point to those topics for which findings from generalized model systems are or are not likely to be appropriate for a particular species.

SPERM STRUCTURE

That the polarized tripartite form of sperm -- head, midpiece, and tail -- is conserved among mammals has been recognized for decades. Since my focus here is the fertilizing function of sperm, I will limit further discussion to the sperm head, this being the "business end" of the cell with regard gamete interaction. This is not to say that other regions of the cell are not utilized during fertilization; the motile apparatus of the tail's principal piece and the mitochondria of the midpiece are essential in translocation and energy production, respectively. But there is no evidence to suggest that molecules which participate in specific interactions with the egg are located in these two latter regions.

In addition to similarity of overall cellular structure, many features of sub-cellular organization, and of specific structural alterations, are constant for mammalian sperm. While their specific dimensions and contours may vary considerably, sperm heads of different species consist of the same homologous regions. For each species, beneath the plasma membrane a secretory granule, the acrosome, is found at the cell's apex situated as a cap over the anterior portion of a rigid, highly condensed nucleus. The acrosome is an organelle required for fertilization and its structural alteration during interaction with the egg represents another general theme recurring throughout the animal kingdom. Recent work suggests that this event, the acrosome reaction (AR), constitutes a classic example of highly regulated exocytosis, and will be discussed further below. During the acrosome reaction, the outer acrosomal membrane fuses with the plasma membrane at various locations, allowing exposure of the acrosomal content to the extracellular environment. One portion of the acrosome that extends laterally to the nucleus, termed the equatorial segment, does not participate in the fusion events of the AR. The hybrid vesicles formed during the AR are lost eventually from the cell surface, with the the inner acrosomal membrane becoming the cell's limiting membrane anteriorly. These structural features of acrosomal exocytosis are constant for all mammalian sperm studied thus far.

COMMON THEMES IN THE CELLULAR SEQUENCE OF GAMETE INTERACTION

Information concerning the molecular organization and composition of the gametes' surfaces is essential for an understanding of their interaction. However, before the molecular basis for the interaction can be examined, it is necessary to establish the cellular sequence, particularly for the sperm cell which is known to undergo dramatic structural re-organization. Determining where and when the AR occurs during gamete interaction, for instance, is important because it defines the membrane surface, either plasma membrane or inner acrosomal membrane, used by the sperm during the steps of fertilization.

While considerable debate and discussion has surrounded definition of the sequence of events during gamete interaction, extensive effort recently has led to findings that suggest a common pattern at the cellular level. The major conclusions that have emerged from that work will be summarized briefly. No attempt is made to include individual references here but they may be found in the reviews cited at the end of this chapter.

I. Capacitation. This final maturation phase of the sperm cell occurs normally in the female reproductive tract, but can be provoked by incubation *in vitro* for a large number of species. Defined operationally as the acquired ability of sperm to fertilize eggs without delay, this reversible process is poorly understood. Species-*dependent* alterations in several parameters, including metabolism, ionic composition, adenylate cyclase activity, and subtleties of acrosomal structure, have been reported, making analysis of the common features of the process difficult. At least two species-independent alterations in the sperm's plasma membrane are now recognized as a function of capacitation: changes in sterol and lipid levels, and the loss of surface-associated components. These alterations may be responsible for the reported increase in membrane fluidity that also occurs during capacitation. In the non-capacitated state, sperm are unable to undergo physiological ARs and capacitation can be considered a prerequisite for this event.

II. Interaction with the cumulus matrix. At the time of fertilization in the oviduct, the oocyte is often surrounded by two extracellular matrices. The inner layer, the zona pellucida (zp), appears to be a secretory product of the egg, and will be discussed further below. The outer layer, in which the cumulus cells are embedded, is not yet well defined but appears to be a product of this population of differentiated granulosa cells. For hamster, mouse, and rabbit sperm, where different populations of sperm cells have been observed, only capacitated, acrosome-intact sperm can pass through the cumulus matrix to the zp; both non-capacitated and acrosome-reacted cells are retained at the outer margin of this layer. Specific components of the sperm surface do not appear responsible for this passage, since a variety of cell types bearing the common feature of motility readily penetrate the hamster cumulus matrix. Rather, it appears that the opposite may be true, that the removal of obstructing factors from the sperm surface is necessary for cumulus matrix penetration. In at least two species, the cow and the pig, it is doubtful whether the cumulus is present at the time of gamete interaction, a finding that also makes it difficult to invoke a specific interactive function for the cumulus matrix. The physiological role of this matrix may be, therefore, to regulate the access of capacitated, acrosome-intact sperm to the vicinity of the egg.

III. Interaction with the zona pellucida. The first specific interaction between the gametes occurs at the level of the zona pellucida (zp). ZP composition has now been studied in several species, with the common finding that it is a relatively simple structure comprised principally of 3-5 glycoproteins. Detailed examination of the interaction of individual mouse zp glycoproteins with homologous sperm has determined that one of the zp glycoproteins, ZP3, serves as the primary ligand for sperm binding. More recent work using pig zp suggests similar findings. Reports of gamete interaction studies with individual zp proteins from other species have not yet appeared.

In initial systematic studies using mouse gametes, and considered controversial for some time, it was found that fertilizing sperm initiate binding to the zp with the plasma membrane in the acrosome-intact state. Subsequently, the same findings have emerged for a large number of mammalian species (Table 1) and can now be considered a generalized occurrence of mammalian gamete interaction.

TABLE 1: Species in which acrosome-intact sperm initiate zp-binding

cow	mouse
guinea pig	pig
hamster	rabbit
human	sheep

Sperm that have penetrated the zp and are located in the perivitelline space are reported universally to be acrosome-reacted. Together with the preceding information on primary binding to the zp, it may be concluded that the location of the AR is the zp surface following primary binding. This has now been observed for several species, as has the induction of ARs by isolated, solubilized zp (Table 2).

TABLE 2: Species demonstrating zp-induced acrosome reactions

cow
hamster
human
mouse
rabbit

The molecular events surrounding the AR are not understood fully, but recent work implicates mechanisms common to other receptor-effector systems, such as hormone- and growth factor-responsive cells. For instance, guanine nucleotide-binding proteins (G-proteins) have not only been identified in sperm from several mammalian species (Table 3), but a G_i-like protein of mouse sperm has been associated specifically with the physiological AR induced by ZP3. Furthermore, recent studies indicate that ZP3 triggers ARs by aggregating the sperm's ZP3 receptors and that the sperm's tyrosine kinase activity is stimulated by ZP3 binding. While still circumstantial and confined primarily to mouse sperm, these findings are striking in their parallel with results from a wide variety of different cell types, and suggest the use of highly conserved extracellular signal transducing mechanisms for acrosomal exocytosis.

TABLE 3: Species with G_i-like proteins in sperm

bull
guinea pig
human
mouse
pig

Once the AR has been triggered, sperm do not dissociate from the zp and binding is maintained. In the mouse, ZP2 is the glycoprotein ligand responsible for this secondary phase of binding with acrosome-reacted sperm. Strong evidence from experiments using boar sperm indicates that sperm employ proacrosin for this interaction. This sperm-specific zymogen and its enzymatic active counterpart, acrosin, are found in all mammalian sperm examined. Indirect evidence in guinea pigs, hamsters, mice and rats supports the use of proacrosin in this role. Further work, particularly with boar sperm, indicates that limited proteolysis together with sperm motility is responsible for sperm penetration through the zp matrix. Freshly exposed acrosin, perhaps with the coordination of other acrosomal enzymes, performs the digestive requirement for zp penetration.

IV. Sperm fusion with the egg plasma membrane. Within the perivitelline space, the acrosome-reacted sperm is in direct contact with the egg plasma membrane, to which it associates and fuses. Ultrastuctural analysis of several different mammalian species indicates that the region of the cell utilized for the initiation of membrane fusion is the plasma membrane overlying the equatorial segment. It should be recalled that, during the AR, this posterior section of the acrosomal cap does not participate in membrane fusion, whereas at this later point during gamete interaction, the equatorial segment initiates the fusion event. This conserved sequence of cellular events implies conserved organization of these subcellular domains at the molecular level. Fusion then proceeds to incorporate the entire sperm into the cytoplasm of the egg, except for the inner acrosomal membrane which does not fuse, but is incorporated into a phagocytic vesicle.

A GENERAL MODEL FOR MAMMALIAN GAMETE INTERACTION

These recent investigations at the cellular and molecular levels provide sufficient data to suggest a sequence of gamete interactions that may apply generally to mammals. Some aspects of this sequence are summarized in Figure 1. It is suggested that the cumulus matrix functions to restrict access of only capacitated sperm to the egg. As a result, under the normal conditions of fertilization, only a select population of capacitated, acrosome-intact sperm will arrive at the zp surface. Binding to the zp is initiated by plasma membrane receptors interacting with ZP3. Primary binding results in receptor aggregation, which itself is the triggering event for acrosomal exocytosis. This signal is transmitted intracellularly via G-proteins and results in regulated fusion between the plasma and outer acrosomal membranes. As a consequence, proacrosin is exposed. Secondary binding between ZP2 and proacrosin of acrosome-reacting/reacted sperm occurs together with pH-dependent autoactivation of proacrosin to acrosin. Sperm motility, together with the action of acrosin to cause limited proteolysis of the zp matrix, permit sperm penetration of the zp. Acrosome-reacted sperm in the perivitelline space associate rapidly with the egg plasma membrane. Fusion between sperm and egg is initiated in the region of the sperm's plasma membrane overlying the equatorial segment.

The many diverse observations made in experiments using different species might lead one to question the generality of this scheme. I believe, however, that many of the apparent differences between species can be accomodated in this scheme if two variable factors, the affinity of the interactions and their kinetics, are considered. Mouse sperm and guinea pig sperm, for example, may represent two extremes of the spectrum for these two parameters. For mouse sperm, it is commonly observed that acrosome-intact sperm initiate zp binding, whereas acrosome-reacted sperm are incapable of initiating binding but will remain associated with the zp if the AR has occurred at the zp surface. In contrast, with guinea pig

FIGURE LEGEND

Figure 1: Proposed general scheme for mammalian gamete interaction.

A) Sperm penetration through the cumulus matrix is achieved by the cell's motility, together with appropriate surface characteristics conferred by capacitation. Specific receptor-ligand interactions do not occur, and enzymatic mechanisms are unnecessary for penetration. The cumulus serves to regulate sperm access to the zp, ensuring that only capacitated, acrosome-intact sperm arrive at the zp surface.

B) At the zp surface, primary binding occurs between receptors on the plasma membrane of the sperm and the zp glycoprotein ZP3.

C) The extracellular signal of zp binding is transmitted across the sperm membrane to trigger the AR in a process involving the aggregation of the sperm's ZP3 receptors by ZP3. This aggregation signal promotes the activity of a tyrosine kinase which is involved in the subsequent activation of G-proteins, and leads to the cascade of events resulting in the AR. Fusion is initiated at multiple locations between the plasma and outer acrosomal membranes while the sperm is bound to ZP3 at the zp surface. Secondary binding between the acrosome-reacting sperm and ZP2 follows directly after the fusion event. Proacrosin serves as at least one of the major sperm components participating in secondary binding.

D) Increased acrosomal pH during the AR leads to autoactivation of proacrosin, resulting in the formation of acrosin. The combined motile force of the sperm and the enzymatic capability of newly exposed acrosin produces strees-induced proteolytic penetration through the zp matrix.

E) A narrow penetration slit through the zp is created due to the limited proteolytic activity of acrosin. Hybrid vesicles bearing sperm receptors for ZP3 remain at the zp surface, and sometimes can be seen as a collar through which the sperm penetrates.

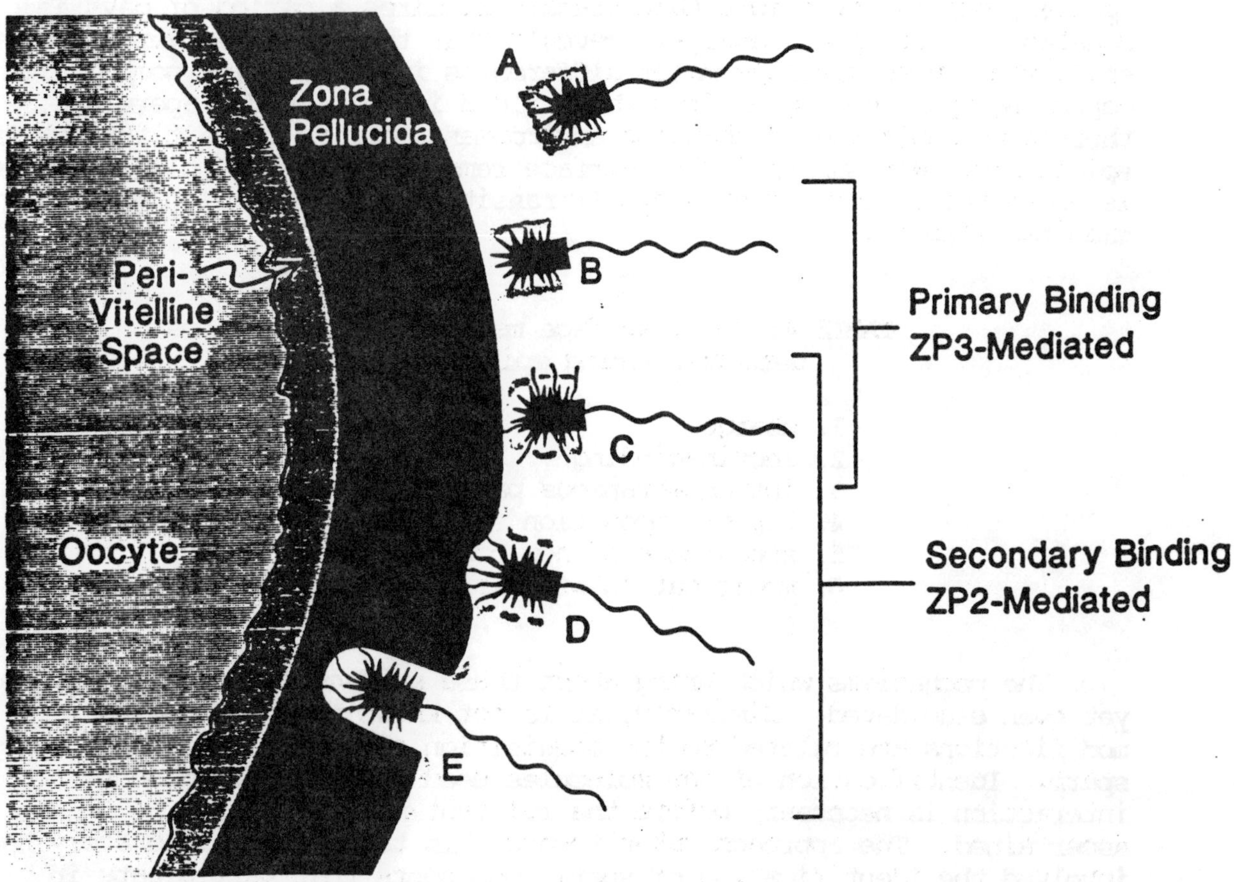

sperm, acrosome-intact sperm associate weakly and in low numbers with the zp, and acrosome-reacted sperm display tenacious interaction with the zp surface. With regard primary binding, the affinity between the sperm's receptors and ZP3 may vary over a wide range for different species. This affinity could be considered high for mouse sperm and low for guinea pig sperm. Opposite affinities may apply to these species for secondary binding. Similarly, with regard kinetics, perhaps ZP3-stimulated ARs occur sufficiently slowly in mouse sperm to permit experimental observation when the sperm is bound to the zp surface, whereas in many other species the event is more difficult to detect due to rapid AR induction at that location. Species variations will be observed, but it seems unlikely that these will involve major differences in fundamental design or mechanism, but rather variations in the affinity or kinetics of individual events.

COMMON THEMES DURING EPIDIDYMAL TRANSIT

Following the differentiation of the haploid spermatozoon, the cell is liberated from the seminiferous epithelium and passes from the testis to the epididymis. Despite the absence of gross morphological changes, sperm acquire the ability to interact successfully with and to fertilize homologous eggs during epididymal transit. Proximal epididymal sperm are weakly motile (at best), cannot recognize the zona pellucida, and cannot undergo physiological ARs; distal epididymal sperm are vigorously motile, bind readily and in large numbers to the zp, and undergo ARs when challenged with zp. Although their lengths may vary, three major subdivisions have been identified cytologically for all species examined: the caput, the corpus and the cauda epididymis. Transit through the organ requires approximately 1-2 weeks; the exact duration varies among species, but the principle that transit requires a period of days is constant. Biochemical analysis reveals that the fluid recovered from the epididymal lumen displays major differences in polypeptide composition depending upon the region from which fluid is collected. Sperm are, therefore, subjected to changing environments during transit through the epididymis. Not surprisingly, surface remodeling at the molecular level is a general feature of epididymal transit for all mammalian species examined (Table 4).

TABLE 4: Sperm surface membrane modifications detected during epididymal transit

1. charge
2. lectin-binding
3. intramembraneous particle distribution
4. lipid composition
5. appearance of new components
6. modification of existing components

The mechanisms which bring about these surface alterations have not yet been elucidated. Similarly, it is not known which of these modifications are related to the acquisition of fertilizing function by sperm. Identification of the molecules used by sperm for gamete interaction is necessary before the relevant mechanisms can be ascertained. The approach taken toward this topic in my laboratory has involved the identification of sperm components that participate in fertilization through the use of anti-sperm monoclonal antibodies (mAbs). These probes have been used in indirect immunofluorescence studies to localize the particular sperm moieties, in functional studies, both *in vitro* and *in vivo* to determine the role of the component in gamete interaction, and in biochemical studies to identify the component involved. We then have the opportunity to examine a specific component as a function of epididymal maturation. One of the sperm components that we have studied using this approach is termed M42 antigen since it was first identified using the M42 mAb. M42 antigen is a high M_r doublet (200/220 kD), located in a restricted region of plasma membrane overlying the acrosome. The M42 mAb blocks fertilization in a concentration dependent manner, both *in vitro* and *in vivo*; the specific event prevented is the ZP3-induced AR. Sperm recovered from each of the three major epididymal regions contain the M42 antigen in an indistinguishable localization pattern despite the finding that caput sperm will not undergo zp-induced ARs. Analysis of the M42 antigen itself reveals that it undergoes structural modification (M_r shift from 220/240 kD in immature sperm to 200/220 kD in mature sperm) coincident with the sperm's acquired ability to undergo zp-induced ARs. M42 antigen's structural modification does not occur as a function of duration of transit, since sperm retained experimentally in the caput epididymis for 2 weeks do not display the mature form of the antigen nor do they undergo zp-induced ARs. These results suggest that it is necessary to expose sperm to particular environmental conditions found in the epididymal lumen distal to the caput to achieve a functional sperm cell.

Considerably more work is required before it can be established that post-translational modification of existing surface proteins, such as that found for M42 antigen, represents a universal mechanism for sperm maturation. Nevertheless, it is apparent that the identification of functional components of the sperm constitutes an essential first step for an understanding of the molecular basis both of gamete interaction and of epididymal maturation.

FOR WHAT TYPES OF STUDIES ARE MODEL SYSTEMS APPROPRIATE?

In the preceeding discussion, I have attempted to illustrate that the cellular and molecular patterns of gamete interaction appear to be highly conserved. This does not imply that all cells will use exactly identical molecules for each event of the process. It is well known from many physiological studies that several aspects of gamete interaction display substantial species specificity. However, the conserved pattern of gamete interaction does imply that, for a species in which experimental material is limited or for one that has not been investigated previously, considerable information concerning the nature of a molecule of interest may be available to facilitate its identification, isolation and investigation. I anticipate that those molecules involved in recognition events, which are likely to be present in low amounts, will be among those that demonstrate species differences. However, the accumulated information derived from several mammalian systems suggests that the most likely location for sperm molecules involved in primary interaction with the zp, for instance, is the plasma membrane overlying the acrosome. With similar reasoning, the plasma membrane overlying the equatorial segment should be examined to identify the sperm component(s) that initiate fusion with the egg plasma membrane. Given the polarized structure of the sperm cell, specification of a component's location is extremely valuable, permitting immediate focus on a restricted area of the cell rather than on general examination of the entire complex structure.

Alternatively, it can also be anticipated that some of the molecules involved in gamete interaction will be very similar among a wide range of species, particularly those involved in cell functions such as ion transport and in extracellular signal transduction. An example of the latter may already be found in the identification of a 41 kD G_i-like protein in the sperm of at least 5 different mammalian species. Other examples may be represented by the sperm's secondary receptor for ZP2, which appears to be proacrosin, and by the sperm's digestive enzyme for zp penetration, acrosin.

For all studies attempting to identify processes or components utilized naturally by gametes in vivo, it is essential that homologous gamete systems be employed under physiological conditions. Spurious results could emerge, and already have emerged, from the use of heterologous gamete interaction studies or the use of non-physiological conditions. Two examples that have arisen in my own lab may be pertinent. The first deals with studies on the mechanism of the AR and the second, with the identification of sperm components involved in fusion with the egg plasma membrane.

Much of our work on the AR in mouse sperm has focussed on the M42 antigen, and inhibition of the process with the M42 mAb, which was discussed earlier. In that work, it was found that M42 mAb inhibited only physiologically-induced (i.e., using zp) ARs, but did not inhibit pharmacologically-induced (i.e., using Ca^{+2} ionophore A23187 or lysophosphatidylcholine) ARs. Subsequent examination of the ability of mouse sperm recovered from the caput, corpus, and cauda epididymis to undergo ARs revealed that all sperm populations could be stimulated to undergo pharmacological ARs, but only mature sperm were stimulated to undergo physiological ARs. Immature caput epididymal sperm do not respond to zp by undergoing ARs. Together, these results suggest that important regulatory aspects of the AR induction mechanism are bypassed during pharmacological AR stimulation, and the results of experiments using pharmacological conditions exclusively to study the AR should be interpreted carefully.

The second example deals with the identification of sperm components that participate in fusion with the egg. We initiated studies on this topic through the use of the M29 mAb. This mAb recognizes a 60 kD mouse sperm protein restricted to the equatorial segment, and inhibits fertilization, both in vitro and in vivo, by blocking homologous sperm fusion with the egg plasma membrane. The mAb is ineffective in the heterologous cross of mouse sperm and zona-free hamster eggs. A converse result has also been found in Paul Primakoff and Diana Myles' labs (University of Connecticut, Farmington). They have identified a mAb against guinea pig sperm that blocks fusion with the egg plasma membrane, but only in a heterologous cross (guinea pig sperm + hamster eggs). The mAb is ineffective in the homologous situation (guinea pig sperm + guinea pig eggs), suggesting its irrelevance as a probe to define physiologically meaningful sperm components. Although opposite findings emerged, the results of these two studies indicate the importance of using homologous gamete systems in attempts to define physiologically relevant components in a particular species.

SUMMARY

In the foregoing discussion, I have tried to summarize evidence to suggest that, despite reported differences, the scheme of gamete interaction among mammals is likely to be more similar than it is different. Species distinctions will be found, but the fundamental design of the system for each species is likely to conform to a common pattern. I anticipate that the molecules involved in gamete recognition events may be homologous, but with varying specificity for each species. Other sperm molecules that are responsible for highly conserved cellular functions, such as ion transport and intracellular signalling, are likely to be much less variable among species. Thus, studies in model systems have provided, and are continuing to provide, an invaluable body of information to understand the cellular and molecular context in which fertilization occurs. There is no substitute for the study of homologous gametes in physiological conditions to identify *individual* molecules that participate in gamete interaction, but extensive studies in model systems can direct attention immediately to the *type* of molecule sought and thereby greatly facilitate investigations in any mammalian species of interest.

Reviews to consult for detailed information and references

Eddy, E.M. 1988. The spermatozoon. Pp. 27-68 in Physiology of Reproduction, Vol.1. E. Knobil and J. Neill, eds. New York: Raven Press.

Fournier-Delpech, S. and M. Courot. 1987. Sperm-zona pellucida binding activity. Oxford Review of Reproductive Biology 9:294-321.

Fraser, L.R. and K.K. Ahuja. 1988. Metabolic and surface events in fertilization. Gamete Res. 20:491-519.

Jones, R. 1989. Membrane remodeling during sperm maturation in the epididymis. Oxford Review of Reproductive Biology, 11:285-337.

Langlais, J. and K.D. Roberts KD. 1985. A molecular membrane model of sperm capacitation and the acrosome reaction of mammalian spermatozoa. Gamete Res. 12:183-224.

Olds-Clarke, P. 1988. Genetic analysis of sperm function in fertilization. Gamete Res. 20:241-264.

Oliphant, G., A.B. Reynolds, and T.S. Thomas. 1985. Sperm surface components involved in the control of the acrosome reaction. Am. J. Anat. 174:269-283.

O'Rand, M.G. 1988. Sperm-egg recognition and barriers to interspecies fertilization. Gamete Res. 19:315-328.

Peterson, R.N. and L.D. Russell. 1985. The mammalian spermatozoon: A model for the study of regional specificity in plasma membrane organization and function. Tissue and Cell 17:769-799.

Robaire, B. and L. Hermo. 1988. Efferent ducts, epididymis, and vas deferens: Structure, functions and their regulation. Pp. 999-1080 in Physiology of Reproduction, Vol.1. E. Knobil and J. Neill, eds. New York: Raven Press.

Saling, P.M. 1989. Mammalian sperm interaction with extracellular matrices of the egg. Oxford Review of Reproductive Biology, 11:339-388.

Setchell, B. and D.E. Brooks. 1988. Anatomy, vasculature, innervation and fluids of the male reproductive tract. Pp. 753-836 in Physiology of Reproduction, Vol.1. E. Knobil and J. Neill, eds. New York: Raven Press.

Topfer-Petersen, E., A.E. Friess, A. Henschen,and W.-B. Schill. 1987. Is a fucose-binding protein involved in sperm-egg interaction in the pig? Pp. 287-296 in New Horizons in Sperm Cell Research. H. Mohri, ed. Tokyo: Japan Scientific Society Press.

Wassarman, P.M. 1987. The biology and chemistry of fertilization. Science 235: 553-560.

Wassarman, P.M. 1987. Early events in mammalian fertilization. Ann. Rev. Cell Biol. 3:109-142.

Wassarman, P.M. 1988. The mammalian ovum. Pp. 69-102 in Physiology of Reproduction, Vol.1. E. Knobil and J. Neill, eds. New York: Raven Press.

Wassarman, P.M. 1988. Zona pellucida glycoproteins. Ann Rev Biochem 57:415-442.

Yanagimachi, R. 1988. Mammalian fertilization. Pp. 135-185 in Physiology of Reproduction, Vol.1. E. Knobil and J. Neill, eds. New York: Raven Press.

MOLECULAR EVENTS PRE- AND POST-FERTILIZATION OF MOUSE EGGS: OOCYTE MATURATION, EGG ACTIVATION, AND POLYSPERMY BLOCK

R. M. Schultz, S. Kurasawa, Y. Endo, and G. S. Kopf

INTRODUCTION

Prior to the fusion of the sperm with the egg, the egg undergoes a series of complex biological processes that allow it to mature and be fertilized. In turn, fertilization initiates a complex series of events termed egg activation. One aspect of egg activation results in cell cleavage and further development. Another aspect of egg activation prevents polyspermy, which leads to aberrant development.

This brief review will first discuss events during oogenesis and oocyte maturation that are involved in production of a fertilizable female gamete. Events comprising the fertilization-induced block to polyspermy will then be addressed. The discussion will be mainly restricted to the mouse, since this is the best characterized system for processes involved in mammalian oocyte maturation and fertilization.

ACQUISITION OF MEIOTIC COMPETENCE

During the period of oocyte growth, which takes about 14 days, mouse oocytes that are arrested in the first meiotic prophase grow from about 15 µm to 80 µm in diameter. Acquisition of meiotic competence is correlated with a specific stage of oocyte growth. In mice, oocytes that are obtained from juvenile mice less than 15 days of age are less than 60 µm in diameter and will not resume meiosis, i.e., undergo meiotic maturation, when placed in a suitable culture medium (Sorensen and Wassarman, 1976). One of the earliest morphological manifestations of meiotic maturation is breakdown of the nuclear membrane, which is called the germinal vesicle. Subsequent to germinal vesicle breakdown (GVBD), a spindle forms. Separation of homologous chromosomes then occurs, with emission of the first polar body and arrest at metaphase II. The frequency at which growing oocytes can resume meiosis increases with increasing diameter, which is a linear function of the age of the donor juvenile mice (Sorensen and Wassarman, 1976).

Stage-specific differences in the spectrum of synthesized polypeptides are correlated with acquisition of meiotic competence (Schultz et al., 1979a). These changes are likely to underlie, in part, the biochemical basis for the acquisition of meiotic competence. Nucleate fragments obtained from fully-grown, meiotically competent oocytes that are less than 60 µm in diameter undergo GVBD, emit the first polar body and arrest at metaphase II (Balakier and Czolowska, 1977; Schultz et al., 1978). Thus, it is likely that the quality of the cytoplasm, and not the amount of cytoplasm, is involved in acquisition of meiotic competence.

Oocyte growth *per se* may not be involved in acquisition of meiotic competence, since acquisition of meiotic competence can be dissociated from oocyte growth (Canipari et al., 1984). Oocytes obtained from juvenile mice about 15 days of age are about 60 µm in diameter. Oocytes obtained from juvenile mice 10 days of age are less than 60 µm in diameter and meiotically incompetent. When these meiotically incompetent oocytes obtained from juvenile mice 10 days of age are cultured for 5 days in a medium that does not support oocyte growth but does support oocyte viability a significant fraction of the oocytes undergo GVBD. Results from a series of similar experiments indicated that a constant amount of combined time that totals 15 days of *in vivo* growth or *in vitro* culture is necessary for acquisition of meiotic competence. To provide a tighter correlation between the changes in the proteins synthesized during oocyte growth and the acquisition of meiotic competence, it should be demonstrated that the changes in protein synthesis that occur during oocyte growth also occur under the conditions of *in vitro* culture that do not sustain oocyte growth but do foster meiotic competence.

OOCYTE MATURATION

Oocytes undergo meiotic maturation, which culminates in the production of an egg; eggs, not oocytes, are capable of being fertilized and giving rise to normal development. Although oocytes that have not reached and arrested at metaphase II can be penetrated by sperm, such oocytes do not develop normally and very quickly degenerate. The subsequent section will discuss briefly some molecular aspects of oocyte maturation.

1. Role of cAMP and Protein Phosphorylation

The follicle exerts an inhibitory influence on oocyte maturation, since oocytes present in preovulatory antral follicles do not resume meiosis, whereas liberation of these oocytes from their follicles results in resumption and completion of meiotic maturation (Pincus and Enzmann, 1935). A substantial body of evidence implicates cAMP in the maintenance of meiotic arrest, although a number of molecules are likely to participate in this process (See Schultz, 1988 and reference therein for a more complete review of this area.).

In vitro maturation is reversibly inhibited by membrane permeable cAMP analogs, (e.g., dibutyryl cAMP;(dbcAMP), or inhibitors of cyclic nucleotide phosphodiesterase (PDE), (e.g., 3-isobutyl-1-methyl xanthine;(IBMX) (Cho et al., 1974; Bornslaeger et al., 1984); the corresponding cGMP analogs do not inhibit maturation *in vitro*. In addition, treatment of oocytes with either the activator of adenylate cyclase, forskolin, or microinjected cAMP inhibits GVBD (Schultz et al., 1983).

The best documented mode of action of cAMP in eukaryotes is to activate a cAMP-dependent protein kinase (protein kinase A, PK-A). Accordingly, it was proposed that cAMP is involved in maintenance of meiotic arrest by activating PK-A. This enzyme phosphorylates, either directly or indirectly, a protein(s) X, which is capable of promoting GVBD (Bornslaeger et al., 1986a). The phosphorylated form, XP, is inactive. Resumption of meiosis *in vitro* is proposed to occur by a decrease in cAMP, which would result in a decrease in PK-A activity. Consequently, a protein phosphatase would shift the equilibrium between XP and X to the dephosphorylated form of X and GVBD would ensue.

Consistent with this model is that microinjection of oocytes, which are incubated in medium containing a concentration of dbcAMP that inhibits maturation, with protein kinase inhibitor (PKI) undergo GVBD

(Bornslaeger et al., 1986a); PKI inhibits PK-A by combining with the free catalytic subunit (C). This result is anticipated, since inactivation of C would result in the conversion of XP to X, and hence maturation would resume. In addition, microinjection of oocytes, which are incubated in a medium that supports oocyte maturation, with the catalytic subunit of PK-A results in inhibiting GVBD. Presumably the excess of C relative to regulatory subunit makes C essentially a cAMP-independent protein kinase. C continues to phosphorylate X and to keep it in its inactive phosphorylated form and accordingly maturation is inhibited.

A decrease in oocyte cAMP does in fact occur prior to GVBD during maturation *in vitro* (Schultz et al., 1983; Vivarelli et al., 1983). This decrease may be causally related to GVBD, since PDE inhibitors, which inhibit GVBD, inhibit this maturation-associated decrease in oocyte cAMP that occurs during a period of time in which the oocytes become committed to resume meiosis (Schultz et al., 1983). Commitment is experimentally defined as follows: After a given period of time in culture, oocytes possessing an intact germinal vesicle are transferred to medium containing either dbcAMP or IBMX and then scored for GVBD at later times. If an oocyte resumes meiosis it is termed "committed". Moreover, microinjection of oocytes, which are incubated in medium containing a concentration of IBMX that inhibits maturation, with purified phosphodiesterase undergo GVBD (Bornslaeger et al., 1986a). Presumably, even though the exogenous microinjected PDE is inhibited by >85% by the IBMX, the excess amount of PDE activity is sufficient to hydrolyze oocyte cAMP. This decrease in cAMP would then occur and GVBD would ensue in the presence of IBMX.

A maturation-associated set of changes in protein phosphorylation occurs during the commitment period (Bornslaeger et al., 1986a) and may be causally related to GVBD. A basic phosphoprotein of M_r 60,000 undergoes an apparent dephosphorylation and a set of phosphoproteins of M_r 24,000, 29,000, and 36,000 exhibit an apparent increase in phosphorylation. These same changes occur in oocytes incubated in medium containing dbcAMP and microinjected with PKI, which induces GVBD. In addition, activators of the calcium, phospholipid- dependent protein kinase, protein kinase C (PK-C) and antagonists of calmodulin, which could regulate a calmodulin-modulated protein kinase, inhibit maturation and the maturation-associated set of changes in protein phosphorylation (Bornslaeger et al., 1986b; Bornslaeger et al., 1984). These agents do not inhibit the maturation-associated decrease in cAMP and microinjection of oocytes incubated in these agents with PKI does not result in GVBD. Thus, the mode of action of these compounds may be distal to that of cAMP.

These changes in protein phosphorylation do not occur in meiotically incompetent oocytes, but do occur in the 10% of oocytes 60 µm in diameter that can undergo GVBD. Injection of meiotically incompetent oocytes with an amount of PKI sufficient to induce maturation in fully-grown oocytes does not result in GVBD in these incompetent oocytes but does elicit the decrease in phosphorylation of the M_r 60,000 phosphoprotein. The increase in apparent phosphorylation of phosphoproteins of M_r 24,000, 29,000 and 36,000 is not observed. It is likely that meiotically competent oocytes lack the ability to phosphorylate these phosphoproteins, since low levels of phosphorylation of these phosphoproteins are detected. Thus, dephosphorylation of the 60,000 M_r protein is not sufficient to induce GVBD and meiotic incompetence may entail deficiencies in the ability of the oocytes to execute phosphorylations that occur downstream to the dephosphorylation of this protein (Bornslaeger et al., 1988). Future work may provide additional insights regarding the proteins involved in this phosphorylation cascade.

2. Maturation Promoting Factor

An activity central to oocyte maturation and subsequent cell cycles is that of Maturation Promoting Factor (MPF)(Masui and Clarke, 1979 and references therein; Kishimoto, 1988 and references therein). MPF is more aptly described as an M phase promoting factor that is involved in the G2 to M transition of the cell cycle (Gerhart et al., 1984). It is directly implicated in chromosome condensation and breakdown of the nuclear membrane and its activity oscillates during the cell cycle. MPF activity shows no evidence for species specificity, since MPF obtained from one species will induce nuclear membrane breakdown when injected into cells of distantly related species (Masui and Clarke, 1979, Kishimoto et al., 1984.; Sorensen et al., 1985). Until recently, MPF has defied purification, due to its instability and cumbersome biological assay (Gerhart et al., 1984). The development of methods for stabilizing the activity and *in vitro* assays that support nuclear membrane breakdown have led to MPF's recent purification from *Xenopus* oocytes (Lohka et al., 1988). The purified protein possesses protein kinase activity, and is itself phosphorylated; it is not known whether phosphorylation regulates its activity. Results of experiments using extracts that contain MPF activity suggest that phosphorylation of MPF, which may be capable of autophosphorylation, may generate the active form of MPF (Cyert and and Kirschner; 1988). Such a mechanism also explains the self-amplification properties of MPF. Interestingly, the *Xenopus* MPF is a CDC2 homologue (Dunphy et al., 1988; Gautier et al., 1988). The CDC2 protein is directly involved in the yeast cell cycle. The protein has an associated protein kinase activity and is itself likely to be regulated by phosphorylation; the level of phosphorylation increases as cells undergo a G2 to M transition (Simanis, V. and Nurse, 1986). Thus, MPF may be an intrinsic component of the cell cycle.

Mouse oocytes appear to possess an MPF-like activity. Fusion of a meiotically competent oocyte that has undergone GVBD with a meiotically incompetent oocyte results in breakdown of the germinal vesicle of the incompetent oocyte (Balakier, 1978). Injection of cytoplasm from mouse oocytes that have undergone GVBD into *Xenopus* (Sorensen et al., 1985) or *Asterina Pectinifera* oocytes (Kishimoto et al., 1984) induces maturation; cytoplasm of mouse oocytes inhibited from maturing does possess MPF activity as deduced by its inability to induce GVBD in the recipient oocytes.

Mouse oocyte MPF activity oscillates during maturation (Hashimoto and Kishimoto, 1988), as is the case in other systems (Gerhart et al., 1984). In a series of elegant experiments, cytoplasm from oocytes at different stages of maturation was injected into *Asterina Pectinifera* oocytes, which were subsequently scored for GVBD. In this manner, the amount of MPF activity was quantified and shown to increase subsequent to GVBD. It reaches a peak at metaphase I, decreases during polar body emission, and then increases again as the oocyte arrests at metaphase II. Inhibiting protein synthesis does not affect the initial appearance of MPF after GVBD, which occurs in the absence of protein synthesis.

MPF generation does not require nuclear contents, since fusion of anucleate fragments of fully-grown oocytes with interphase blastomeres derived from 2-cell mouse embryos results in almost immediate chromosome condensation in the interphase nuclei (Balakier and Czolowska, 1977); the anucleate fragments were used at a time when the nucleate fragments had undergone GVBD.

Meiotically incompetent mouse oocytes also possess an "anti-MPF" activity. Fusion of a meiotically competent oocyte with an intact GV with a meiotically incompetent oocyte preserves the integrity of each nucleus, which resides in the common cytoplasm (Fulka et al., 1985). The loss of this anti-MPF activity may be required for acquisition of meiotic competence, and further characterization of this activity is required.

3. Acquisition and Maintenance of Sperm Decondensing Activity

The nucleus of the fertilizing sperm is highly condensed due to chromatin-associated basic proteins called protamines, which contain high amounts of cysteine (Yanagimachi, 1988 and references therein). These cysteine residues are oxidized during sperm maturation and their reduction during fertilization is necessary for the sperm nucleus to decondense in the egg cytoplasm and thus be transformed into the male pronucleus.

A pivotal role for glutathione in the process of sperm nuclear decondensation is likely, since glutathione is the major biological thiol reducing agent and treatment of eggs with diamide, which is an antioxidant of glutathione, reversibly inhibits the decondensation of microinjected hamster sperm nuclei (Perreault et al., 1984). In addition, treatment of mice *in vivo* with an inhibitor of glutathione synthesis, L-buthionine-S,R-sulfoximine (BSO) inhibits sperm decondensation of mouse sperm nuclei following *in vitro* fertilization (Calvin et al., 1986). The inability of the sperm nucleus to decondense is apparently due to inadequate amounts of reducing power present in the egg, since disulfide-poor sperm nuclei decondense when microinjected into either GV intact oocytes or eggs (Perreault et al., 1984; Zirkin et al., 1985).

The ability of sperm nuclei to decondense in egg cytoplasm is dependent on the maturational state of the oocyte (Perreault et al., 1988). Sperm microinjection experiments indicate that decondensing activity is maximal in metaphase II arrested eggs but is barely detectable in GV intact oocytes. Correlated with these differences in these decondensing potentials is a maturation-associated increase in the amount of glutathione. Inhibiting this maturation-associated increase in glutathione content with BSO inhibits the ability of the matured oocyte to decondense microinjected sperm nuclei.

Sperm nuclei in a GV-intact oocyte will not decondense, whereas sperm nuclei in an oocyte that has undergone GVBD will decondense. Moreover, since dithiothreitol-treated hamster sperm nuclei will decondense in the cytoplasm of GV-intact oocytes (Perreault, et al., 1984), the contents of the GV may be necessary for the development of this decondensing activity. The interaction of the nucleoplasm with the cytoplasm may result in generating the sperm decondensing activity by providing the cytoplasm with (1) the activity (2) a "co-factor" necessary for activity, (3) an activity that "activates" the sperm-decondensing activity, or (4) an activity that inhibits a sperm-decondensing inhibitory activity. Last, it should be noted that GVBD is apparently required for development of cytoplasmic activities that control male and female pronuclei formation (Balakier and Tarkowski, 1980; Yanagimachi, 1988 and references therein).

FERTILIZATION AND EGG ACTIVATION

Although fertilization of mouse eggs induces a small increase in the absolute rate of protein synthesis (Schultz et al., 1979b), it initiates a dramatic series of changes in the pattern of protein synthesis (Schultz et al., 1979b; Cullen et al., 1980). Although the one-cell embryo supports a low level of transcription (Clegg and Piko, 1982), transcription is not necessary for these changes in protein synthesis, since they occur in either physically enucleated zygotes or zygotes incubated in the presence of a-amanitin (Petzoldt et al., 1980; Braude et al., 1979). Most of these changes are due to post-translational modifications of existing proteins (Van Blerkom, 1981) or recruitment of maternal mRNA (Cascio and Wassarman, 1982); In addition, there is a small subset of changes that appear autonomous of fertilization and are apparently initiated by oocyte maturation (Howlett and Bolton, 1985). Both the fertilization-induced and fertilization-independent changes in protein synthesis are likely involved, at least in part, in the onset of DNA synthesis in the pronuclei, cleavage to the 2-cell stage, and transition from a nonproliferative to a proliferative state with the concomitant conversion of a meiotic cell cycle to a mitotic one. What controls these events is not known, but perturbing protein phosphorylation during the first cell cycle can inhibit both cleavage to the two-cell embryo, as well as activation of transcription of the zygotic genome, which also occurs in the two-cell embryo (Poueymirou and Schultz, 1987). This result is consistent with protein phosphorylation being a major type of protein modification thata occurs during the first cell cycle (Van Blerkom, 1981).

1. Block to Polyspermy

A major response of the egg to the fertilizing sperm is the block to polyspermy. All species apparently have mechanisms to block polyspermy. For example, sea urchins possess a fast electrical block that operates at the level of the plasma membrane (Jaffe and Gould, 1983 and references therein).

Although mouse eggs generate a plasma membrane block to polyspermy (Wolf, 1978; Stewart-Savage, and Bavister, 1988 for hamster), there is no evidence to support the existence of a fast electrical block to polyspermy at the level of the plasma membrane (Jaffe et al., 1983). The major mechanism for the polyspermy block in mice appears to be an egg-induced modification of the *zona pellucids* (ZP). The ZP is an extracellular coat that surrounds the oocyte and is responsible for species-specific binding of sperm and induction of the acrosome reaction (Wassarman, 1987, 1988 and references therein). It should be noted, however, that rabbits do not possess a *zona* block to polyspermy; their primary block apparently resides at the plasma membrane (Yanagimachi, 1988). Moreover, dispermic mouse eggs can be restored to the monospermic condition by sperm loss due to cytoplasmic blebbing (Yu, and Wolf, 1981); this would constitute a third line of defense against polyspermy. The mechanism of this sperm loss is poorly understood, although microfilaments are likely involved, since the process is inhibited by cytochalasin B (Yu, and Wolf, 1981).

The mouse egg's *zona pellucida* is composed of three sulfated glycoproteins called ZP1, ZP2, and ZP3 (Wassarman, 1988 and references therein that pertain to structural and functional aspects; Bleil and Wassarman, 1980a; Shimizu et al., 1983). ZP1, which has an apparent molecular weight of about 200,000 daltons (Bleil and Wassarman, 1980a), is a dimer connected by intermolecular disulfide bonds and is likely to serve as a cross-linker responsible for maintaining the three dimensional structure of the *zona pellucida* (Greve and Wassarman, 1985). ZP2 isolated from oocytes or unfertilized eggs has an apparent molecular weight of 120,000 daltons, which is observed under either non-reducing or reducing conditions of gel electrophoresis (Bleil and Wassarman, 1980a). Fertilization results in the modification of ZP2 to a form called $ZP2_f$. Under nonreducing conditions, $ZP2_f$ has an apparent molecular weight of 120,000, whereas under reducing conditions it has an apparent molecular weight of 90,000 daltons (Bleil et al., 1981). This modification is a proteolytic cleavage that results in two fragments held together by disulfide bonds. Since ZP2 can bind to acrosome reacted sperm, whereas $ZP2_f$ cannot, ZP2 may mediate sperm binding subsequent to the acrosome reaction (Bleil and Wassarman, 1986). ZP3 has an apparent molecular weight of 83,000 daltons and the O-linked carbohydrate portion accounts for the sperm receptor activity of ZP3, which is lost after fertilization (Bleil and Wassarman, 1980b; Florman et al., 1984). ZP3 also possesses all of the sperm acrosome reaction-inducing activity of the ZP, and can do so to the same extent as that induced by the ionophore A23187 (Bleil and Wassarman, 1983).

Based on the aforementioned properties of the mouse *zona* and that only acrosome-intact sperm bind to the *zona*—acrosome-reacted sperm do not bind— the following sequence of events has been proposed for sperm-*zona* interactions and fertilization of mouse eggs. Acrosome-intact sperm bind to the *zona pellucida* (Saling et al., 1979a; Florman and Storey, 1982); binding is species-specific and mediated by ZP3. Sperm bound to ZP3 then undergo the acrosome reaction, which is also mediated by ZP3. ZP2 mediates the binding of acrosome-reacted sperm (Bleil and Wassarman, 1986), which then penetrate the *zona pellucida* and gain access to the perivitelline space. The acrosome-reacted sperm then bind to and fuse with the egg's plasma membrane.

In response to fertilization, the egg undergoes the "cortical granule reaction" (Szollsi, 1967; Barros and Yanagimachi, 1971, 1972; Wolf and Hamada, 1977). Cortical granules subjacent to the plasma membrane are thought to fuse with the plasma membrane and release into the perivitelline space enzymes that convert ZP2 to $ZP2_f$ and modify ZP3 such that it no longer possesses either sperm receptor activity or the ability to induce an acrosome reaction. Thus, acrosome-intact sperm can no longer bind to the *zona* and acrosome-reacted sperm bound to the *zona* can no longer interact with and penetrate the *zona*, since these sperm do not interact with $ZP2_f$ (Bleil and Wassarman, 1986). This series of events is proposed to comprise the *zona* reaction or the *zona* block to polyspermy. A membrane block to polyspermy, however, also develops with time.

Although much is known about the cortical granule reaction in lower species and how the contents of the granules modify the extracellular coats surrounding eggs in these species, little is known about the mammalian cortical granule reaction (Gulyas, 1980). This is in large part due to the difficulty in obtaining large amounts of biological material from mammalian species. In turn, this has prevented generation of molecular markers for mammalian cortical granules. Moreover, until recently, the only way to assess accurately the status and distribution of cortical granules was by electron microscopy, which is a time consuming process and difficult to quantify easily.

Polyphosphatidylinositol turnover, which has been implicated as a response of the egg to fertilization in lower species (Turner et al., 1984, 1986; Whitaker and Irvine, 1984; Swann and Whitaker, 1986), generates two second messengers, i.e., sn 1,2-diacylglycerol, which activates the calcium- and phospholipid-dependent protein kinase, protein kinase C (PK-C), and inositol-1,4,5-trisphosphate (IP_3), which releases calcium from intracellular stores (Berridge, 1984; Berridge and Irvine, 1984). Protein phosphorylation(s) catalyzed by PK-C is implicated in regulating exocytotic processes (Nishizuka, 1984; Takai et al., 1984). Since the cortical granule reaction in mammalian eggs involves an exocytotic process, the role of mouse egg PK-C in the early events of the fertilization process was examined (Endo et al., 1987 b,c).

Treatment of eggs with biologically active phorbol diesters or a diacylglycerol, compounds that activate PK-C, inhibits both sperm penetration and fertilization (Endo et al., 1987c). Biologically inactive phorbol diesters, which do not activate PK-C, do not inhibit either sperm penetration or fertilization. This inhibition is due to an egg-induced modification of the *zona pellucida,* such that ZP2 is converted to $ZP2_f$, while ZP3 retains its sperm receptor activity. This latter observation accounts for the finding that sperm binding is not reduced in eggs treated with PK-C activators. The inhibition of fertilization is due to the inability of ZP3 to induce a complete acrosome reaction, which was determined by using an assay that monitors of the progression of the acrosome reaction.

The progression of capacitated, acrosome-intact sperm to acrosome-reacted sperm can be monitored by changes in staining patterns using the fluorescent probe chlortetracycline (Saling and Storey, 1979b). Three major fluorescent staining patterns have been characterized with this assay. The B-pattern of capacitated sperm is correlated with acrosome intact sperm, as assessed by transmission electron microscopy (Florman and Storey, 1982). The S-pattern represents an intermediate stage and appears prior to completion of the acrosome reaction. This pattern correlates with loss of the ability of sperm to maintain a transmembrane pH gradient (Lee and Storey, 1985). The AR-pattern corresponds to acrosome-reacted sperm, as determined by transmission electron microscopy (Saling et al., 1979b).

ZP3 isolated from untreated eggs possesses the ability to induce the B to S to AR transitions. In contrast, ZP3 isolated from eggs treated with PK-C activators can induce the B to S transition, but not the S to AR transition. Accordingly, sperm treated with ZP3 isolated from these eggs treated with PK-C activators accumulate in the S pattern. Although previous studies indicated that ZP3 isolated from 2-cell embryos does not induce the acrosome reaction, the methods used in these studies assayed an end point, i.e., the completion of the acrosome reaction, and would not detect intermediates in this process (Bleil and Wassarman, 1983). When tested with the chlortetracycline assay, sperm incubated with ZP3 isolated from 2-cell embryos do not even undergo the B to S transition.

Sperm arrested in the S pattern by *zonae* isolated from PK-C activator-treated eggs can be induced to undergo the S to AR transition by treatment with either ionophore A 23187 or solubilized *zonae* from untreated eggs; solubilized *zonae* from 2-cell embryos do not induce this transition (Kligman, Storey, and Kopf, unpublished results). Thus, the S pattern in which the sperm accumulate by treatment with ZP3 isolated from PK-C activator-treated eggs represents an intermediate stage of the acrosome reaction that can be induced to undergo subsequent steps and complete the acrosome reaction.

These studies demonstrate that treatment of eggs with activators of protein kinase C results in an egg-induced modification of the *zona* such that there is a dissociation of the sperm receptor activity from the acrosome reaction-inducing activity of ZP3. In contrast, fertilization results in the loss of both the sperm receptor and acrosome reaction-inducing activities of ZP3. Presumably, differences in ZP3 obtained from untreated and phorbol diester-treated eggs reflect those portions of the molecules that participate in the acrosome reaction. Studies examining biochemical differences in ZP3 obtained from untreated and phorbol diester-treated eggs, therefore, should facilitate analysis of those portions of ZP3 that are involved in inducing the acrosome reaction. The use of *zonae* from phorbol diester-treated eggs should also facilitate studies on the mechanism(s) of the acrosome reaction, since it is now possible to use these *zonae* to study independently the B to S transition from the S to AR transition. Such an experimental system may be of great value in determining the biochemical correlates of the S pattern, which has characteristics of an intermediate stage prior to the completion of the acrosome reaction.

The effects of IP_3 microinjected into mouse eggs have also been examined with respect to its effect on egg activation, *zona* modifications, and sperm receptor activities of the *zona* (unpublished results). A low percentage (15%) of eggs microinjected with IP_3 at a final concentration of 4 μM become activated, as evidenced by second polar body emission within 1.5 h. In contrast, eggs injected with the vehicle do not activate. *Zonae* from the IP_3-microinjected eggs that activate always show the loss of ZP2 and its conversion to $ZP2_f$. Although IP_3-injected eggs do not activate, 85% of these eggs reveal a conversion of ZP2 to $ZP2_f$. In about 70% of these cases the conversion is total and in the other 30% about 50% of the ZP2 is modified. Of the vehicle injected eggs, about 30% reveal a modification of ZP2, which is usually only partially modified.

The half-maximal concentration of IP_3 necessary to elicit the change in ZP3 is about 5 nM, and this corresponds well to that necessary to induce the cortical granule-mediated elevation of the fertilization envelope in sea urchin eggs, as well as calcium release from intracellular stores in other systems (Whitaker and Irvine, 1984; Swann and Whitaker, 1986). Extracellular calcium is not required for mouse eggs injected with IP_3 to display the ZP2 modification, and this is consistent with release of calcium from intracellular stores in the egg. Microinjection of either $I(1,4)P_2$, $I(2,4,5)P_3$, or $I(1,3,4)P_3$, each of which does not release intracellular calcium, to a final concentration of 4 μM fails to induce a modification of ZP2. Moreover, microinjection of inositol 1,3,4,5-tetrakisphosphate, which is implicated in regulating calcium channels in the plasma membrane of eggs of lower species (Irvine and Moor, 1986, 1987), does not induce a conversion of ZP2 to $ZP2_f$.

IP_3 injection of mouse eggs may also modify ZP3, since IP_3-injected eggs bind fewer sperm than vehicle injected eggs. The lower extent of binding could represent the inability of acrosome-reacted bound sperm to establish a secondary binding with ZP2, for the following reason: If ZP2 mediates the binding of acrosome-reacted sperm and $ZP2_f$ cannot interact with acrosome-reacted sperm, then although the sperm can bind to IP_3-injected eggs and undergo the acrosome reaction, they cannot establish the secondary binding with ZP2, which has been modified to $ZP2_f$. Thus, the bound sperm will dissociate. This explanation for the lower level of binding is made less likely since the sperm used in these experiments are treated with pertussis toxin, which prevents the sperm from undergoing the acrosome reaction (Endo et al., 1987a). Thus, their interaction with ZP2 is prevented and their interaction with the *zona* is restricted to ZP3. The lower level of sperm binding to IP3-injected eggs is therefore likely to be due to a reduced level of ZP3 sperm receptor activity.

Fertilization is associated with a characteristic set of changes in the pattern of protein synthesis. Although IP_3 injected eggs display a modification in ZP2, they do not reveal the changes in the pattern of protein synthesis associated with fertilization. This is consistent with the very low level of egg activation, as assessed by pronuclear formation. Thus, although IP_3 can apparently bring about an egg-induced modification of the *zona*, it does not elicit a full egg activation response, but rather a subprogram of events that occurs during egg activation.

The mechanism of IP_3-induced modifications of the *zona* is likely to be via an IP_3-stimulated release of intracellular calcium, which somehow is involved in cortical granule exocytosis. Periodic increases in intracellular free calcium concentration occur following fertilization in hamster eggs (Miyazaki et al., 1986). Injection of IP_3 to a final concentration of 80 nM induces a transient increase in intracellular free calcium that spreads over the entire egg within a second (Miyazaki, 1988). In addition, IP_3 also stimulates a hyperpolarization of the membrane potential. These changes are very similar to those that normally occur after fertilization.

G proteins are a family of guanine nucleotide binding proteins that are activated by GTP and serve to couple various extracellular signals to their intercellular effectors, which can be involved in generation of second messengers (Gilman, 1987 and references therein; Neer and Clapham, 1988, and references therein). Generation of sn 1,2-diacylglycerol and IP_3 is believed to be mediated by a G protein stimulated phospholipase C, and in sea urchins, fertilization is correlated with a rapid turnover of phosphatidylinositol bisphosphate (Turner et al., 1984). In addition, activation of G proteins is believed to result in cortical granule exocytosis in sea urchins (Turner et al., 1986).

If mammalian sperm initiate a signal transduction sequence that is mediated by a G protein, a cascade of fertilization-induced events should be triggered by microinjected GTP. Consistent with this hypothesis is the observation that microinjection of GTPgS into hamster oocytes also triggers transient increases in intracellular free calcium and hyperpolarization of the membrane potential, and GDPbS inhibited this GTPgS-induced response (Myasaki, 1988).

Although these data are consistent with a fertilization-induced, G protein coupled, PK-C/IP_3 mediated stimulation of a cortical granule reaction, which in turn effects the modifications of the *zona* that constitute the *zona* block to polyspermy, experiments have not yet correlated the biological and biochemical changes with changes in the number of cortical granules in response to these agents. *Lens culinaris* agglutinin is a lectin that apparently stains cortical granules, and accordingly provides a convenient marker to monitor the cortical granule reaction (Cherr et al., 1988). Other lectins may also provide potential markers for cortical granules (Lee et al., 1988). Results of recent experiments indicate that fertilized or ionophore-activated eggs have dramatically reduced numbers of these lectin staining granules (Ducibella, personal communication). Moreover, eggs treated with PK-C activators have a partially reduced number of these granules (Ducibella, Kopf, and Schultz, unpublished observations).

Future studies are required to ascertain effects of IP_3 and GTP on the release of these granules, as well as characterizing the cortical granules with respect to their enzymatic contents. The development of micro-assay procedures will facilitate these studies, which may reveal the types of enzymes involved in the modification(s) of both ZP2 and ZP3. In addition, the development of cortical granule probes may be used to study cortical granule biogenesis and to reveal if there is a heterogeneity in the cortical granule population. Electron microscopy studies reveal the existence of light and dark staining populations of cortical granules in mouse eggs (Nicosia et al., 1977). This may reflect granules at different stages of packaging their internal contents, or could reflect heterogeneity of mature granules. Such heterogeneity exists in lower species. For example, in sea urchins, although the cortical granules appear homogeneous in the transmission electron microscope, only about 20% of them contain a cortical granule antigen, as determined by immunoelectronmicroscopy (Anstrom et al., 1988). The partial modification of ZP3 coupled with the partial reduction in the number of LCA-staining granules in response to PK-C activators is consistent with cortical granule heterogeneity in mammalian eggs, but further studies are clearly required to support this conjecture.

UNRESOLVED QUESTIONS AND FUTURE DIRECTIONS

Little is known regarding the molecular basis for the acquisition of meiotic competence. In the future, subtraction hybridization of cDNA libraries generated from meiotically competent and incompetent oocytes may allow the cloning of cDNAs specific to meiotically competent oocytes. Analysis of such clones may provide insights regarding how oocytes develop and differentiate. This may be of extreme importance, since an understanding of factors involved in acquisition of meiotic competence may lead to improved systems that support oocyte growth and acquisition of meiotic competence *in vitro*.

Although protein phosphorylation is implicated in regulating meiotic maturation, we know very little concerning the sequence of events that comprise meiotic maturation and the proteins involved in this process. Meiotic maturation entails a G2 to M transition in the cell cycle, and therefore is a problem in cell cycle regulation. Specific cellular oncogenes, some of which are protein kinases, are implicated in cell cycle regulation in other systems. Interestingly, recent studies show changes in the temporal patterns of expression of specific oncogenes, e.g., c-mos, which is a serine/threonine protein kinase, during oocyte growth and maturation (Propst et al., 1988). Future studies addressing functional aspects of these gene products, coupled with the recent purification and identification of maturation promoting factor as a homolog of the fission yeast cell cycle control protein encoded by the $cdc2^+$ gene will undoubtedly shed light on the process of meiotic maturation at the molecular level and define more clearly the role of protein phosphorylation in regulating this process. Again, an understanding of the molecular mechanisms of the process of meiotic maturation may lead to improvements in culture systems that support maturation *in vitro*.

Not all oocytes that mature into eggs during maturation are capable of being fertilized and giving rise to normal development. This type of maturation is termed "cytoplasmic" maturation. An example of cytoplasmic maturation was discussed above, namely, the acquired ability of the egg during maturation to decondense the sperm nucleus. We still do not understand at the molecular level what constitutes "cytoplasmic" maturation and its regulation. A major problem in medically assisted conception is to identify eggs that are capable of being fertilized and giving rise to normal development. Although matured eggs may appear similar on morphological grounds, they are likely to possess profound differences with respect to their state of cytoplasmic maturation. These differences may compromise their ability to be fertilized and develop.

The sequence of events from sperm fusion with the egg's plasma membrane to egg activation and the *zona* polyspermy block is still not well defined. Although it is likely that this process is mediated by a G protein(s), the biochemical nature of the sperm receptor on the egg's plasma membrane, the molecular identity of the G protein(s) involved, and the coupling mechanism of the putative receptor with the G protein(s) and the consequence of this interaction are still unknown. For example, does fertilisation of mammalian eggs result in activation of a phospholipase C with the subsequent production of diacylglycerol and IP_3? Analysis at the biochemical and molecular levels should provide basic and essential information regarding these issues. Such an understanding may lead to refinement of conditions that foster fertilization.

The molecular basis for the mammalian cortical granule reaction in response to fertilization is still unknown. Although studies in lower species and initial studies in the mouse and hamster implicate calcium, much more work is required to understand this exocytotic event at the molecular level. The biochemical composition of mammalian cortical granules is still undefined and how the contents of these granules modify the *zona* to elicit the *zona* block to polyspermy remains to be determined. Moreover, the actual biochemical changes that occur in the *zona* proteins following egg activation needs to be elucidated. Work directed at isolating and characterizing mammalian cortical granules will help resolve many of these issues. In addition, studies that focus on first determining the biochemical identity of determinants on ZP3 and ZP2 that are involved in sperm binding and the acrosome reaction will pave the way for subsequent studies that address the nature of the changes in these proteins that result in loss of these biological activities associated with these *zona* proteins. Results of such studies should establish the biochemical basis for the block to polyspermy. Such knowledge may result in *in vitro* fertilization culture systems in which the incidence of polyspermy is further reduced.

ACKNOWLEDGMENTS

Research performed by the authors was supported by grants from the National Institutes of Health (HD 180604 to R.M.S. and 19096 to G.S.K. and HD 22732 to G.S.K. and R.M.S.) and grants from the Mellon Foundation (G.S.K.) and the University Research Fund (R.M.S.). S.K. was supported by the Rockefeller Foundation. G.S.K. and R.M.S. would like to thank Philip Hugo for assistance with some of the experiments described above and Jeff Bleil for stimulating discussions about the role of ZP2 in sperm binding.

REFERENCES

Anstrom, J.A., Chin, J.E., Leaf, D.S., Parks, A.L., and Raff, R.A. (1988). Immunocytochemical evidence suggesting heterogeneity in the population of sea urchin egg cortical granules. *Dev. Biol.* **125**: 1-7.

Balakier, H. (1978). Induction of maturation in small oocytes from sexually immature mice by fusion with meiotic or mitotic cells. *Exp. Cell Res.* **112**: 137-141.

Balakier, H., and Czolowska, R. (1977). Cytoplasmic control of nuclear maturation in mouse oocytes. *Exp. Cell Res.* **110**: 466-469.

Balakier, H., and Tarkowski, A.K. (1980). The role of germinal vesicle karyoplasm in the development of the male pronucleus in the mouse. *Exp. Cell Res.* **128**: 79-86.

Barros, C., and Yanagimachi, R. (1971). Induction of the *zona* reaction in golden hamster eggs by cortical granule material. *Nature (London)* **233**; 268-269.

Barros, C., and Yanagimachi, R. (1972). Polyspermy-preventing mechanisms in the golden hamster egg. *J. Exp. Zool.* **180**: 251-266.

Berridge, M.J. (1984). Inositol triphosphate and diacylglycerol as second messengers. *Biochem. J.* **220**: 345-360.

Berridge, M.J., and Irvine, R.F. (1984). Inositol triphosphate, a novel second messenger in cellular signal transduction. *Nature (London)* **312**: 315-321.

Bleil, J.D., and Wassarman, P.M. (1980a). Structure and function of the *zona pellucida*: Identification and characterization of the proteins of three mouse oocyte's *zona pellucida*. *Dev. Biol.* **76**: 185-203.

Bleil, J.D., and Wassarman, P.M. (1980b). Mammalian sperm-egg interaction: Identification of a glycoprotein in mouse egg *zonae pellucidae* possessing receptor activity for sperm. *Cell* **20**: 873-882.

Bleil, J.D., and Wassarman, P.M. (1981). Mammalian sperm-egg interaction: Fertilization of mouse eggs triggers modification of the major *zona pellucida* glycoprotein, ZP2. *Dev. Biol.* **86**: 189-197.

Bleil, J.D., and Wassarman, P.M. (1983). Sperm egg interactions in the mouse: Sequence of events and induction of the acrosome reaction by a *zona pellucida* glycoprotein. *Dev. Biol.* **95**: 317-324.

Bleil, J.D., and Wassarman, P.M. (1986). Autoradiographic visualization of the mouse egg's sperm receptor bound to sperm. *J. Cell Biol.* **102**: 1363-1371.

Bornslaeger, E.A., Mattei, P.M., and Schultz, R.M. (1986a). Involvement of cAMP-dependent protein kinase and protein phosphorylation in regulation of mouse oocyte maturation. *Dev. Biol.* **114**: 453-462.

Bornslaeger, E.A., Mattei, P.M., and Schultz, R.M. (1988). Protein phosphorylation in meiotically competent and incompetent mouse oocytes. *Mol. Reprod. Dev.* **1**, 19-25.

Bornslaeger, E.A., Poueymirou, W.T., Mattei, P., and Schultz, R.M. (1986b). Effects of protein kinase C activators on germinal vesicle breakdown and polar body emission of mouse oocytes. *Exp. Cell Res.* **165**: 507-517.

Bornslaeger, E.A., Wilde, M.W., and Schultz, R.M. (1984). Regulation of mouse oocyte maturation: Involvement of cyclic AMP phosphodiesterase and calmodulin. *Dev. Biol.* **105**: 488-499.

Braude, P., Pelham, H., Flach, G., and Lobatta, R. (1979). Post-transcriptional control in the early mouse embryo. *Nature (London)* **282**: 102-105.

Calvin, H.I., Grosshans, K., and Blake, E. (1986). Estimation and manipulation of glutathione levels in prepubertal mouse ovaries and ova: Relevance of sperm nucleus transformations in the fertilized egg. *Gamete Res.* **14**: 265-275.

Canipari, R., Palombi, F., Riminucci, M., and Mangia, F. (1984). Early programming of maturation competence in mouse oogenesis. *Dev. Biol.* **102**: 519-524.

Cascio, S.M., and Wassarman, P.M. (1982). Program of early development in the mammal: Post-transcriptional control of a class of proteins synthesized by mouse oocytes and early embryos. *Dev. Biol.* **89**: 397-408

Cherr, G.N.,Drobnis, E.Z., and Katz, D.F. (1988). Localization of cortical granule constituents before and after exocytosis in the hamster egg. *J. Exp. Zool.* **246**: 81-93.

Cho, W.K., Stern, S., and Biggers, J.D. (1974). Inhibitory effect of dibutyryl cAMP on mouse maturation *in vitro*. *J. Exp. Zool.* **187**: 383-386.

Clegg, K.B., and Piko, L. (1982). RNA synthesis and cytoplasmic polyadenylation in the one-cell mouse embryo. *Nature (London)* **295**: 342-344.

Cullen, B.R., Emigholz, K., and Monahan, J.J. (1980). The transient appearance of specific proteins in one-cell mouse embryos. *Dev. Biol.* **76**: 215-221.

Cyert, M.S., and Kirschner, M.W. (1988) Regulation of MPF activity *in vitro*. *Cell* **53**: 185-195.

Dunphy, W.G., Brizuela, L., Beach, D., and Newport, J.W. (1988). The Xenopus cdc2 protein is a component of MPF, a cytoplasmic regulator of mitosis. Cell **54**: 423-431.

Endo, Y., Lee, M.A., and Kopf, G.S. (1987a). Evidence for the role of a guanine nucleotide-binding regulatory protein in the *zona pellucida*-induced mouse sperm acrosome reaction. *Dev. Biol.* **119**: 210-216.

Endo, Y., Mattei, P., Kopf, G.S., and Schultz, R.M. (1987b). Effects of a phorbol ester on mouse eggs: Dissociation of sperm receptor activity from acrosome reaction-inducing activity of the mouse *zona pellucida* protein, ZP3. *Dev. Biol.* **123**: 574-577.

Endo, Y., Schultz, R.M., and Kopf, G.S. (1987c). Effects of phorbol esters and a diacylglycerol on mouse eggs: Inhibition of fertilization and modification of the *zona pellucida*. *Dev. Biol.* **119**: 199-209.

Florman, H.M., and Storey, B.T. (1982). Mouse gamete interactions: The *zona pellucida* is the site of the acrosome reaction leading to fertilization *in vitro*. *Dev. Biol.* **91**: 121-130.

Florman, H.M., and Wassarman, P.M. (1985). O-linked oligosaccharides of mouse egg ZP3 account for its sperm receptor activity. *Cell* **41**: 313-324.

Florman, H.M., Bechtol, K.B., and Wassarman, P.M. (1984). Enzymatic dissection of the functions of the mouse egg's receptor for sperm. *Dev. Biol.* **106**: 243-255.

Fulka, , J., Jr., Motlik, J., Fulka, J., and Crozet, N. (1985). Inhibition of nuclear maturation in fully grown porcine and mouse oocytes after their fusion with growing porcine oocytes. *J. Exp. Zool.* **235**: 255-259.

Gautier, J., Norbury, C., Lohka, M., Nurse, P., and Maller, J. (1988). Purified maturation-promoting factor contains the product of a *Xenopus* homolog of the fission yeast cell cycle control gene $cdc2^{+}$. **Cell** 54: 433-439.

Gerhart, J., Wu, M., and Kirschner, M. (1984). Cell cycle dynamics of an M-phase-specific cytoplasmic factor in *Xenopus laevis* oocytes and eggs. *J. Cell Biol.* **98**: 1247-1255.

Gilman, A.G. (1987). G proteins: Transducers of receptor-generated signals. *Ann. Rev. Biochem.* **56**: 615-649.

Greve, J.M., and Wassarman, P.M. (1985). Mouse egg extracellular coat is a matrix of interconnected filaments possessing a structural repeat. *J. Mol. Biol.* **181**: 253-264.

Gulyas, B.J. (1980). Cortical granules of mammalian eggs. *Int. Rev. Cytol.* **63**: 357-392.

Hashimoto, N., and Kishimoto, T. (1988). Regulation of meiotic metaphase by a cytoplasmic maturation-promoting factor during mouse oocyte maturation. *Dev. Biol.* **126**: 242-252.

Howlett, S.K., and Bolton, V.N. (1985). Sequence and regulation of morphological and molecular events during the first cell cycle of mouse embryogenesis. *J. Embryol. Exp. Morphol.* **52**: 209-225.

Irvine, R.F., and Moor, R.M. (1986). Micro-injection of inositol 1,3,4,5 tetrakisphosphate activates sea urchin eggs by a mechanism dependent on external Ca^{2+}. *Biochem. J.* **240**: 917-920.

Irvine, R.F., and Moor, R.M. (1987). Inositol (1,3,4,5) tetrakisphosphate-induced activation of sea urchin eggs requires the presence of inositol trisphosphate. *Biochem. Biophys. Res. Commun.* **146**: 284-290.

Jaffe, L.A., and Gould, M. (1985). Polyspermy-preventing mechanisms. In "Biology of Fertilization", Metz, C.B., and Monroy, A., eds., vol. 3, pp. 223-250, Academic Press, New York.

Jaffe, L.A., Sharp, A.P., and Wolf, D.P. (1983). Absence of an electrical polyspermy block in the mouse. *Dev. Biol.* **96**: 317-323.

Kishimoto, T. (1988). Regulation of metaphase by a maturation-promoting factor. *Develop. Growth and Differ.* **30**: 105-115.

Kishimoto, T., Yamazaki, K., Kato, Y., Koide, S.S., and Kanatani, H. (1984). Induction of starfish oocyte maturation by maturation-promoting factor of mouse and surf clam oocytes. *J. Exp. Zool.* **231**: 293-295.

Lee, M.A., and Storey, B.T. (1985). Evidence for plasma membrane impermeability to small ions in acrosome-intact mouse spermatozoa bound to mouse *zonae pellucidae*, using an aminoacridine fluorescent probe: Time course of the *zona*-induced acrosome reaction monitored by both chlortetracycline and pH probe fluorescence. *Biol. Reprod.* **33**: 235-246.

Lee, S.H., Ahuja, K.K., Gilburt, D.J., and Whitingham, D.G. (1988). The appearance of glycoconjugates associated with cortical granule release during mouse fertilization. *Development* **102**: 595-604.

Lohka, M.J., Hayes, M.K., and Maller, J.L. (1988). Purification of maturation-promoting factor, an intracellular regulator of early mitotic events. *Proc. Natl. Acad. Sci., U.S.A.* **85**: 3009-3013.

Masui, Y., and Clarke, H.J. (1979). Regulation of oocyte maturation. *Int. Rev. Cytol.* **57**: 185-282.

Miyazaki, S. (1988). Inositol 1,4,5-trisphosphate-induced calcium release and guanine-nucleotide binding protein-mediated periodic calcium rises in golden hamster eggs. *J. Cell Biol.* **106**: 345-353.

Miyazaki, S., Hashimoto, N., Yoshimoto, Y., Kishimoto, T., Igusa, Y., and Hiramoto, Y. (1986). Temporal and spatial dynamics of the periodic increase in intracellular free calcium at fertilization in the golden hamster. *Dev. Biol.* 118: **259**-267.

Neer, E.J., and Clapham, D.E. (1988). Roles of G protein subunits in transmembrane signalling. *Nature (London)* **333**: 129-134.

Nicosia, S.V., Wolf, D.P., and Masato, I. (1977). Cortical granule distribution and cell surface characteristics in mouse eggs. *Dev. Biol.* **57**: 56-74.

Nishizuka, Y. (1984). The role of protein kinase C in cell surface signal transduction and tumour promotion. *Nature (London)* **308**: 693-698.

Perreault, S.D., Barbee, R.R., and Slott, V.L. (1988). Importance of glutathione in the acquisition and maintenance of sperm nuclear decondensing activity in maturing hamster oocytes. *Dev. Biol.* **125**: 181-186.

Perreault, S.D., Wolff, R.A., and Zirkin, B.R. (1984). The role of disulfide bond reduction during mammalian sperm nuclear decondensation in vivo. *Dev. Biol.* **101**: 160-167.

Petzoldt, U., Hoppe, P.C., and Illmensee, K. (1980). Protein synthesis in enucleated fertilized and unfertilized mouse eggs. *Wilhelm Roux's Arch. Dev. Biol.* **189**: 215-219.

Pincus, G., and Enzmann, E.V. (1935). The comparative behaviour of mammalian eggs *in vivo* and *in vitro*. *J. Exp. Med.* **62**: 665-675.

Poueymirou, W.T., and Schultz, R.M. (1987). Differential effects of activators of cAMP-dependent protein kinase and protein kinase C on cleavage of one-cell mouse embryos and protein synthesis and phosphorylation in one- and two-cell embryos. *Dev. Biol.* **121**: 489-498.

Propst, F., Rosenbery, M.P., and Vande Woude, G.F. (1988). Proto-oncogene expression in germ cell development. *Trends in Genetics* **4**: 183-187.

Saling, P.M., and Storey, B.T. (1979). Mouse gamete interactions during fertilization *in vitro*: Chlortetracycline as fluorescent probe for moue sperm acrosome reaction. *J. Cell Biol.* **83**: 544-555.

Saling, P.M., Sowinski, J., and Storey, B.T. (1979). An ultrastructural study of epididymal mouse spermatozoa binding to *zonae pellucidae in vitro*: Sequential relationship to the acrosome reaction. *J. Exp. Zool.* **209**: 229-238.

Schultz, R.M. (1988). Molecular aspects of oogenesis. In "Experimental Approaches to Mammalian Embryonic Development", Rossant, J., and Pedersen, R.A., eds. pp. 195-237, Cambridge University Press, New York.

Schultz, R.M., Letourneau, G.E., and Wassarman, P.M. (1978). Meiotic maturation of mouse oocytes *in vitro*: Protein synthesis in nucleate and anucleate oocyte fragments. *J. Cell Sci.* **30**: 251-264.

Schultz, R.M., Letourneau, G.E., and Wassarman, P.M. (1979a). Program of early development in the mammal: Changes in the patterns and absolute rates of tubulin and total protein synthesis during oocyte growth in the mouse. *Dev. Biol.* **73**: 120-133.

Schultz, R.M., Letourneau, G.E., and Wassarman, P.M. (1979b). Program of early development in the mammal: Changes in the patterns and absolute rates of tubulin and total protein synthesis during oogenesis and early embryogenesis in the mouse. *Dev. Biol.* **68**: 341-359.

Schultz, R.M., Montgomery, R.R., and Belanoff, J.R. (1983). Regulation of mouse oocyte maturation: Implication of a decrease in oocyte cAMP and protein dephosphorylation in commitment to resume meiosis. *Dev. Biol.* **97**: 264-273.

Shimizu, S., Tsuji, M., and Dean, J. (1983). *In vitro* biosynthesis of three sulfated glycoproteins of murine zonae pellucidae by oocytes grown in follicle culture. *J. Biol. Chem.* **258**: 5858-5863.

Simanis, V., and Nurse, P. (1986) The cell cycle control gene cdc2+ of fission yeast encodes a protein kinase potentially regulated by phosphorylation. *Cell* **45**: 261-268.

Sorensen, R.A., and Wassarman, P.M. (1976). Relationship between growth and meiotic maturation of the mouse oocyte. *Dev. Biol.* **50**: 531-536.

Sorensen, R.A., Cyert, M.S.,and Pedersen, R.A. (1985). Active maturation-promoting factor is present in mature mouse oocytes. *J. Cell Biol.* **100**: 1637-1640.

Stewart-Savage, J., and Bavister, B.D. (1988). A cell surface block to polyspermy occurs in golden hamster eggs. *Dev. Biol.* **128**; 150-157.

Swann, K., and Whitaker, M. (1986). The part played by inositol trisphosphate and calcium in the propagation of the fertilization wave in sea urchin eggs. *J. Cell Biol.* **103**: 2333-2342.

Szollsi, D. (1967). Development of cortical granules and the cortical reaction in rat and hamster eggs. *Anat. Rec.* **159**: 431-446.

Takai, Y., Kikkawa, U., Kaibuchi, K., and Nishizuka, Y. (1984). Membrane phospholipid metabolism and signal transduction for protein phosphorylation. *Adv. Cyclic Nuc. Prot. Phos. Res.* **18**: 119-154.

Turner, P.R., Jaffe, L.A., and Fein, A. (1986). Regulation of cortical granule exocytosis in sea urchin eggs by inositol 1,4,5-trisphosphate and GTP-binding protein. *J. Cell Biol.* **102**: 70-76.

Turner, P.R., Sheetz, M.P., and Jaffe, L.A. (1984). Fertilization increases the polyphosphoinositide content of sea urchin eggs. *Nature (London)* **310**: 414-415.

Van Blerkom, J. (1981). Structural relationship and post-translational modification of stage-specific proteins synthesized during early preimplantation development in the mouse. *Proc. Natl. Acad. Sci. U.S.A.* **78**: 7629-7633.

Vivarelli, E., Conti, M. De Felici, M., and Siracusa, G. (1983). Meiotic resumption and intracellular cAMP levels in mouse oocytes treated with compounds which act on cAMP metabolism. *Cell Diff.* **12**: 271-276.

Wassarman, P.M. (1987). Early events in mammalian fertilization. *Ann. Rev. Cell Biol.* **3**: 109-142.

Wassarman, P.M. (1988). *Zona pellucida* glycoproteins. *Ann. Rev. Biochem.* **57**: 414-442.

Whitaker, M., and Irvine, R.F. (1984). Inositol 1,4,5-trisphosphate microinjection activates sea urchin eggs. *Nature (London)* **312**: 636-639.

Wolf, D.P. (1978). The block to sperm penetration in zona-free mouse eggs. *Dev. Biol.* **64**: 1-10.

Wolf, D.P., and Hamada, M. (1977). Induction of zonal and oolemmal blocks to sperm penetration in mouse eggs with cortical granule exudate. *Biol. Reprod.* **17**: 350-354.

Yanagimachi, R. (1988). Mammalian fertilization. In "The Physiology of Reproduction", Knobil, E., Neill, J., et al., eds., Vol. 1, pp. 135-185, Raven Press, Ltd., New York.

Yu, S.-F., and Wolf, D.P. (1981). Polyspermic mouse eggs can dispose of supernumerary sperm. *Dev. Biol.* **82**: 203-210.

Zirkin, B.R., Soucek, D.A., Chang, T.S.K., and Perreault, S.D. (1985). *In vitro* and *in vivo* studies of mammalian sperm nuclear decondensation. *Gam. Res.* **11**: 349-365.

GLYCOLYTIC PATHWAY IN PREIMPLANTATION MAMMALIAN EMBRYOS

John D. Biggers

INTRODUCTION

A highly integrated network of chemical reactions called metabolism enables cells to extract energy and reducing power from their environments and to synthesize the building blocks of their macromolecules. The role of metabolism in the generative process is therefore fundamental. It provides the energy whereby a new organism can develop following the blueprint determined by its genetic endowment.

In aerobic organisms metabolism is subserved ubiquitously by three linked systems - the glycolytic pathway, the citric acid cycle and the oxidative respiratory chain. Intuitively it seems trivial to study these highly conserved systems in detail in preimplantation mammalian embryos, since they are more readily studied in other systems. Observations that began to accumulate over 30 years ago (see Biggers, 1987, for a review), however, show that such a conservative view cannot be sustained. Pioneering studies by Whitten (1956, 1957) on media for the culture of preimplantation mouse embryos found that 2-cell stages that would not develop if glucose alone was the only energy source would develop if lactate was present. Subsequent work by several investigators led to the conclusion that during oogenesis in the mouse the metabolic process is restricted and that it is restored at about the 8-cell stage of development (Brinster, 1965a,b; Brinster and Thompson, 1966; Biggers et al., 1967). A key finding was that metabolism is restricted in the maturing oocyte stage and in the zygote to the extent that only pyruvate or oxaloacetate could support development (Biggers et al., 1967). These early observations are the origin of the widely assumed view that pyruvate is an important source of energy in early mammalian development and should be incorporated in all embryo culture media.

These observations also led to an extensive literature on the metabolism of preimplantation embryos (see Biggers and Stern, 1973; Biggers, 1976; Biggers and Borland, 1976; Wales, 1975; Pike, 1981; Weitlauf and Nieder, 1984; Kaye, 1986; Biggers et al., 1989, for reviews). In this paper I wish to revisit particularly the glycolytic pathway, whose function seems to account for the original observations on the nutritional requirements of the preimplantation mouse embryo. I suggest that three areas can be usefully discussed at the present time to provide a basis for future research. These areas are: (1) What is the evidence that metabolic pathways are restricted in preimplantation development? (2) What overall metabolic fluxes are functional during preimplantation development? (3) How is the glycolytic pathway controlled during early mammalian development?

EVIDENCE THAT METABOLIC PATHWAYS ARE RESTRICTED IN OOGENESIS AND EARLY MAMMALIAN DEVELOPMENT

Four types of evidence suggest that metabolic pathways are restricted during oogenesis and early mammalian development. The evidence comes from changes in the morphology of the mitochondria, which can be correlated with the metabolic state of the embryo, nutritional studies with chemically defined media, the uptake of metabolic substrates, and the metabolic fate of substrates. The morphological evidence is important since it is the only evidence that does not involve the manipulation of embryos *in vitro*.

Morphology of the Mitochondria during Oogenesis and Preimplantation Development

A pioneer study with the electron microscope on oocyte matuation described the pleiomorphic stucture of cristae in the mitochondria of the mouse (Yamada et al., 1957). Some of the cristae were oriented transversely across the organelle, while others were arranged concentrically inside the plasma membrane. Later Mazanec and Dvorak (1964) described mitochondria with concentric cristae in the early cleavage stages of the rat, and the presence of these were soon confirmed in the mouse (Sackler, 1968; Calarco and Brown, 1969; Hillman and Tasca, 1969; Reinius, 1969). This type of mitochondrion has been reported also in oocytes of the guinea-pig, hamster, rabbit, cow, monkey and human, and in the cleavage stages of the rabbit [see Stern et al.(1971), for a review]. Carefully timed studies showed that after a critical time of development almost all mitochondria with concentric cristae disappeared, after which time all mitochondria have transverse cristae. In the mouse this maturation in mitochondrion morphology is complete by the 8-cell stage (Stern et al., 1971) and in the rabbit by the morula stage (Anderson et al.,1970). Of importance is the fact that this maturation is complete by the time the oxygen consumption of the mouse embryo (Mills and Brinster, 1967) and the rabbit embryo (Fridhandler et al., 1957) increases suddenly. Moreover, the mitochondrial maturation is associated in the mouse with the restoration of the ability of the embryos to utilize malate, citrate and 2-oxoglutarate (Kramen and Biggers, 1971). Although the genesis of these morphological changes in mitochondrial morphology is unknown, one possible explanation is that they reflect the energy state of the blastomeres, as has been described in other cell types (Hackenbrock, 1968).

Nutritional studies with chemically defined media

The early results obtained on the nutritional requirements of the preimplantation mouse embryo are summarized in Fig. 1. Meiotic maturation of the oocyte and cleavage through the first division after fertilization require the presence of pyruvate or oxaloacetate. Development of the 2-cell stage can only be supported by pyruvate, oxaloacetate, phosphoenolpyruvate and lactate. Glucose alone is unable to support development of the preimplantation mouse embryo until after the 8-cell stage.

Comparable studies on the nutritional requirements of the rabbit have led to conflicting results. The design of definitive experiments is complicated for two reasons [see Kane (1987a) for a review]. First, the large rabbit ovum, with a volume about five times that of the mouse, has considerable energy reserves which allow it to undergo several cleavage divisions without the incorporation of any energy source in the medium (Kane, 1972). Second, rabbit preimplantation embryos are able to use as energy sources a variety of long- and short-chain fatty acids that contaminate many samples of bovine serum albumin (Kane, 1979). Using a complex modification of Ham's F10 without a macromolecule, Daniel (1967) reported that pyruvate, phosphoenolpyruvate and lactate improved the development of rabbit embryos. These results conflict with those obtained by Kane (1987b), who used a basic protein-free medium, containing polyvinylalcohol as a macromolecule. He found that pyruvate and glucose can separately increase the number of cell divisions of one-cell rabbit embryos over a 48h period (Fig. 2), while phosphoenolphosphate, oxaloacetate and lactate had no significant effects. Thus there may be an advantage to including pyruvate in the medium, though the requirement is not absolute, as in the mouse.

The need for pyruvate and lactate in the culture of pig preimplantation embryos has been studied [see Davis (1985), for a review]. Davis and Day (1978) and Stone et al (1984) have reported that the incorporation of pyruvate and lactate is detrimental to development. Petters (personal communication) has not found these substrates detrimental, but has failed to show thay have any beneficial effects on the development of 1-cell and 4-cell pig embryos.

In vitro studies using chemically defined media with other species are far less complete. Pyruvate is often incorporated into media for the culture of non-human primate (Bavister et al., 1983) and human embryos (Edwards et al., 1970), but the practice does not seem to be based on any critical observations.

To summarize, the *in vitro* studies on the development of the preimplantation stages of three mammalian species (mouse, rabbit, pig) suggest the requirements for energy substrates may differ widely.

Uptake of metabolic substrates

Recently, by using very sensitive ultramicrofluorometric methods of biochemical analysis (Mroz amd Lechene, 1980), it has become possible to analyze the uptake of pyruvate and glucose by single preimplantation embryos (Leese et al., 1984; Butler et al., 1988). The results obtained by Leese and Barton (1984), using embryos flushed directly from the genital tract, are shown in Fig. 3. Prior to the 8-cell stage the embryos preferentially take up pyruvate, while older embryos take up glucose. The change-over in relative uptakes occurs roughly at the time of compaction. Similar results have been obtained using embryos produced *in vitro* (Gardner and Leese, 1986). Both of these sets of results are consistent with the earlier nutritional studies on the mouse which suggested that metabolic fluxes of preimplantation mouse embryos change with development.

Recently it has been possible to measure the uptake of pyruvate by single human oocytes and spare human embryos obtained from an IVFET Program (Leese et al., 1986). The results are shown in Fig. 4. The fresh oocytes took up an average of 36 pmol/oocyte/h, while normal, cleaving embryos took up less. The uptake was diminished still further by the blastocyst stage. These very preliminary observations suggest that the utilization of pyruvate by the human preimplantation embryo follows a developmental pattern similar to that observed in the mouse.

Metabolic Fate of Substrates

A physiological test of the utilization of glucose by the glycolytic pathway is to compare the amount of CO_2 produced from exogenous glucose and pyruvate respectively. Such comparisons were done some years ago on the mouse preimplantation embryo by Brinster (1967). The results of Brinster are summarized in Fig. 5. At the one- and two-cell stages relatively little glucose is metabolized to CO_2, whereas considerably more pyruvate gives rise to CO_2. After the 8-cell stage, however, relatively more glucose is metabolized to CO_2. In another study Brinster and Harstad (1977) showed that the preferential utilization of pyruvate over glucose occurs in primordial germ cells isolated from the germinal ridge of 15-day-old mouse fetuses. These results suggest that in the mouse relatively little glucose is metabolized through the glycolytic cycle during oogenesis, and that glycolysis is not restored until after the 8-cell stage. Similar results on the production of CO_2 from glucose have recently been reported by Wales (1986) using single mouse preimplantation embryos.

Comparable studies on the rabbit (Brinster, 1968, 1969) also show that the ability of the mature oocyte and one-cell rabbit embryo to oxidize glucose to CO_2 is very restricted, whereas pyruvate can be readily oxidized. Further studies on the oocyte of the monkey (Brinster, 1971) and cow (Rushmer and Brinster, 1973) have also shown that these cells oxidize pyruvate to carbon dioxide much more readily than glucose.

Recently Rieger and Guay (1988) have examined the metabolism of glucose and pyruvate in the seven-day-old cow blastocyst. They found that pyruvate was readily oxidized to CO_2 and that the rate of this process could be significantly stimulated by dinitrophenol, a decoupler of oxidative phosphorylation. Glucose, on the other hand, seemed to be converted in quantity to phosphoenolpyruvate. The inhibitor, however, had no stimulatory effect. These results suggest that in the cow full glycolysis is not restored until after the seventh day of development.

A Caveat - Be Alert for Artifacts

There is always the possibility that the metabolic effects that have been observed in preimplantation embryos are due to the *in vitro* conditions under which the embryos are observed. An example has recently come to light in studies on the breakdown of glycogen by mouse preimplantation embryos. The concentrations of glycogen in the mouse preimplantation embryo analyzed immediately after flushing from the female genital tract are shown in Fig. 6. There is little glycogen in the one-cell stage. At the two-cell stage considerable quantities of glycogen are synthesized, reaching a peak concentration at the 8-cell stage (Stern and Biggers, 1968; Ozias and Stern, 1973). Synthesis of glycogen also occurs in early mouse preimplantation embryos developing *in vitro* (Ozias and Stern, 1973). At the blastocyst stage the amount of glycogen diminishes *in utero*. In contrast the glycogen content remains elevated in blastocysts raised *in vitro*. Recent work has shown that the amount of glycogen present in mouse blastocysts that developed in the uterus is regulated by the rate of degradation, which is low in blastocysts produced under *in vitro* conditions (Edirisinghe and Wales, 1984). Very recently Khurana (1987) has obtained evidence that the pO^2 controls the rate of degradation of glycogen in mouse blastocysts. The results shown in Fig. 7 demonstrate that 20% oxygen almost completely inhibits glycogen breakdown and that it is still 50% inhibited at an oxygen concentration of 2.5%. It is well established that an oxygen concentration of 20% is not optimal for the culture of mouse preimplantation embryos (Whitten, 1971; Quinn and Harlow, 1978; Harlow and Quinn, 1979). There is also evidence that a low oxygen tension is beneficial for the culture of sheep and cattle preimplantation embryos (Tervit et al., 1972; Wright et al., 1976). Measurements of the oxygen tension in the uterus of the rabbit (Bishop, 1956) and the rhesus monkey (Mass et al., 1976) suggest that the oxygen concentration may be less than 5 percent, particularly if the organ is under the influence of progesterone. Thus the recent work of Khurana (1987), showing that oxygen tension affects the rate of degradation of glycogen, is possibly the first demonstration of a biochemical artefact in metabolism produced by conditions commonly used in embryo culture.

METABOLIC FATE OF GLUCOSE

An analysis of the fate of exogenously supplied glucose to an organism *in vitro* is complex. The possible pathways are shown in Fig. 8. The glucose is first transported through the plasma membrane of the cells into the glucose pool. Then it can be used through either the glycolytic pathway, the pentose phosphate pathway or the glycogen synthesis/degradation shuttle.

A simplistic interpretation of the experiments on the fate of glucose and pyruvate at different stages of preimplantation development in the mouse is that the glycolytic pathway is shut down during oogenesis while the citric acid cycle and the respiratory pathways are left intact. As development proceeds the glycolytic pathway becomes functional. Such an interpretation could be explained in terms of the absence of an enzyme in the glycolytic pathway in the early stages, which is remedied by the activation of the appropriate gene at a critical time in preimplantation development.

In a critical review of earlier studies (Biggers, 1976) it was argued that the total shutdown of glycolytic pathways in the mouse does not occur at any stage. The evidence, in fact, suggested that the glycolytic pathway in the mouse oocyte and preimplantation embryo operates at a low level throughout development, and that the citric acid cycle is also not operating to full capacity. Under such circumstances the metabolic pathways should be considered intact in the sense that all enzymes and their substrates are present. Appropriate analyses of the metabolic networks are in terms of the fluxes passing at any given time through the component pathways [see Reich and Sel'Kov (1981), for a thorough theoretical review]. Such studies have barely begun in the investigation of preimplantation development. So far two approaches have been used. One is an examination of the relative metabolic fluxes originating with glucose through the glycolytic pathway and the pentose-phosphate pathway in the rabbit. The other involves the identification of specific rate-limiting steps in a metabolic pathway using starvation and replacement feeding experimental strategies.

The first attempts to estimate the relative amounts of glucose metabolized through the glycolytic pathway and the pentose-phosphate pathway were done on the rabbit (Fridhandler, 1961; Brinster, 1968) and mouse (Brinster, 1967) by measuring the amounts of $^{14}CO_2$ produced from glucose labelled in the C-1 and C-6 positions repectively. The results suggested that during the early cleavage stages flux throught the pentose-phosphate pathway is high in the rabbit and low in the mouse. The interpretation of these experiments, however, is equivocal (Biggers and Stern, 1973). A more reliable assessment of the relative fluxes through the two pathways can be obtained only if the total glucose utilization is also known (Katz et al., 1966). Recently O'Fallon and Wright (1986) have re-examined the relative fluxes of the glycolytic pathway and the pentose-phosphate pathway in the preimplantation mouse embryo using the methods described by Katz et al. in their studies of rat adipose tissue. They showed that glucose is metabolized to CO_2 through both the glycolytic and pentose pathways from the 2-cell to the late blastocyst. Evidence was presented which suggests that the relative activities of these two pathways fluctuate during the preimplantation period, with peak relative fluxes through the pentose pathway at the 2-cell and compacted morula stages and a low relative flux at the late blastocyst. A very recent study on the relative avtivities of the glycolytic pathway and the pentose-phosphate pathway in the pig suggests that the rate of glucose utilization is low up to the time of compaction, after which it rapidly increases. Prior to compaction the glucose is used predominantly through the pentose-phosphate pathway, while after compaction it is used almost exclusively by glycolysis.

The technique of starvation and refeeding to locate rate-limiting steps in the metabolic pathways in preimplantation mouse embryos was introduced by Barbehenn et al. (1974). The principle of this method is to compare the levels of certain metabolites in the glycolytic and citric acid pathways in single embryos at different stages of development in the presence and absence of glucose or pyruvate. Fig. 9 compares the effect of starvation and refeeding glucose on the concentrations of glucose-6-phosphate and fructose-1,6-bisphosphate in the 2-cell, 8-cell, morula and blastocyst stages of mouse preimplantation development. Following starvation the concentrations of glucose-6-phosphate fell in all developmental stages. On refeeding glucose the concentration of the metabolite was restored to basal levels or above within five minutes. In contrast starvation had little effect on the concentrations of fructose-1,6-bisphosphate at all developmental stages. Refeeding glucose had no effect on the concentration of metabolite at the 2- and 8-cell stages. However, at the morula and blastocyst stages the concentration of the metabolite increased significantly. The prompt rise in the concentration of glucose-6-phosphate on refeeding glucose at the 2-cell stages suggests that neither the transport of glucose into the blastomeres nor the activity of hexokinase are rate-limiting steps that prevent glucose from supporting development of this early stage of development. The observations that the concentration of fructose-1,6-bisphosphate is unaffected by starvation or refeeding of glucose suggest that the activity of the enzyme phosphofructokinase is minimal. The fact that the concentration of the metabolite increases on refeeding glucose in the 8-cell, morula and blastocyst stages suggests that the activity of the enzyme increases after the two-cell stage of development. The fact that the response to refeeding increases in amount with development provides evidence that the activity of the enzyme is increasing. These results provide strong evidence that phosphofructokinase is a rate-limiting enzyme that controls the utilization of glucose by the preimplantation mouse embryo.

In a further study, Barbehenn et al. (1978) compared the effect of starvation and refeeding glucose and glucose plus pyruvate on seven metabolites in single 2-cell, 8-cell, morula and blastocyst stages of mouse development. The metabolites were glucose-6-phosphate, fructose-6-phosphate, fructose-1,6-bisphosphate, citrate, isocitrate, alpha oxoglutarate and malate. The results are shown in Fig. 10. Two major effects of pyruvate were observed. First, glucose caused an increase in fructose-1,6-bisphosphate above resting levels; particularly from the 8-cell stage onwards the addition of pyruvate strongly inhibited the accumulation of the metabolite. Clearly the presence of pyruvate strongly inhibits the conversion of fructose-6-phosphate to fructose-1,6-bisphosphate. Secondly, glucose had no effect on the concentrations of citrate, isocitrate, alpha-oxoglutarate and malate. When pyruvate was also present, the concentrations of citrate, isocitrate and alpha-oxoglutarate were raised at all four stages of development. The concentration of malate was not raised by the presence of pyruvate. Barbehenn et al. (1978) suggest that there is a rate-limiting step between alpha-oxoglutarate and malate which regulates fluxes through the citric acid cycle.

CONTROL OF GLYCOLYSIS

From an evolutionary point of view the glycolytic pathway is one of the ancient, conserved metabolic pathways (Boiteux and Hess, 1981; Fothergill-Gilmore, 1986), being used by both aerobic and anaerobic organisms. The pathway consists of a chain of three classes of enzyme that serve regulatory functions (Fig. 11). These classes are interconvertible enzymes, cooperative enzymes and Michaelis-Menten enzymes [see Boiteux and Hess (1981) for a review].

The two interconvertible enzymes, glycogen phosphorylase and pyruvate dehydrogenase, are placed at the beginning and end of the chain. The activities of these enzymes are controlled by chemical interconversion to different molecular species by processes which occur at relatively slow rates. The enzymes control the movement of metabolites into and out of the pathway.

Glycogen phosphorylase controls the entry of glucose-6-phosphate from glycogen stores. Pyruvate dehydrogenase controls the channeling of acetycoenzyme A to the citric acid cycle or the fatty acid cycle. Two cooperative enzymes - phosphofructokinase and pyruvate kinase - are also located at either end of the glycolytic pathway. The activites of these cooperative enzymes, in contrast to the interconvertible enzymes, are modified at fast rates through allosteric control mechanisms. They also regulate the exit of metabolites from the glycolytic pathway. Phosphofructokinase controls the transfer of metabolites from the glycolytic pathway into lipid synthesis pathways, through the production of dihydroxyacetone phosphate. Pyruvate kinase regulates the exit of substrates from the glycolytic pathway into pathways that result in the synthesis of amino acids and related compounds. The Michaelis-Menten enzymes, if they are in a simple chain, control each other sequentially through substrate concentrations (Crabtree and Newsholme, 1987). However, if they are present at an intersection with another metabolic pathway, they may also participate in a control point. As an example Boiteux and Hess (1981) use the enzyme glceraldehyde phosphate dehydrogenase, which is located at the intersection of the glycolytic and pentose-phosphate pathways. Fluxes through this control point and their direction will depend on the concentrations of two substrates and the redox potential of the cofactor nicotinamide adenine dinucleotide.

Our knowledge about the control points of the glycolytic pathway in preimplantation mammalian embryos is still sparse. Among the interconvertible enzymes there is evidence that the amount of glycogen phosphorylase is low until the morula stage in the mouse (Hsieh et al., 1979). These investigators, however, assumed that their assay method for the enzyme measures both the active a and inactive b forms of the enzyme. If by any chance the b form in the embryo resembles the b form found in the liver, the total amount of enzyme would be grossly underestimated. We do not know, therefore, whether this enzyme is controlled by being switched from an inactive to active form at the morula stage. By this stage there is also a marked drop in the concentration of glycogen synthetase (Stern, 1970), and glycogen concntrations in the embryo fall (Stern and Biggers, 1968) (Fig. 12). At present we have no information on the regulation of the other interconvertible enzyme, pyruvate dehydrogenase, in preimplantation embryos.

Most of our information on the regulation of the glycolytic pathway in the preimplantation embryo concerns the cooperative enzyme phosphofructokinase. This enzyme is controlled allosterically by several factors - concentrations of adenylates, citrate, protons and phosphate (Wu and Davis, 1981). The role of the adenylates (ATP, ADP, AMP) in the regulation of the glycolytic pathway in mouse preimplantation embryos has been discussed in terms of Atkinson's energy charge (Atkinson, 1968; Biggers, 1976; Biggers and Borland, 1976). The energy charge of the adenylate pool is defined as half of the number of anhydride-bound phosphate groups per adenine moiety, which is given by;

$$\text{Energy charge} = ([\text{ATP}] + 0.5[\text{ADP}])/([\text{ATP}] + [\text{ADP}] + [\text{AMP}])$$

If all of the adenylate pool is in the form of ATP, the energy charge equals one, and if all of the pool is in the form of AMP, the energy charge is zero. A high energy charge favors the biosynthetic pathways, while a low energy charge favors energy-generating pathways that maintain the levels of ATP. Further details on Atkinson's energy charge are given by Reich and Sel'kov (1981). It has been known since the work of Quinn and Wales (1971) that the mouse zygote has a high ATP/ADP ratio and that this ratio falls as cleavage proceeds. As a result of developing a method to estimate ATP, ADP and AMP simultaneously on single embryos, Leese et al. (1984) have been able to estimate both the ATP/ADP ratio and the energy charge at different stages of mouse preimplantation development (Table I). The results show both a high ATP/ADP ratio and a high energy charge in the 1-cell mouse embryo and that both parameters fall with development. The high energy charge at the beginning of development may allosterically inhibit phosphofructokinase and cause a restriction in the use of glucose by early mouse embryos. It is also known that citrate is present in high concentrations in preimplantation mouse embryos (Barbehenn et al., 1974). These high concentrations of citrate could also reinforce the inhibition of phosphofructokinase. However, the effects of citrate on phosphofructokinase depend on the pH and P_i concentration. Until the interactions between the effects of these factors is studied, the physiological role of phosphofructokinase in the regulation of glycolysis in preimplantation mouse embryos will be obscure. Little is known about the other cooperative enzyme in the glycolytic pathway - pyruvate kinase. However it has been suggested recently that this enzyme is rate limiting in the use of glucose by the cow blastocyst (Rieger and Guay, 1988).

CONCLUSION

The evidence assembled in this review establishes that the metabolism of the mature oocyte differs from that of adult cells. The modifications, which are presumably produced during oogenesis, are later restored by the blastocyst stage. In particular the glycolytic pathway seems affected, so that glucose utilization is restricted over a period spanning fertilization. As yet our understanding of the rate-limiting processes which are operative in the oocyte and early cleavage stages is limited and explanations are sought in terms of the well known control points. These explanations, however, may not be the final answer, since recent work has emphasized other mechanisms which may control glycolysis. These involve the role of the cellular structure of the cytoplasm, such as the cytoskeleton (Swezey and Epel, 1986; Masters et al., 1987), and metabolite-modulated dynamic enzyme associations (Ovadi, 1988). Future research should keep these alternative possibilities in mind.

REFERENCES

Anderson, E., Condon, W. and Sharp, D. 1970. A study of oogenesis and early embryogenesis in the rabbit, Oryctolagus cuniculus, with special reference to the structural changes of mitochondria. J. Morph. 130: 67-92.

Atkinson, D.E. 1968. The energy charge of the adenylate pool as a regulatory parameter. Interaction with feedback modifiers. Biochem. 7:4030-4034.

Barbehenn, E.K., Wales, R.G. and Lowry, O.H. 1974. The explanation for the blockade of glycolysis in early mouse embryos. Proc. Nat. Acad. Sci. U.S.A. 71: 1056-1060.

Barbehenn, E.K., Wales, R.G. and Lowry, O.H. 1978. Measurement of metabolites in single preimplantation embryos: a new means to study metabolic control in early embryos. J. Embryol. exp. Morph. 43: 29-46.

Bavister, B.D., Boatman, D.E., Leibfried, M.L., Loose, M. and Vernon, M.W. 1983. Fertilization and cleavage of rhesus monkey oocytes in vitro. Biol. Reprod. 28: 983-999.

Biggers, J.D. 1976. Bioenergetic aspects of fertilization and embryonic development of the mouse. In: Ebling, F.J.G. and Henderson, I.W. eds. Biological and Clinical Aspects of Reproduction. Excerpta-Medica, Amsterdam-Oxford, pp 128-133.

Biggers, J.D. 1987. Pioneering mammalian embryo culture. In: The Mammalian Preimplantation Embryo. ed. B.D. Bavister, pp.1-22. Plenum Press, New York.

Biggers, J.D. and Borland, R.M. 1976. Physiological spects of growth and development of the preimplantation mammalian embryo. Ann. Rev. Physiol. 38: 95-119.

Biggers, J.D. and Stern, S. 1973. Metabolism of the preimplantation embryo. Adv. Reproductive Physiology 6: 1-59.

Biggers, J.D., Gardner, D.K., and Leese, H.J. 1989. Control of carbohydrate metabolism in preimplantation mammalian embryos. In: Regulation of Growth in Development, eds. I.Y.Rosenblum and S. Heyner, CRC Press, Boca Raton, FL (in press).

Biggers, J.D., Whittingham, D.G. and Donahue, R.P. 1967. The pattern of energy metabolism in the mouse oocyte and zygote. Proc. Nat. Acad. Sci., USA, 58: 560-567.

Bishop, D.W. 1956. Oxygen concentrations in the rabbit genital tract. Proc. 3rd. Int. Congress Animal Reprod. Cambridge. Section 1, 53-55.

Boiteux, A. and Hess, B. 1981. Design of glycolysis. Phil. Trans. R. Soc. Lond. B 293: 5-22.

Brinster, R.L. 1965a. Studies on the development of mouse embryos *in vitro*. II. The effect of energy source. J. Exp. Zool. 158: 59-

Brinster, R.L. 1965b. Studies on the development of mouse embryos *in vitro*. IV. Interaction of energy sources. J. Reprod. Fertil., 10: 227-240.

Brinster, R.L. 1967. Carbon dioxide production from glucose by the preimplantation mouse embryo. Exp. Cell Res. 47: 271-277.

Brinster, R.L. 1968. Carbon dioxide production from glucose by the preimplantation rabbit embryo. Exp. Cell Res. 51: 330-334.

Brinster, R.L. 1969. Radioactive carbon dioxide production from pyruvate and lactate by the preimplantation rabbit embryo. Exp. Cell Res. 54: 205-209.

Brinster, R.L. 1971. Oxidation of pyruvate and glucose by oocytes of the mouse and Rhesus monkey. J. Reprod. Fert. 24:187-191.

Brinster, R.L. and Harstad, H. 1977. Energy metabolism in primordial germ cells of the mouse. Exp. Cell Res. 109: 111-117.

Brinster, R.L. and Thomson, J.L. 1966. Development of eight-cell mouse embryos *in vitro*. Exp. Cell Res. 42: 308-

Butler, J.E., Lechene, C., and Biggers, J.D. 1988. Noninvasive measurement of glucose uptake by two populations of murine embryos. Biol. Reprod. 39:779-786.

Calarco, P.G. and Brown, E.H. 1969. An ultrastructural and cytological study of preimplantation development of the mouse. J. exp. Zool., 171: 253-283.

Crabtree, B. and Newsholme, E.A. 1987. A systematic approach to describing and analysing metabolic control systems. TIBS 12, 6-12.

Daniel, J.C. Jr. 1967. The pattern of utilization of respiratory metabolic intermediates by preimplantation mammalian rabbit embryos *in vitro*. Exp. Cell Res. 47: 619-624.

Davis, D.L. 1985. Culture and storage of pig embryos. J. Reprod. Fert., Suppl. 33: 115-124.

Davis, D.L. and Day, B.N. 1978. Cleavage and blastocyst formation by pig eggs *in vitro*. J. Anim. Sci., 46: 1043-1053.

Dvorak, M. and Tesarik, J. 1985. Differentiation of mitochondria in the human preimplantation embryo grown *in vitro*. Scripta Medica 58: 161-170.

Edirisinghe, W.R. and Wales, R.G. 1984. Effect of parental administration of oestrogen and progesterone on the glycogen metabolism of mouse morulae-early blastocysts *in vivo*. J. Reprod. Fert. 72: 67-73.

Edwards, R.G., Steptoe, P.C. and Purdy, J.M. 1970. Fertilization and cleavage *in vitro* of preovulator human oocytes. Nature 227:1307-1309.

Flood, M.R. and Wiebold, J.L. 1988. Glucose metabolism by preimplantation pig embryos. J. Reprod. Fert. 84: 7-12.

Fothergill-Gilmore, L.A. 1986. The evolution of the glycolytic pathway. TIBS 11: 47-51.

Fridhandler, L. 1961. Pathways of glucose metabolism in fertilized rabbit ova at various pre-implantation stages. Exp. Cell Res. 22: 303-316.

Fridhandler, L., Hafez, E.S.E. and Pincus, G. 1957. Developmental changes in the respiratory activity of rabbit ova. Exp. Cell Res. 13: 132-139.

Gardner, D.K and Leese, H.J. 1986. Non-invasive measurement of nutrient uptake by single cultured pre-implantation mouse embryos. Human Reproduction 1: 25-27.

Hackenbrock, C.R. 1968. Ultrastructural bases for metabolically linked mechanical activity in mitochondria. II. Electron transport-linked ultrastructural transformations in mitochondria. J. Cell Biol. 37: 345-369.

Harlow, G.M. and Quinn, P. 1979. Foetala and placental growth in the mouse after preimplantation development *in vitro* under oxygen concentrations of 5 and 20 percent. Aust. J. Biol. Sci., 32: 363-369.

Hillman, N. and Tasca, R.J. 1969. Ultrastructural and autoradiographic studies of mouse cleavage stages. Amer. J. Anat. 126:151-174.

Hsieh, B., Chi, M.M.-Y., Knor, J. and Lowry, O.H. 1979. Enzymes of glycogen metabolism and related metabolites in preimplantation mouse embryos. Devel. Biol. 72: 342-349.

Kane, M.T. 1972. Energy substrates and culture of single cell rabbit ova to blastocysts. Nature 238: 468-469.

Kane, M.T. 1979. Fatty acids as energy sources for culture of one-cell rabbit ova to viable morulae. Biol. Reprod. 20: 323-332.

Kane, M.T. 1987a. *In Vitro* growth of preimplantation rabbit embryos. In: The Mammalian Preimplantation Embryo. Ed. B.D. Bavister, pp. 193-217. Plenum Press, New York.

Kane, M.T. 1987b. Minimal nutrient requirements for culture of one-cell rabbit embryos. Biol. Reprod. 37:775-778.

Katz, J., Landau, B.R. and Bartsch, G.E. 1966. The pentose cycle, triose phosphate isomerization, and lipogenesis in rat adipose tissue. J. Biol. Chem. 241: 727-740.

Kaye, P.L. 1986. Metabolic aspects of the physiology of the preimplantation embryo. In Rossant, J. and Pedersen, R.A., eds. Experimental approaches to mammalian embryonic development. Cambridge University Press, Cambridge, pp. 267-292.

Khurana, N.K. 1987. Interactions between the embryo and its environment during the early stages of pregnancy. Ph.D. Thesis, Murdoch University, Western Australia.

Kramen, M.A. and Biggers, J.D. 1971. Uptake of tricarboxylic acid cycle intermediates by preimplantation mouse embryos *in vitro*. Proc. Nat. Acad. Sci.,U.S.A. 68: 2656-2659.

Leese, H.J. and Barton, A.M. 1984. Pyruvate and glucose uptake by mouse ova and preimplantation embryos. J. Reprod. Fert. 72: 9-13.

Leese, H.J., Biggers, J.D., Mroz, E.A., and Lechene, C. 1984. Nucleotides in a single mammalian ovum or preimplantation embryo. Anal. Biochem. 140:443-448.

Leese, H.J., Hooper, M.A.K., Edwards, R.G. and Ashwood-Smith, M.J. 1986. Uptake of pyruvate by early human embryos determined by a non-invasive technique. Human Reproduction 1: 181-182.

Mass, D.H.A., Storey, B.T. and Mastroianni, L. Jr. 1976. Oxygen tension in the oviduct of the rhesus monkey *Macaca mulatta*.. Fert. Steril., 27: 1312-1317.

Masters, C.J., Reid, S. and Don, M. 1987. Glycolysis - new concepts in an old pathway. Molecular and Cellular Biochem. 76: 3-14.

Mazanec, K. and Dvorak, M. 1964. On the submicroscopical changes of the segmenting ovum in the albino rat. Morfologie 2: 103-108.

Mills, R.M. and Brinster, R.L. 1967. Oxygen consumption of preimplantation mouse embryos. Exp. Cell Res. 47: 337-344.

Mroz, E.A., and Lechene, C. 1980. Fluorescence analysis of picolitre samples. Anal. Biochem. 102:90-96.

O'Fallon, J.V. and Wright, R.W. 1986. Quantitative determination of the pentose phosphate pathway in preimplantation mouse embryos. Biol. Reprod. 34: 58-64.

Ovadi, J. 1988. Old pathway - new concept: control of glycolysis by metabolite-modulated dynamic enzyme associations. TIBS 13: 486-490.

Ozias, C.B. and Stern, S. 1973. Glycogen levels of preimplantation mouse embryos developing in vitro. Biol. Reprod. 8: 467-472.

Pike, I.L. 1981. Comparative studies of embryo metabolism in early pregnancy. J. Reprod. Fert., Suppl. 29: 203-213.

Quinn, P. and Harlow, G.M. 1978. The effect of oxygen on the development of preimplantation mouse embryos in vitro. J. exp. Zool. 206: 73-80.

Quinn, P. and Wales, R.G. 1971. Adenosine triphosphate content of preimplantation mouse embryos. J. Reprod. Fert. 25: 133-135.

Reich, J.G. and Sel'Kov, E.E. 1981. Energy Metabolism of the Cell. Academic Press, New York.

Reinius, S. 1969. Morphology of oviduct, gametes and zygotes as a basis of oviductal function in the mouse. RK-tryck, Uppsala.

Rieger, D. and Guay, P. 1988. Measurement of the metabolism of energy substrates in indiviual bovine blastocysts. J. Reprod. Fert. 83: 585-591.

Rushmer, R.A. and Brinster, R.L. 1973. Carbon dioxide production from pyruvate and glucose by bovine oocytes. Exp. Cell Res. 82: 252-254.

Sackler, M.L. 1968. Morphological changes in mitochondria and associated organelles during development of the preimplantation mouse embryo. J. Cell. Biol. 39: 117a.

Stern, S. 1970. The activity of glycogen synthetase in the cleaving mouse embryo. Proc. Soc. Stud. Reprod., Columbus, OH, Third Annual Meeting: p.3 (Abstr.).

Stern, S. and Biggers, J.D. 1968. Enzymatic estimation of glycogen in the cleaving mouse embryo. J. exp. Zool. 168: 61-66.

Stern, S., Biggers, J.D. and Anderson, E. 1971. Mitochondria and early development of the mouse. J. exp. Zool. 176: 179-192.

Stone, B.A., Quinn, P. and Seamark, R.F. 1984. Energy and protein sources for development of pig embryos cultured beyond hatching in vitro. Anim. Reprod. Sci. 7:405.

Swezey, R.R. and Epel, D. 1986. Regulation of glucose-6-phosphate dehydrogenase in sea urchin eggs by reversible association with cell structural elements. J. Cell Biol. 103: 1509-1515.

Tervit, H.R., Whittingham, D.G. and Rowson, L.E.A. 1972. Successful culture in vitro of sheep and cattle ova. J. Reprod. Fert. 30: 493-497.

Wales, R.G. 1975. Maturation of the mammalian embryo: biochemical aspects. Biol. Reprod. 12:66-81.

Wales, R.G. 1986. Measurement of metabolic turnover in single mouse embryos. J. Reprod. Fert. 76: 717-725.

Weitlauf, H.M. and Nieder, G.L. 1984. Metabolism in preimplantation embryos. J. Biosci. 6, Suppl. 2, 33-42.

Whitten, W.K. 1956. Culture of tubal mouse ova. Nature 177: 96-

Whitten, W.K. 1957. Culture of tubal ova. Nature 179: 1081-1082.

Whitten, W.K. 1971. Nutrient requirements for the culture of preimplantation embryos *in vitro*. In: Advances in the Biosciences Ed. G. Raspe. Vol.6 pp. 129-141. Pergamon Press, Oxford.

Wright, R.W. Jr., Anderson, G.B., Cupps, P.T. and Drost, M. 1976. Successful culture *in vitro* of bovine embryos to the blastocyst stage. Biol. Reprod. 14: 157-162.

Wu, T.L. and Davis, E.J. 1981. Regulation of glycolytic flux in an energetically controlled cell-free system: the effects of adenine nucleotide ratios, inorganic phosphate, pH and citrate. Arch. Biochem. Biophys. 209: 85-99.

Yamada, E., Muta, A., Motomura, A. and Koga, H. 1957. The fine structure of the oocyte in the mouse ovary studied with electron microscope. Kurume Med. J. 4: 148-171.

REGULATIVE POTENTIAL OF MICROMANIPULATED EMBRYOS

V.E. Papaioannou

Introduction

Uses of Micromanipulation

Spurred on by the promise of biotechnology for improvements in health and reproductive performance in humans and animals, scientists from diverse fields have recently been applying manipulative experimental techniques to the embryos of a wide range of species. At both molecular and cellular levels, material is being added, subtracted, or altered for the accomplishment of purposes as diverse as studying embryonic induction and producing genetically superior agricultural animals. The growth of molecular biology has added a new dimension to the venerable practice of embryo manipulation, making possible the introduction of specific genetic material or the perturbation of endogenous gene activity in the developing embryo. The power of this approach will be realized only in combination with a consideration of developmental processes on the level of the organism as a whole.

The manipulations to be described in this paper are of a cellular rather than a nuclear or genetic variety. Although the object of the experiments was not always primarily embryonic regulation, together the studies provide a compendium of information on the capacity of the organism to compensate for cellular perturbations during preimplantation stages of development. The emphasis will be on the upward regulation of cell number, in other words, the ability of an embryo to compensate for a decrease in cell number at different embryonic stages.

Types of Manipulation

The indeterminate cleavage of mammals means that cell fates are not precisely fixed from the outset but rather that cells retain a high degree of developmental plasticity or regulative capacity throughout the early stages of embryogenesis. Experimental evidence supports the contention that developmental potential of cells is gradually narrowed as development progresses and cells become established in particular pathways. When the direction of differentiation of a cell or region of the embryo has become irrevocably established, its fate is the same as its potential and it can be said to be committed or determined. The mechanism of this cellular determination remains elusive.

Mechanical manipulation is a cornerstone of experimental embryology. In addition to the investigation of the fate and potential of individual cells, the organization of the embryo as a whole can also be considered using micromanipulation techniques. The interactions between cells or groups of cells, the influence of parts on the whole, and the role of morphogenetic movements in embryogenesis can all be approached using cellular micromanipulation. Isolation of parts of an embryo to examine their properties separated from the rest of the embryo either in culture, in ectopic sites or in the uterus is one possible procedure. This method tests the potential of a cell or tissue for independent development by altering cell and tissue interactions and, in explants, by removing the mediating influences of the maternal environment.

Alternatively, part of an embryo can be destroyed or removed and the embryo grown *in vitro* or in the uterus. The further development of the embryo will provide evidence as to the importance of the missing part to the development of the whole and the capacity of the embryo to compensate for damage. Cells or tissues may also be recombined by transplantation or exchanged between different embryos. In this type of experiment, genetic markers allow for the identification of the contribution of each component in the composite, chimeric embryo. Finally, cells or parts may be added to an otherwise complete embryo, and in an extreme form of this procedure, whole embryos can be aggregated together. The successful development of chimeric animals following the aggregation of two or more complete embryos is a testament to the regulative capacity of the early mammalian embryo.

Regulation in Mammalian Embryos

Experimental embryologists have made good use of the innate capacity of the mammalian embryo to adjust its developmental program to compensate for perturbations. From the embryo's point of view, this feature is presumably advantageous, providing a measure of flexibility to overcome and compensate for adverse conditions that might occur during gestation. In an animal as complex as the mammal, with a dynamic maternal-embryonic and maternal-fetal interaction during gestation, this degree of flexibility provides some insurance for the considerable reproductive investment of each pregnancy.

As a practical consideration, particularly in the manipulation of human embryos, this regulative ability allows some leeway in the various techniques associated with artificially assisted conception. The fact that the embryo will be able to tolerate less than ideal conditions and still develop into a normal individual broadens the range of manipulations that can ethically be applied in overcoming infertility in humans. In agricultural animals it will define the limits of economically viable procedures. Thus, it is all the more important to determine the limits of the regulative capacity at all stages of early embryogenesis and to determine differences between different species. In doing so we will be learning about the basic processes of development as well as establishing a firm and reasonable basis for intervention in the reproductive process.

Use of Micromanipulation in Studying Regulation in Laboratory and Domestic Species

Preimplantation Development

The mammalian embryo travels down the oviduct as it cleaves and usually enters the uterus toward the end of this process, although the stage at which it does so varies among mammals (Table 1). Late in cleavage, cells of the embryo which up to this time have been similar, loosely associated spheres, begin to flatten onto one another and compact as a prelude to formation of the blastocyst. With blastocyst formation comes the first overt sign of morphological differentiation of two cell types, the inner cell mass (ICM) and the outer layer of trophectoderm form as the result of a divergence in gene expression in the two cell populations. This first differentiation into distinct cell types also heralds the earliest commitment of cells to specific cell lineages and results from the restriction in potential of the previously totipotent or at least multipotent blastomeres. Following hatching from the zona pellucida, the blastocysts in some species, such as the horse, mouse and human, remain spherical, whereas in others, such as pig and cow, it expands and elongates considerably before implantation. The time and stage at implantation varies in different species, as do maternal-fetal interactions resulting in distinctively different placentation types.

These early events of preimplantation development have been well studied in the mouse from both morphological and experimental perspectives and a reasonably full picture has been drawn. But for other species of mammals such information is still sketchy. Morphologically, the embryos of most eutherian mammals resemble the mouse, although the timing of events varies (Table 1). It is becoming evident from experimental work that differences in timing are not the only distinctions and that embryos of different species with the same cell numbers or at the same morphological stage may be fundamentally different in the degree of determination that has taken place and the regulative potential of blastomeres.

Regulation and Cell Potential During Cleavage and Blastulation

The most stringent test of cell potential is whether an isolated embryonic cell is totipotent, that is, whether it is capable of regulating its development such that it can compensate for an overall reduction in embryonic cell number and produce a viable offspring. It must be determined if any or all of the cells of an early embryo have this capacity in order to fully define regulation. For example, development of one blastomere from a 2-cell embryo is an indication of individual blastomere totipotency, but both blastomeres must be shown to develop normally before totipotency of 2-cell blastomeres in general can be claimed. Likewise, the production of even a single pair of identical twins from bisection of an embryo is proof that two halves can be totipotent, but the plane of bisection with respect to any asymmetry in the embryo must be considered. For the polarized blastocyst, it is evident that only bisection of the ICM, giving balanced half-embryos, can result in successful twinning (Gardner 1972; 1974) and so totipotency of half-blastocysts in general cannot be claimed.

1) Laboratory species: In the mouse, the potential of single isolated blastomeres to form embryos appears to be restricted relatively early (Table 2). Single blastomeres from the 2-cell stage, but not later stages, are capable of developing into complete, viable offspring (Tsunoda & McLaren 1983). Although individual blastomeres from embryos up to at least the 8-cell stage appear to be totipotent in terms of the tissues they can produce in a chimera (Kelly 1975), they are not capable of organizing into viable fetuses in isolation (Rossant 1976). In the rat, a complete egg cylinder was reported to have developed from one blastomere of the 2-cell stage, although development stopped shortly thereafter (Nicholas & Hall 1942). In more recent experiments, identical twin offspring have been produced from 2-cell blastomeres of the rat (Matsumoto et al, 1989). Single blastomeres from 4- and even 8-cell stages of the rabbit have shown totipotency (Seidel 1952, 1956; Moore, Adams & Rowson 1968) indicating a greater regulative capacity of cleaving blastomeres in this species than in the mouse.

The survival of half mouse embryos can be as high as 65 percent for the 2-cell stage (Tsunoda & McLaren 1983), and over 45 percent for the 4-cell (Rossant 1976), morula (Tsunoda & McLaren 1983; Nagashima et al. 1984), and blastocyst (Gardner 1974). However, survival of both halves of the same embryo has been demonstrated only for the 2-cell stage and the 8- to 16-cell stage (Mousatafa & Hahn 1978; Gartner & Braunack 1981; Tsunoda & McLaren 1983; Nagashima et al. 1984; Table 2). A recent study of late morula/early blastocyst stages of rabbit embryos indicates that half embryo survival and the survival of both halves is possible in this species as well (Yang & Foote 1987).

2) Domestic species: Work in domestic species has been less systematic but there has been considerable success in the production of identical twins of several species. Totipotency of isolated, individual sheep blastomeres is maintained up to the 8-cell stage (Trounson & Moore 1974; Willadsen 1981), but no more than three of the blastomeres from a 4-cell embryo and one of the blastomeres from an 8-cell embryo have yet been proved totipotent. This may be a technical failure but it could also represent a biological restriction in the potential of some of the blastomeres by these stages and indicate that the morphologically similar blastomeres have already undergone restriction in potential. In another experiment, 4 identical quadruplets were born from a single sheep embryo that had been quartered at the 8-cell stage (Willadsen 1981). If these quarters consisted of daughter-cell pairs from the 4-cell stage, as seem likely from the procedure used (Willadsen 1980), then this result might argue that all 4 blastomeres at the 4-cell stage do indeed retain totipotency. A high rate of survival of half embryos to term has been demonstrated for the 2-cell stage through the blastocyst stage in sheep, and both halves of all of these stages have produced identical twin offspring (Willadsen 1979, 1980; Gatica et al., 1984).

The regulative potential of single blastomeres of cow embryos has not been determined, but quarter embryos from the 8-cell stage have produced a set of triplets (Willadsen & Polge 1981) and from the 32-64-cell morula stage, 4 sets of twins (Willadsen et al. 1981), indicating that at least some subsets of blastomeres at these stages are totipotent. Bisection of cow embryos has resulted in successful production of identical twins at the early morula stage through the blastocyst stage (Ozil et al 1982; Lambeth et al. 1983; Ozil 1983; Williams et al. 1984) and in high rates of half embryo survival. Reported losses of embryos late in gestation, however, clearly indicate that not all embryos that implant are capable of complete development. A study of the causes of fetal mortality in these cases might well provide insight into possible restrictions of potential or regulative capacity.

Little work has been published on manipulation of other domestic species. In the horse, 3 offspring have been reported from bisected 2-8-cell stages, although no twin pairs, and 4 offspring comprising 2 sets of twins have been obtained from isolated 4-cell blastomeres (Allen 1982; Allen & Pashen 1984). Thus, at least two of the blastomeres at the 4-cell stage are totipotent. In the pig, single blastomeres of 2-8-cell stages have been isolated and grown in vitro to the blastocyst stage (Moore et al. 1969; Menino & Wright 1983). Recently, several abstracts have appeared indicating that up to 50% of half embryos at the morula to blastocyst stage can regulate and develop into offspring but as yet there is no proof that two halves of the same embryo have this capacity (Rorie et al 1985; Nagashima et al 1987). In the goat, there is a report that twins can develop from halved blastocysts (Iritani, 1987).

3) Summary: From this somewhat sketchy picture of domestic and laboratory species, it appears that the capacity for independent development of a single blastomere is retained to a later stage in cleavage in rabbit, sheep and horse embryos than in the most frequently used model, the mouse. Survival of half embryos may be similar or better in sheep and cows than in mice, and the survival of both halves or even quarters to produce identical offspring has been successful in domestic species with significantly fewer attempts, indicating a greater regulative capacity of partial embryos in these species compared with the mouse. The occurrence of blastulation at a later cleavage division and thus a higher total cell number in some species, including humans, may allow greater flexibility in embryonic organization. If there is a critical number of ICM cells below which normal development of the fetus will not occur, embryos that blastulate with a higher total cell number and thus a higher proportion of ICM cells (Buehr & McLaren 1974) may be able to tolerate a proportionately greater decrease in cell number. However, the relationship between cell number and embryonic viability needs to be determined to test this idea, and other factors such as later determination of specific cell lineages may also play an important role in species differences.

The available information from these and other types of manipulation studies indicates that mammalian embryos share with the mouse a considerable flexibility in regulative development and may also share mechanisms of determination. It is important, however to critically evaluate each species to avoid unwarranted extrapolation between species. One need only remember some exceptional animals such as the nine-banded armadillo, which regularly produces monozygous quadruplets, to realize that smaller but significant variations among more closely related species might well exist.

How is Regulation Accomplished?

1) Regulation in the early embryo: From the variation seen in the experiments described above, it appears that regulation does not always occur even in situations where the embryo or blastomere has the potential to reprogram its development. Technical factors affecting regulation, how completely, how quickly and through what mechanisms it is accomplished are questions that must be approached experimentally. Does effective regulation involve only cell number or is embryo volume important? Is it necessary that all tissues be regulated to the same degree? These are questions that bear directly on manipulations that are applied to human embryos and thus are a matter for urgent consideration. We have recently been studying these questions using the mouse as a model and the following is a summary of unpublished work done in collaboration with Dr. K.M. Ebert and Dr. J.D. Biggers. We have examined the response of the mouse embryo to a 50% reduction by destroying one blastomere at the 2-cell stage and then following the development of the resulting half embryos.

Tarkowski (1959a, 1959b) found that half embryos surviving until the 12th day of gestation had reached the same stage of development as control embryos and that the birth weight of live young derived from half embryos was not significantly different. Other studies have provided evidence that mouse half embryos adjust to approximately normal size by the 7th day of gestation (Lewis & Rossant 1982). In contrast, 18 day fetuses derived from half embryos have been shown to weigh significantly less than those from controls (Tsunoda & McLaren 1983). Another study found that half embryos produced by destruction of one blastomere at the 2-cell stage regulate their size between 7.5 and 10.5 days of development although the half embryos may become smaller than controls again by 13.5 days (Rands 1986).

Our approach has been to examine the immediate results of halving 2-cell embryos on the formation of the blastocyst and to correlate this with developmental potential. We have examined the effects of culture *in vitro* versus return of the embryos to the reproductive tract, and also examined a surrogate *in vivo* environment, the immature mouse oviduct, for its ability to support development of half embryos. Embryos were first recovered at the 2-cell stage, 1 blastomere was destroyed by lysis, they were then either cultured *in vitro* or transferred to pseudopregnant or immature oviducts for 2 days. Following recovery, they were assessed for morphology, and cell number in the ICM and trophectoderm, and some were retransferred to one day asynchronous foster mothers to determine their further developmental potential.

a) Morphology. *In vitro* culture results in a delay in morphogenetic development, but in our study half embryos were at similar morphological stages as intact controls, indicating that morphogenesis of the blastocyst does not depend on cell number (Smith & McLaren, 1977; Fernandez & Izquierdo, 1980). There was evidence of a much higher percentage of normal development *in vivo* than *in vitro*, although it is possible that retarded and degenerating embryos were more difficult to recover.

b) Cell Number and Viability. Within each culture condition, cell number in half embryos was approximately half that of controls as expected from other studies indicating that upward cell number regulation does not occur until after implantation (Rands 1986). Our results indicate that this relationship holds under a variety of culture conditions. When the embryos were transferred to foster mothers as the most stringent measure of viability, it was found that cell number of the preimplantation embryos correlated with viability and that half embryos from all culture conditions had lower viability than controls (Figure 1).

This does not present the whole picture, however. Figure 2 provides a different representation which subdivides the reproductive loss into implantation failure as distinct from failure to develop after implantation. Among the intact control embryos, the loss in viability, which is correlated with a lower cell number, can be seen to affect both of these parameters. The half embryos, with their correspondingly lower cell numbers and viability, however, present a different picture. The proportion of failure due to postimplantation losses, i.e. failure to develop normally after evocation of a decidual response, increases dramatically, indicating development of the trophectoderm at the expense of the inner cell mass. Since blastulation occurs at a particular time regardless of the number of cells, and since the proportion of inner cells is predicted to be lower in embryos of low cell number at the time of blastulation, this result implicates ICM ratio at that time as being a causative factor in the lowered viability of these embryos.

c) ICM Ratio. The proportion of ICM cells to total cells was determined for embryos of each group using a differential dye method to distinguish between ICM and trophectoderm (Handyside & Hunter 1984). A preliminary comparison between half and control embryos for a given culture condition fits with the prediction that embryos with lower cell number at the time of blastulation will have a lower ICM cell ratio and also with the prediction that a low ICM ratio will decrease viability. The results are incomplete as yet but the indications are that *in vivo* culture, even in a nonpregnant uterus, promotes the formation of a relatively large ICM and that the opposite is true of *in vitro* culture (V.E. Papaioannou and K.M. Ebert, unpublished data).

2) Effect of Regulation on Later Development: In a separate study, half embryos were produced in the same way but were immediately transferred to foster mothers and allowed to complete their development to

term. The object of this study was to assess phenotypic variation among mice produced from half embryos and to compare them with genetically identical mice developing from intact embryos. In other words, half embryos were examined for evidence of a greater phenotypic variation that might indicate incomplete of inconsistent regulation. In addition to intact control embryos that were recovered from the mothers and retransferred (transferred controls), an additional control was to compare mice developing from undisturbed pregnancies (controls).

a) Reproductive Success. Preimplantation, fetal and neonatal losses in the different groups of animals are indicated in Figure 3. The lower success of pregnancy among recipients that received half embryos compared with control embryos argues in favor of a failure that is embryo-related and is consistent with the results of the previous study. Implantation failure and failure during gestation could not be distinguished in this study since implantation sites were not counted, but during gestation, half embryos were again less successful than controls, and this indicates a limitation to regulation.

An interesting finding was a higher neonatal mortality in mice that developed from half embryos. It is known that neonatal loss is associated with an increase in the duration of pregnancy and that the duration of pregnancy is increased by small litter size (McLaren 1970). In our study, embryo transfer, regardless of the embryos transferred, resulted in small litters and the duration of pregnancy was increased in inverse proportion to the litter size. Thus the increased neonatal loss can be explained as an indirect reproductive effect and not as a reflection of a decreased survival potential of mice developing from half embryos.

b) Variation After Birth. A detailed analysis of phenotypic variation among the surviving mice developing from half embryos is underway (V.E. Papaioannou, J. Mkandawire & J.D. Biggers, unpublished data). Preliminary results have shown little variation and no differences that could be attributed to incomplete regulation. The sex ratio among controls, transferred controls and half embryo mice was not significantly different. Maturation events such as age at eye opening and age at vaginal opening were similar for all three groups, as were growth curves for body weight and tail length.

Although a number of measures have yet to be analyzed, it appears that complete regulation has taken place in all aspects of the animal by the time of birth in mice developing from embryos halved at the 2-cell stage. The considerable variation seen among developing half embryos during preimplantation and postimplantation development, however, is an indication that there are critical points or crises in development that serve to eliminate embryos that have not regulated. Implantation is certainly one of these critical points since a failure here dooms the embryo. The narrow window of mutual compatibility between mother and embryo for successful implantation is a stringent requirement for a particular level of embryonic development so that development of function must have taken place by this time even if cell number is not regulated. The timing of blastocyst formation is an indication that this occurs although preimplantation losses associated with half embryos in our study indicate that sometimes the half embryo is still incapable of eliciting a decidual response.

If implantation is successfully initiated, the relationship between the number of cells in the ICM and in the trophectoderm of the blastocyst may present another critical developmental point. Although it is not known whether there is a minimum necessary number of ICM cells, it is known that these cells stimulate proliferation of the trophectoderm (Gardner et al 1973) and that the entire fetus as well as some extraembryonic membranes are derived from the ICM (Papaioannou 1982). A low ICM cell number at the time of blastocyst formation could create problems at specific stages (a threshold effect) or put embryos at a disadvantage throughout development. More experimental work will need to be done to further illuminate the mechanisms of regulation, not only the regulation of cell number, but also regulation in the proportion of differentiated cell types and regulation of function. Study of the forces acting on embryos to eliminate defective embryos or those that have not successfully regulated will provide insight into possible improvements in techniques for handling embryos of any species.

REFERENCES

Allen, W.R. 1982. Embryo transfer in the horse. In: "Mammalian Egg Transfer", C.E. Adams, Ed., CRC Press, Inc., Boca Raton, Florida, pp 135-154.

Allen, W.R., and R.L. Pashen. 1984. Production of monozygotic (identical) horse twins by embryo micromanipulation. J. Reprod. Fert. 71:607-613.

Brackett, B.G. 1978. Experimentation involving primate embryos. In: "Methods in Mammalian Reproduction", J.C. Daniel, Jr., Ed., Academic Press, Inc., New York, London, S.F., pp. 333-357.

Buehr, M., and A. McLaren. 1974. Size regulation in chimaeric mouse embryos. J. Embryol. exp. Morph. 31:229-234.

Davis, J., and H. Hesseldahl. 1971. Comparative embryology of mammalian blastocysts. In: "The Biology of the Blastocyst", R.J. Blandau, Ed., University of Chicago Press, pp. 27-48.

Enders, A.C., and S. Schlafke. 1981. Differentiation of the blastocyst of the Rhesus Monkey. Am. J. Anat. 162:1-21.

Fernandez, M.S., and L. Izquierdo. 1980. Blastocoel formation in half and double mouse embryos. Anat. Embryol. 160:77-81.

Gardner, R.L. 1972. An investigation of inner cell mass and trophoblast tissues following their isolation from the mouse blastocyst. J. Embryol. exp. Morph. 28:279-312.

Gardner, R.L. 1974. Microsurgical approaches to the study of early mammalian development. In: "Birth Defects and Fetal Development, Endocrine and Metabolic Factors" K.S. Moghissi, Ed., Thomas, Illinois, pp. 212-233.

Gardner, R.L., V.E. Papaioannou, and S.C. Barton. 1973. Origin of the ectoplacental cone and secondary giant cells in mouse blastocysts reconstituted from isolated trophoblast and inner cell mass. J. Embryol. exp. Morph. 30:561-572.

Gartner, K., and E. Baunack. 1981. Is the similarity of monozygotic twins due to genetic factors alone? Nature 292:646-647.

Gatica, R., M.P. Boland, T.F. Crosby, and I. Gordon. 1984. Micromanipulation of sheep morulae to produce monozygotic twins. Theriogenology 21:555-560.

Hamilton, W.J., and F.T. Day. 1945. Cleavage stages of the ova of the horse, with notes on ovulation. J. Anat. 79:127-130.

Handyside, A.H., and S. Hunter. 1984. A rapid procedure for visualising the inner cell mass and trophectoderm nuclei of mouse blastocysts *in situ* using polynucleotide-specific fluorochromes. J. Exp. Zool. 231:429-434.

Heuser, C.H., and G.L. Streeter. 1941. Development of the Macaque Embryo. In: "Contributions to Embryo, No. 181, vol 29", Carnegie Inst. of Wash., Publication 525, Wash. DC, pp. 15-55.

Iritani, A., 1987. A recent advance in embryology and embryo manipulation and their potential contribution to agriculture. Proc. IVth AAAP Animal Science Congress, Hamilton, New Zealand, pp. 126-129

Kelly, S.J. 1975. Studies of the potency of the early cleavage blastomeres of the mouse. In: "The early development of mammals", British Society for Developmental Biology Symposium 2, M. Balls, and A.E. Wild, Eds., Cambridge University Press, pp. 97-105.

King, G.J., B.A. Atkinson, and H.A. Robertson. 1982. Implantation and early placentation in domestic ungulates. J. Reprod. Fert., Suppl. 31:17-30.

Lambeth, V.A., C.R. Looney, S.A. Voelkel, D.A. Jackson, K.G. Hill, and R.A. Godke. 1983. Microsurgery on bovine embryos at the morula stage to produce monozygotic twin calves. Theriogenology 20:85-95.

Lewis, N.E., and J. Rossant. 1982. Mechanism of size regulation in mouse embryo aggregates. J. Embryol. exp. Morph. 72:169-181.

Lewis, W.H., and C.G. Hartman. 1941. Tubal ova of the rhesus monkey. In "Contributions to Embryology, No. 180 Vol XXIX" Carnegie Institution of Washington Publication 525, Wash. D.C., pp. 7-15.

Matsumoto, K., M. Miyake, K. Utsumi, and A. Iritani. 1989. Production of identical twins by seperating two-cell rat embryos. Gamete Res. 22:257-263

Mayer, J.F., and H.I. Fritz. 1974. The culture of preimplantation rat embryos and the production of allophenic rats. J. Reprod. Fert. 39:1-9.

McLaren, A. 1970. The fate of very small litters produced by egg transfer in mice. J. Endocr. 47:87-94.

Menino, A.R., Jr., and R.W. Wright, Jr. 1983. Effects of pronase treatment, microdissection and zona pellucida removal on the development of procine embryos and blastomeres <u>in vitro</u>. Biol Reprod. 28:433-446.

Moore, N.W., C.E. Adams, and L.E.A. Rowson. 1968. Developmental potential of single blastomeres of the rabbit egg. J. Reprod. Fert. 17:527-531.

Moore, N.W., C. Polge, and L.E.A. Rowson. 1969. The survival of single blastomeres of pig eggs transferred to recipient gilts. Aust. J. Biol. Sci. 22:979-982.

Moustafa, V.L.A., and J. Hahn. 1978. Experimentelle erzengung von identischen mausezwillingen. Dtsch. Tierarztl. Wschr. 85:242-244.

Nagashima, H., Y. Katoh, K. Shibata, and S. Ogawa. 1987. Development of half-embryos produced from moruale and blastocysts in pigs. Therio. 27:262.

Nagashima, H., K. Matsui, T. Sawaski, and Y. Kano. 1984. Production of monozygotic mouse twins from microsurgically bisected morulae. J. Reprod. Fert. 70:357-362.

Nicholas, J.S., and B.V. Hall. 1942. Experiments on developing rats. II. The development of isolated blastomeres and fused eggs. J. Exp. Zool. 90:441-459.

Ozil, J.P. 1983. Poduction of identical twins by bisection of blastocysts in the cow. J. Reprod. Fert. 69:463-468.

Ozil, J.P., Y. Heyman, and J.-P. Renard. 1982. Production of monozygotic twins by micromanipulation and cervical transfer in the cow. Vet. Rec. 110:126-127.

Papaioannou, V.E. 1982. Lineage analysis of inner cell mass and trophectoderm using microsurgically reconstituted mouse blastocysts. J. Embryol. exp. Morph. 68:199-209.

Papaioannou, V.E., and K.M. Ebert. 1988. Comparative aspects of embryo manipulation in mammals. In: "Experimental Approaches to Mammalian Development". J. Rossant and R. Pederson, eds. Cambridge University Press, pp. 67-96.

Perry, J.S., and I.W. Rowlands. 1962. Early pregnancy in the pig. J. Reprod. Fert. 4:175-188.

Rands, G.F. 1986. Size regulation in the mouse embryo. II. The development of half embryos. J. Embryol. exp. Morph. 98:209-217.

Rorie, R.W., S.A. Voelkel, C.W. McFarland, L.L. Southern, and R.A. Godke. 1985. Micromanipulation of day-6 procine embryos to produce split-embryo piglets. Therio. 23:225.

Rossant, J. 1976. Postimplantation development of blastomeres isolated from 4- and 8-cell mouse eggs. J. Embryol. exp. Morph. 36:283-290.

Seidel, F. 1952. The development potential of an isolated blastomere in the 2-cell stage mammalian egg (In German). Die Naturwissenschaften 39:355-356.

Seidel, F. 1956. Nachweis eines zentrums zur bildung der keimscheibe im saugetierel. Naturwissenschaften 43:306-307.

Smith, R., and A. McLaren. 1977. Factors affecting the time of formation of the mouse blastocoele. J. Embryol. exp. Morph. 41:79-92.

Steven, D., and G. Morris. 1975. Development of foetal membranes. In: "Comparative Placentation". D.H. Steven, ed. N.Y. Academic Press, p. 58-86.

Tarkowski, A.K. 1959a. Experimental studies on regulation in the development of isolated blastomeres of mouse eggs. Acta Theriologica 3:191-267.

Tarkowski, A.K. 1959b. Experiments on the development of isolated blastomeres of mouse eggs. Nature 184:1286-1287.

Trounson, A.O., and N.W. Moore. 1974. Attempts to produce indentical offspring in the sheep by mechanical division of the ovum. Aust. J. Biol. Sci. 27:505.

Tsunoda, Y., and A. McLaren. 1983. Effect of various procedures on the viability of mouse embryos containing half the normal number of blastomeres. J. Reprod. Fert. 69:315-322.

Willadsen, S.M. 1979. A method for culture of micromanipulated sheep embryos and its use to produce monozygotic twins. Nature 277:298-300.

Willadsen, S.M. 1980. The viability of early cleavage stages containing half the normal number of blastomeres in the sheep. J. Reprod. Fert. 59:357-362.

Willadsen, S.M. 1981. The developmental capacity of blastomeres from 4- and 8-cell sheep embryos. J. Embryol. exp. Morph. 65:165-172.

Willadsen, S.M., H. Lehn-Jensen, C.B. Feheilly, and R. Newcomb. 1981. The production of monozygotic twins of preselected parentage by micromanipulation of non-surgically collected cow embryos. Theriogenology 15:23-29.

Willadsen, S.M., and C. Polge. 1981. Attempts to produce monozygotic quadruplets in cattle by blastomere separation. Vet. Rec. 108:211-213.

Williams, T.J., R.P. Elsden, and G.E. Seidel, Jr. 1984. Pregnancy rates with bisected bovine embryos. Theriogenology 22:521-531.

Yang, X., and R.H. Foote. 1987. Production of identical twin rabbits by micromanipulation of embryos. Biol. Reprod. 37:1007-1014.

TABLE 1: Comparative Cell Numbers and Timing (Days After Fertilization) of Developme

(Modified from Papaioannou and Ebert, 1986)

	Cleavage		Compaction of morula		Blastulation		Entry into uteru
	2c (days)	8c (days)		Cell		Cell	
no.	(days)	Days	Days Stage[a]	no.	Days ference[c]	–	
Laboratory Species							
Mouse	1	2	2.5	8-16	3	32	3
Rat	1.5	3.5			4.5		3.5
Rabbit	0.5	1.5	2	16-32	3	128	3
Domestic Species							
Pig	0.5-1	2-3	3.5	8-16	3.5-5	16	2.5-3
Sheep	1	2	3		4.5	64	3
Cow	1	2-4	4-5	16	5-6		3
Horse	1	3	4-5	>15	7-8		5-6
Primates							
Rhesus Monkey	1	2-3	>4	>26	By 7		4
Baboon			5		5-8		4-5
Human	1.5-2.5	2.5-3[b]	3-5	16	4.5-5[d]	64-107	3-4

NOTE:

a ES = early somite; BC = blastocyst; 16 mm = crown-rump length.

b *In vitro*.

c 1, Mayer and Fritz (1974); 2, Davis and Hesseldahl (1971); 3, Perry and Rowlands Robertson (1982); 5, Willadsen (1980);, 6, Hamilton and Day (1945); Steven and M Heuser and Streeter (1941); Lewis and Hartman (1941); Enders and Schlafke (1981) (1978).

TABLE 2: Percentage Normal Development of Manipulated Partial Embry Viable Fetuses or Offspring (Modified from Papaioannou & Ebert, 1

	Single blastomere from			Half-embryos from				Both halv		
	2c	4c	8c	4c	8-16c	Compact Morula	BC	2c	4c	8
Mouse	65	0	0	46	30	42	46	40	0	
Rabbit	30	19	11	—[a]	—	30		—	—	
Sheep	52	16	6	100	80	58		31	>0[b]	
Cow	—	—	—	—	16	46	48	—	—	
Horse	—	36	—		16	—	—	—	—	

NOTE: Results from different studies are combined (see text for references).

a Not done.

b Two pairs of twins have been produced following freezing of one half of each pair

c This figure represents only two twin pairs (Willadsen, 1980).

Uterine Receptivity, Maternal Recognition of Pregnancy and Early Embryonic Loss

R. Michael Roberts

SUMMARY

Much early embryonic loss may be of genetically normal embryos that are lagging behind the uterus in their development. Embryo transfer experiments have shown in a variety of species that some degree of synchrony must be observed between the embryo donor and the recipient. Embryos transferred to advanced recipients are particularly prone to loss. Since asynchrony may occur as a result of retarded ovulation, time of fertilization and low rates of cleavage, a certain amount of natural embryonic wastage is anticipated. Possible reasons for such embryonic loss are 1) the changing uterine milieu is only narrowly permissive, 2) embryos fail to signal their presence in time to prevent a return to ovarian cyclicity, 3) the embryos are rejected immunologically. The above are discussed in relation to the steroid control of uterine secretory activity and the production of luteotrophic and antiluteolytic substances by the conceptus. The recent discovery that interferons are major products of preimplantation ungulate conceptuses raises the possibility that these compounds may be important in modulating immune responses towards the fetal allograft. Their delayed production might, therefore, contribute to embryonic loss.

Magnitude of Embryonic Losses

Considerable prenatal mortality probably occurs in all mammals. In the human, a species that exhibits extremely high rates of embryonic loss, a successful pregnancy occurs only in about 1 in 5 menstrual cycles in women having regular intercourse and not practicing contraception (1). Unfortunately it is not clear to what extent fertilization failure contributes to this low success rate, although two studies have indicated that a conceptus was present at some time between days 6 and 18 after ovulation in about 60% of such cycles (2,3). Of these, between one-third (2) and one-half (3) came to term. The results strongly suggest that most embryonic losses occur early in pregnancy, probably soon after implantation.

In sheep (4) and cattle (5), where fertilization rates are very high, between 60 and 80% of ovulations give rise to live births. Most of these losses appear to occur within the first two or three weeks of pregnancy. With sheep, it is also clear that when more than a single ovulation occurs on one ovary, embryonic mortality increases significantly (6). In litter-bearing animals such as the pig, between 30 and 40% of conceptuses are lost at sometime during gestation, whereas fertilization failure is rare (see 7). For example, a mature sow sheds 15 to 16 ova, all of which generally become fertilized. About 30% of these are lost before day 30 of pregnancy and a further 10% before day 100. Thus only 9 to 10 piglets are born.

Reproductive failure is one of the most costly and limiting production factors in the livestock industry. The economic loss to the livestock industry due to reproductive failure is difficult to estimate with a high degree of certainty due to incomplete records and undefined effects on other production costs. Losses due to conception failure and embryonic mortality in swine alone have been estimated to total approximately $200 million annually in the United States (8). In the case of the human repeated failure of pregnancy constitutes a health topic of considerable concern and often of personal economic cost. In the next section we shall briefly review some of the causes of such early pregnancy failure.

Causes of Embryonic Loss

In the case of farm animals, husbandry practices and environmental factors such as heat stress, high nutritional plane, severe underfeeding, specific nutritional deficiencies and intrauterine infections have been found to increase embryonic mortality. However, in well managed herds these factors are not generally held to be important, and other explanations must be sought for the majority of pregnancy losses.

Genotypic abnormalities in the fertilized egg can undoubtedly be lethal. Two major types of genetic lesions can be anticipated. First, lethal genes may be expressed in embryo early development upon activation of the embryonic genome in pigs. Evidence for embryonic loss due to the presence of lethal genes includes sire differences as they relate to embryonic mortality rate (9) and a marked decrease in embryonic survival rate associated with increased inbreeding (10). Parallel observations have been made in other species including the mouse, and anecdotally, in man. The second type of abnormality is in chromosome number and gross structure. Certainly in humans there is a high incidence of chromosomal abnormalities associated with spontaneous abortions in the first and second trimester (11). However, information is sketchy and conflicting for very early embryos when losses are highest. Chromosomal analyses of such embryos in a variety of species have revealed that chromosomal abnormalities, particularly polyploidy, exist but occur at a relatively low rate (see 12).

Although Bishop (13) has argued that much of the "basal" prenatal loss is of genetically abnormal embryos, it seems fair to assume that a significant fraction of embryos are lost not because they are genetically abnormal but for some other reason. In species where only single young are normally born this would lead to termination of the pregnancy, while in litter-bearing species the number of young would be reduced overall.

Among other reasons that have been proposed to account for early pregnancy loss we shall consider only those that may be related to the fact that the uterus and conceptus may become out of phase with each other. Specifically we shall discuss the possibilities that a) the uterine environment can be inadequate or even embryotoxic, b) the conceptus fails to signal its presence to the mother so that the pregnancy is not recognized, and c) the conceptus is rejected immunologically. Although these possibilities have been partitioned for purposes of discussion, it will become quite clear that each of these subjects overlap.

Implantation and Placentation

Before proceeding further with this discussion it should be recognized that the sequence of physiological and morphological events that precede implantation in different groups of mammals is bewilderingly diverse (14). Consequently it is difficult to conjure up generalities that will apply to all species. Implantation and the establishment of a placenta are clearly crucial for continued development of the embryo. Similarly, the pregnancy must be "recognized" by the mother if she is to respond appropriately to the conceptus. However, the manner in which these processes come about in different species is by no means uniform. This point is illustrated in Fig. 1 where the range of events that precedes implantation in eutherian species is presented in diagramatic form. The first cleavage division usually takes at least 24 h and subsequent divisions 12 h or more. The conceptus enters the uterus on about days 3-4 and blastocyst formation begins. It is around this stage that different species begin to show diversity in the way the conceptus develops. In man and in a number of rodent species the blastocyst in an appropriately receptive uterus hatches quite early, attaches to the uterine epithelium and subsequently begins to implant. In other species, notably the rabbit, the cells of the blastocyst undergo further rounds of division, the blastocyst enlarges, and only then does the process of implantation begin.

Implantation is a complex and variable process that begins with the apposition of the trophectoderm to the surface epithelium at an appropriate and usually favored site for nidation. In many species, the initial attachment is followed by limited spreading of the trophectoderm over the surface of the implantation site. Often the conceptus becomes partially or almost completely enclosed by the endometrium in response to a synchronously occurring decidual response. These stages are followed by penetration of trophoblast cells through the epithelium and the eventual establishment of the placenta. Placentation has been classified on the basis of the numbers of cell layers that separate the blood supplies of the mother and conceptus (15). In the case of the human and rodents the absorptive surface of the trophoblast quickly comes into direct contact with maternal blood and is said to be hemochorial. Thus a reliance on the uterine milieu for nutritional or even endocrine support may be relatively short-lived in such species. The blastocysts of cattle, sheep and pigs hatch relatively late but then continue to enlarge without implanting or even attaching firmly (Fig. 1). In the sheep, for example, the hatched blastocyst at days 11-12 is approximately 2 mm in diameter. By day 13 it has enlarged and begins to elongate into a thread-like form. At day 14 the elongated blastocyst measures 3 to 5 cm in length, whereas by day 16 it can reach up to 20 cm and extend into both uterine horns. True implantation with placentome formation begins much later, between days 20 and 30 (15). Events in the cow follow a similar sequence but occur slightly later in real time, with elongation beginning around days 15 to 16. The situation in the pig is even more curious. Here spherical conceptuses at day 11 average about 8-10 mm in diameter but can reach over one meter in length three days later (16). In this polytoceous species, elongated conceptuses arrange themselves end to end, each occupying not more than 10 to 20 cm of uterus as they follow the contours of the villous folds of the endometrium. Moreover, the uterine epithelium is never eroded throughout pregnancy. All the substances required for fetal and placental sustenance must, therefore, pass across this intact epithelium. It is for this reason that uterine secretions (or histotrophe) are thought to play a vital role in pregnancy of the pig and in other species where an invasive type of implantation either does not occur or occurs relatively late in pregnancy (17).

Asynchrony of Embryo and Mother is Related to Early Embryonic Loss

It has become clear from embryo transfer experiments in all species examined that asynchrony between the transferred embryo and the stage of the development of the recipient's uterus can lead to embryonic death. In general, embryos transferred to less advanced recipients are tolerated better than the reverse. However, because there are some interesting differences between species and because the effects of asynchrony on the immediate development of embryos can also differ, it is worth examining some experimental results in more detail.

In rodents, 24 h advanced embryos survive as well as synchronously developing embryos. They remain in delay until the uterus catches up. However, 24 h retarded embryos survive poorly (see Ref. 18). Gates (18) noted that mouse conceptuses that were cleaving more slowly in vitro had a poorer survival rate after transfer than ones that had been dividing rapidly. Losses of such embryos transferred at the blastocyst stage seemed to occur at the immediate post-implantation period. These observations have been extended by Warner and her colleagues (19) who have defined a specific gene (the ped gene) that appears to have an influence on controlling the rate of cell cleavage in preimplantation mouse embryos. There have been indications that an analogous gene exists in other species as well as the mouse.

Rabbit conceptuses also tolerate asynchrony poorly (20). This has been shown both by transfer to pseudopregnant does or by either delaying or accelerating the passage of the embryo through the fallopian tube of pregnant does. As with the majority of species, rabbit conceptuses held in the oviduct do not implant or advance beyond the blastocyst stage.

Asynchrony has been studied intensively over the past two decades in farm animals because of the economic importance of embryo transfer. In sheep, conception rate is little changed if the donor and recipient ewes are in estrous within 48 h of each other (21,22). However, when day 4 or day 9 embryos are transferred to day 7 or to day 6 recipients respectively, complete failure occurs even when a synchronously transferred embryo is developing within the same uterus (23). Observations of embryos transferred 48 h out of phase with recipient ewes have indicated that the development of the more advanced embryos tends to slow while that of the delayed embryos accelerates (12). In this way some early effects of asynchrony are probably minimized. Sheep embryos which are either 48 h-advanced or 48 h-delayed relative to the onset of estrus of the recipient ewe can also survive together in opposite horns of the same uterus (24). This observation is of particular interest since advanced embryos seem to cause the demise of retarded embryos in the pig (25). It would appear that there is greater tolerance to asynchrony in the sheep than in the pig but this tolerance might decrease as time after estrus of the recipient is increased (23).

One final observation pertinent to embryo transfer in the sheep has come from the experiments in which progesterone was injected into ewes on days 0-3 of the estrous cycle to cause an early, although artificial, luteal phase (26). The uteri of these ewes at day 6 provided a suitable environment for the survival of 10-day old embryos. Thus these aresults are indicative of the likely importance of uterine secretions in selecting against asynchronous embryos.

Unfortunately, in the majority of experiments testing asynchrony in the sheep, the precise time and stage of embryonic development at which death occurs has not been accurately established. For example, it is of considerable interest whether the embryos die before the period of maternal recognition of pregnancy or whether they continue to develop but fail to produce an adequate signal. Alternatively, they may signal their presence to the maternal system but die at a later stage.

Until recently there have been relatively few studies on the effects of asynchrony in pigs. The results of such studies have generally confirmed observations made in other species, namely that older embryos transferred to younger uteri often thrive even if the separation in "age" is as long as 48 h (25,27). However, the reverse situation is again poorly tolerated. Pregnancy rates were significantly reduced when embryos were transferred to recipients that were only 24 h ahead of the donors. In addition, only one of 22 animals became pregnant of those that came into heat two days before the donors (27). In this particular animal the total embryonic survival was also low. This study gave similar results whether donor embryos were tested from the early stages or at the hatched blastocyst stage. However, it was not determined exactly when the embryos died.

Attempts have been made in pigs to advance the uterine environment during early pregnancy (28). However, injection of pregnant gilts with progesterone in the early luteal phase increased circulating levels of progesterone prematurely, but did not induce an advanced embryocidal uterine environment. Rather, embryonic mortality was unchanged.

On the other hand, it has become clear that pig embryos themselves cause a major alteration of the uterine milieu at about day 11 when they first begin to synthesize estrogen (29). This effect can be mimicked in nonpregnant gilts by injecting exogenous estrogen (30) and leads to a marked increase in the total quantity and a change in the quality of proteins in the uterine secretions. Calcium and prostaglandin levels also become elevated. Pope and First (7) reported that exogenous estrogen administered on days 12 and 13, the stage when embryos are synthesizing large amounts of estrogen, did not interfere with pregnancy. However, estradiol 17-β treatment on days 9 and 10 was embryocidal. Further studies by Morgan et al. (31) have confirmed that estrogen treatment at day 9 or days 9 and 10 advanced uterine secretion of calcium and various proteins and was highly embryocidal. Recent work from the same laboratory has shown surprisingly that embryonic death occurs after elongation at about day 14 and not at a time immediately coincident with the estrogen rise (R.D. Geisert, unpublished results). Diekman and Long (32) have reported similar embryotoxic effects from feeding the estrogenic compound, zearalenone, to pregnant gilts prior to blastocyst elongation. In an earlier study Pope et al. (25) showed that when day 5 embryos were transferred into one uterine horn and day 7 embryos into the other uterine horn of recipient gilts only 6 days after their previous estrus (day 0), a high proportion of the advanced embryos survived while the majority of the retarded embryos died. The loss also occurred sometime after day 11 but was not accurately pinpointed. Presumably, the premature estrogen production by the more advanced embryos led to the demise of those that were less developed.

It is clear, therefore, that the asynchronous embryo, particularly one that is delayed relative to the uterus, has a poorer chance of survival than one in synchrony. What the basis of its death might be is unknown but could well be due to multiple factors, some of which are discussed in the subsequent three sections of this paper.

The Necessity of the Uterine Environment

It has been well documented that the epithelium which lines the surface and glands of the uterus is active in secretion, that these secretions contain protein components different from those in plasma, and that the temporal patterns of protein synthesis are correlated with the circulating steroid hormone concentrations of the mother. However, only a limited number of these luminal secretory proteins have been purified. Here a few of the better characterized progesterone and estrogen responsive compounds will be discussed in order to illustrate their range of functions.

Endometrial protein 15 (EP 15, "pregnancy associated 2-globulin" or "pregnancy specific protein 14", depending upon author) of the human is a product of the uterine epithelium during the secretory phase of the cycle and during the first trimester of pregnancy. EP 15 has significant sequence similarity to proteins of the β-lactoglobulin gene family which include components that bind retinol and a variety of the nonpolar ligands (34). However, no specific function for EP 15 has yet become evident.

Uteroglobin, historically the first progesterone-induced protein to be characterized in mammals, is secreted by the uterine epithelium of the rabbit in early pregnancy and pseudopregnancy (see 35). Interestingly, it is an exocrine product of several organs, including the lung where control by corticosteroid rather than progesterone is evident. A variety of functions have been ascribed to uteroglobin. The most recent is as an anti-inflammatory agent which acts to inhibit phospholipase C and the generation of arachidonic acid (35).

The mouse glandular epithelium has recently been shown to be a source of growth factors, including epidermal growth factor and colony stimulating factor-1 (CSF-1). The latter is synthesized in response to progesterone and has been postulated to be important in controlling proliferation of the placenta (36).

The pig, whose uterine secretions are probably the best characterized of all mammals, possesses a diffuse epitheliochorial type of placentation where the uterine epithelium is never invaded and where the blood supplies of the mother and conceptus remain separated by several cell layers. The conceptus is believed to have a long-term reliance on material released as secretions into the uterine luminal milieu. The first secretory protein of the pig to be studied in detail was uteroferrin which, although an acid phosphatase, clearly delivers iron to the conceptus (37). Recently a family of retinol-binding proteins have also been described (see 17). Like uteroferrin, these proteins are synthesized in response to progesterone and appear to be responsible for transplacental transport of a water insoluble nutrient. Other abundant proteins found in porcine uterine secretions during pregnancy include several plasmin/trypsin inhibitors which may act to protect the uterine mucosal surface from the action of conceptus proteases, and lysozyme, an enzyme long recognized as having antimicrobial activity (17). As in the mouse, porcine uterine secretions have been shown to be a source of growth factors (38).

In species where the conceptus quickly invades the uterine wall any reliance upon epithelial secretions may be shorter lived than in the pig. In the rodent and human, the invading trophoblast quickly becomes surrounded by decidual tissue. A number of progesterone-responsive secretory products of the human and baboon deciduum have been described. Perhaps the most interesting is EP 14, a protein which is identical to the IGF-1 binding protein of plasma (39), and is likely to be identical with PP12, PEP, α-2-PEG and several other acronyms (Ref. 33).

There are several uterine secretory proteins that respond to or are produced in response to estrogen (see 40-43). Among these are murine lactoferrin (40) and the so-called CUPED glycoprotein (41), a high molecular weight glycoprotein in the cat, whose function remains unknown. In addition, there is now evidence that a range of carrier proteins for water soluble vitamins such as riboflavin, biotin and thiamin are produced by the liver, and possibly by the uterus, in response to estrogen (43). Although these vitamins are water soluble, the presence of such carrier molecules may be important in transporting such vitamins to the conceptus. Active or passive immunization of females against these carrier molecules appears sufficient to cause abortion but has no nutritionally deleterious effect on the mother. Adiga et al. (43) have speculated that the need for estrogen in implantation of some mammals may be in part due to a requirement for such carrier proteins for nutritional support of the embryo. What is particularly intriguing is that these proteins have been remarkably conserved in evolution from oviparity to viviparity and have been recently identified in the human.

It thus appears that the major proteins secreted by the endometrium during early pregnancy which are responsive to progesterone and estrogen fall into a number of categories. These categories include nutrient and hormone transport molecules, growth factors, protease inhibitors, antimicrobial components and immunosuppressive agents. Others may be involved in promoting implantation or attachment. Many of these proposed roles are at best speculative and may have to be revised. Moreover, the possibility should not be ignored that there may be a general requirement for protein in the local environment of the embryo to fulfill nonspecific ligand binding, osmoregulatory and buffering roles similar to the functions of albumin in plasma. Certainly in those species exhibiting an invasive type of placentation, any nutritional reliance upon uterine secretions might be expected to be short-lived, since the trophoblast quickly comes into close contact with the maternal blood supply. Only in species such as the pig where the trophoblast invades late, or not at all, is it anticipated that uterine secretions have any long-term function in support of the embryo.

Even though so little is known of the protein milieu, even less is known about the concentrations of major ions, the amounts of sugar, lipid, amino acid and micronutrients, the levels of dissolved gases and even the pH in the uterine fluids bathing the early conceptus. It seems likely that low molecular weight components are present at concentrations that are different from those in plasma, and that their relative compositions, like those of the proteins, fluctuate in response to maternal hormones. Our lack of knowledge in these areas is reflected in current uncertainty about the most appropriate media and culture conditions that should be used to support early embryos in vitro.

Although our knowledge of the changing composition of the uterine environment is very limited, it is clear that abnormalities could easily occur as a result of ovarian dysfunction or abnormal uterine response to steroids that would place the conceptus at odds with the mother. The loss of asynchronous embryos might well result from an inappropriate quality, quantity or even toxic nature of uterine secretions. Even when temporal synchrony exists between an embryo and the mother, a hormonal profile similar to that observed during a normal pregnancy must be maintained if conceptuses are to survive in a mother whose ovaries have been removed. Although the timing of the rise in luteal progesterone is variable and the ultimate levels reached not uniform, the pattern of progesterone release is clearly important for pregnancy success. Also an association of lower progesterone concentrations with a failure to conceive has been noted in sheep as well as cattle. So far, however, few comprehensive studies in any species have shown that administration of progesterone in the luteal phase can provide a significant improvement in fertility (see 12). It is possible that some of the differences in progesterone concentration between pregnant and bred but nonpregnant animals may be related to the luteotrophic effects of some embryonic products.

Maternal Recognition of Pregnancy and Pregnancy Failure

Physiological recognition of the presence of an embryo by the mother results from the production of biochemical mediators by the embryo. These mediators critically intervene in the normal progression of the maternal estrous or menstrual cycle such that the embryo is maintained in an environment conducive to continued growth and development. In most mammalian species physiological recognition of pregnancy is associated with extended luteal function, and this process must occur either by direct luteotropic support in primate species where luteal function is promoted by the production of chorionic gonadotropin (44) by prevention of luteolysis as is seen in domestic ruminants and swine (45) or by a combination of such events. In swine and domestic ruminants, the uterine production or release of a luteolytic substance is reduced.

In domestic sheep, the release of the embryonic signal(s) critical for maternal recognition of pregnancy first occurs between days 12 and 13 post-mating (45). The active substance has been shown to be proteinaceous, produced for only a limited period during pregnancy (46), and is thought to be a low molecular weight acidic component known as ovine trophoblast protein-1 (or oTP-1). oTP-1 appears to act as a paracrine hormone directed to the uterus (45). Recently oTP-1 has been shown to be closely related structurally to interferons of the so-called II class (47,48), a result that is of particular interest since interferons not only are known to affect prostaglandin metabolism in their target tissues but also to have potent immunomodulatory properties. The implications of the latter phenomenon will be addressed further in the later discussion of immunorejection.

Based on similar types of experiments, an homologous protein identified as bovine trophoblast protein-1 (bTP-1) has been inferred as a critical factor in maternal recognition of pregnancy the bovine. bTP-1 has also been shown to be interferon-like in both its structure and properties (49).

The active component produced by porcine embryos believed to promote protection of corpora lutea from luteolysis has long been proposed to be estrogen (see 45). Certainly administered to gilts between days 11 and 14 of the estrous cycle leads to prolonged luteal lifespan. However, it is of interest that porcine embryos also begin to secrete interferons during this same period (J. Cross, R.M. Roberts, unpublished results).

In both primates and in the domestic species discussed above, the timing of the release of the active substance, whether it be a gonadotropin, an interferon or an estrogen, begins very close to the time when corpora lutea begin their normal cyclical loss of function. Rescue occurs just prior to when the corpora lutea are irreversibly committed to regression. Clearly embryos delayed in their development or producing the active component in low amounts might fail to signal their presence such that they are lost as luteolysis proceeds. Whether this inability to communicate adequately with the mother contributes to pregnancy failure is not known, but it clearly must be regarded as potentially important in view of the earlier discussion on the importance of synchrony between embryo and mother.

Immunological Rejection of Embryos

In a normal outbred population it is improbable that any randomly mated male or female or any embryo or its mother will share the same transplantation antigens since the genes controlling these loci are highly polymorphic. The embryo, therefore, can be considered to be an allograft, and, like any organ or tissue graft, should normally be rejected. In order to survive, allografts must either be placed at sites that are not easily penetrated by cells of the recipient's immune system or else must be protected by treating the recipient with compounds that are broadly immunosuppressive. Because the uterus is not an immune privileged site and because immune functions of the mother is not generally depressed during pregnancy, the fetal hemiallograft must either behave or be organized differently than a surgically placed allograft (see 50,51). Why most embryos escape immunological attack and whether or not immune rejection contributes extensively to embryonic losses, particularly in early pregnancy, remain unanswered questions. Unfortunately, that limited knowledge that currently exists has largely been derived from the mouse, a species which may not be an appropriate model for mammals that do not have an invasive type of placentation.

The rejection of embryos by the maternal immune system as a cause of infertility is not a well documented phenomenon since it is difficult to tell whether the infiltration of lymphocytes into the "graft" is the basis or consequence of embryonic failure. It has been postulated to occur during rejection of interspecific hybrids and chimeras (52) and in cases of spontaneous abortion in mice (53) and in humans (see 51,52), but even in these instances the data are not clear cut.

The multiple histocompatibility antigens, which are believed to play the major but probably not the sole role in governing the acceptance or rejection of grafts, exist as two classes of cell surface proteins, each products of closely linked genes. Only the class I type are known to be present in placental tissue, and there is considerable controversy as to how early in pregnancy they are expressed (19,50). Although class I antigens have been reported absent on blastocysts and on the early invasive trophoblast of the mouse, no data are available for the domestic animal species. Nevertheless, since other antigens may contribute to the "foreignness" of a graft (54), the early embryo may not be immunologically inert. Certainly in most species the blastocyst seems capable of initiating the dilation and proliferation of blood capillaries and the accumulation of fluid and serum proteins close to the zone of attachment. Whether there is attraction of maternal immune cells to this site is unclear.

Embryos certainly do not remain immunologically neutral for very long. Foreign antigens are expressed and are known to be recognized by the mother (50,51). Consequently, considerable research has gone into defining "immunosuppressive" substances produced either by the embryo itself or by the endometrium in response to progesterone that might blunt maternal immune activity at the fetal-maternal interface. Paradoxically, there is also limited, and admittedly somewhat anecdotal, evidence that some sort of immune response is necessary for a pregnancy to succeed (see 50). If these reports are correct, the fetal allograft on the one hand must protect itself against immune attack yet on the other be sufficiently foreign that its presence be recognized. This recognition may serve to generate necessary lymphokines or growth factors at the maternal interface. In this regard, a particularly exciting result from our laboratory has been the discovery that early embryos of domestic ungulates produce alpha-interferons in substantial quantities (47,49). We have suggested that these substances are responsible for maternal recognition of pregnancy. However, these proteins are also potent modulators of the immune system and can delay the rejection of artificial grafts (53,54). Their effects on cells of the immune system are complex; they are not simply immunosuppressive in a broad sense. For example, interferons do have the ability to inhibit the proliferation of T-lymphocytes in response to a challenge by antigen and consequently inhibit the incorporation of [^{3}H]-thymidine into mixed lymphocyte cultures or into lymphocytes stimulated with mitogen (55), two common and probably overused tests for immunosuppression. On the other hand, they strongly promote the expression of class I MHC genes and have been reported to activate natural killer cells.

The embryonic interferons of ruminants are first produced by the trophectoderm coincident with the time that the spherical blastocysts begin to expand (45,46). IFN's are released by pig embryos at an equivalent period of their development, and we have preliminary evidence that the mouse embryo also produces IFN around the time of implantation (J. Cross, C. Farin and R.M. Roberts, unpublished results). There are several other reports of later stage placental tissue containing IFN's (see 48). It is tempting to suggest that the production of IFN's by the trophectoderm or later stage placental tissue may provide one mechanism whereby the embryo evades destruction as a graft by limited local adjustment of the maternal immune system. We also speculate that a failure to produce IFN's at an appropriate time or in insufficient quantities might compromise that embryo. Thus, late developing or slow growing embryos will again be placed at a considerable disadvantage in the face of an agressive immunological challenge.

Natural Asynchrony and Embryonic Loss

It seems inescapable that the uterus is only narrowly permissive to the early preimplantation conceptus. Even before the pregnancy has been recognized the conceptus is advancing rapidly in its development and the uterine milieu is simultaneously undergoing rapid change as the endometrium responds to maternal steroids and to the local presence of a conceptus. Conceptuses developing slowly or otherwise out of phase with the mother may not be capable of coping with these changes and will be lost for a variety of reasons as discussed earlier. The uterine environment seems likely to place strong selective pressures on embryos such that only those keeping pace with the endometrium will survive. This requirement for synchrony may be the means whereby genetically abnormal embryos, which cannot keep pace with events with the uterus, are discarded. However, retardation of embryos may also occur naturally as a result of at least three different causes (18). First, the timing of the first meiotic division may lead to some oocytes lagging in their maturation. Certainly ovulation in multiparous species is generally not a synchronous process and can occur over several hours. Pope et al. (56,57) have provided strong evidence that the variability in embryonic development seen at days 11 and 12 in Western breeds of pig is the result of late developing eggs giving rise to less developed embryos. It is these immature embryos that subsequently appear to be lost. A second natural cause of retardation in embryos may relate to the time of fertilization and the completion of the second meiotic division. In mice, for example, fertilization of all shed ova requires several hours (see 18). Finally, as mentioned earlier, in relation to the ped gene (19), embryos cleave at different rates. Even within inbred mice strains, however, development of all embryos is not equal. At day 3 1/2 mouse embryos have usually undergone anywhere from 4 to 6 cleavages, resulting in a theoretical developmental separation of 24 h (18). There is even greater variability following superovulation. Clearly all the above events could lead to wastage if the uterus selects against the less advanced but otherwise normal embryos.

It is perhaps worth commenting upon the fact that embryos from many species, once they have been coaxed past an early developmental block (itself probably the result of inappropriate culture conditions), can thrive in culture and develop to the blastocyst stage and beyond in media only slightly modified from ones that were developed primarily for fibroblast growth. Such observations suggest that uterine milieu, rather than constituting a rich embryotrophic medium, provides instead a narrowly permissive environment promoting the survival of only those embryos which can develop at a rate that keeps them appropriately in phase with the mother.

Acknowledgements. Support was provided by grants from the National Institutes of Health (HD21896) and the United States Department of Agriculture (87-CRCR-1-2543).

References

1. Short, R.V. 1984. Species differences in reproductive mechanisms, pp. 24-61. in Reproduction in Mammals: 4. Reproductive Fitness, C.R. Austin and R.V. Short, eds. Cambridge University Press.

2. Edmonds, D.K., K.S. Lindsay, J.F. Miller, E. Williamson and P.J. Wood. 1982. Early embryonic mortality in women. Fert. Steril. 38:447-453

3. Smart, Y.C., L.S. Fraser, T.K. Roberts, R.L. Clancy and A.W. Cripps. 1982. Fertilization and early pregnancy loss in healthy women attempting conception. Clin. Reprod. Fert. 1:177-184

4. Edey, T.W. 1979. Embryo mortality, pp. 315-325. in Sheep Breeding, G.J. Tomes, D.E. Robertson, R.J. Lightfoot and W. Haresign, eds., Butterworth, London.

5. Ayglon, W. 1978. A review of embryonic mortality in cattle. J. Reprod. Fertil. 54:483-493

6. Doney, J.M., R.G. Gunn and W.F. Smith. 1973. Transuterine migration and embryo survival in sheep. J. Reprod. Fertil. 34:363-367.

7. Pope, W.F. and N.L. First. 1985. Factors affecting the survival of pig embryos. Theriogenology 23:91-105.

8. Gerrits, R.J., T.H. Blosser, H.G. Purchase, C.E. Terrell and E.J. Warwick. 1979. Economics of improving reproductive efficiency in farm animals, p. 413. in Beltsville Symposia in Agricultural Research. 3. Animal Reproduction, H.W. Hawk, ed., John Wiley & Sons, New York.

9. Perry, J.S. and I.W. Rowlands. 1962. Early pregnancy in the pig. J. Reprod. Fertil. 4:175-181.

10. Squires, C.D., G.E. Dickerson and D.T. Mayer. 1952. Influence of inbreeding, age and growth rate on sow's sexual maturity, rate of ovulation, fertilzation and embryonic survival. Res. Bull. Mo. Agric. Exp. Stn., No. 494.

11. Hassold, T., N.J. Chen, J. Funkhouser, T. Joos, B. Manuel, J. Matsmura, A. Matsuyarma, C. Wilson, J.A. Yamane and D.A. Jacobs. 1980. A cytogenetic study of 1000 spontaneous abortions. Annls. Hum. Genet. 44:151-178.

12. Wilmut, I., D.I. Sates and C.J. Ashworth. 1980. Maternal and embryonic factors associated with prenatal loss in mammals. J. Reprod. Fertil. 76:851-864.

13. Bishop, M. 1964. Paternal contribution to embryonic death. J. Reprod. Fertil. 7:383-396.

14. McClaren, A. 1982. The embryo, pp. 1-25. in Reproduction in Mammals, Vol. 2: Embryonic and fetal development, C.R. Austin and R.V. Short, eds., Cambridge, Cambridge University Press.

15. King, G.J., B.A. Atkinson and H.A. Robertson. 1982. Implantation and early placentation in domestic ungulates. J. Reprod. Fertil. (Suppl)31:17-30.

16. Anderson, L.L. 1978. Growth, protein content and distribution of early embryos. Anat. Rec. 190:143-154.

17. Roberts, R.M. and F.W. Bazer. 1988. The function of uterine secretions. J. Reprod. Fertil. 82:875-892.

18. Gates, A.H. 1965. Rate of ovular development as a factor in embryonic survival, pp. 270-288. in Preimplantation stages of Pregnancy, G.E. Wolstenholme and G.E.W. O'Connor, eds., CIBA Foundation Symposium, J.A. Churchill Ltd., London.

19. Warner, C.M., M.S. Brownell and M.A. Ewoldsen. 1988. Why aren't embryos immunologically rejected by their mothers? Biol. Reprod. 38:17-29.

20. Adams, C.E. 1965. The influence of maternal environment on preimplantation stages of pregnancy in the rabbit, pp. 345-373. in Preimplantation Stages of Pregnancy, G.E.W. Wolstenholme and M.O. O'Connor, eds., CIBA Foundation Symposium, J.A. Churchill Ltd., London.

21. Moore, N.W. and J.B. Shelton. 1964. Egg transfer in sheep. Effect of degree of synchronization between donor and recipient, age of egg and site of transfer on the survival of transferred eggs. J. Reprod. Fert. 7:145-152.

22. Rowson, L.A. and R.M. Moor. 1966. Embryo transfer in the sheep: the significance of synchronising oestrus in the donor and recipient animal. J. Reprod. Fert. 11:207-212.

23. Wilmut, I. and Sales, D.I. 1981. Effect of asynchronous environment on embryonic development in sheep. J. Reprod. Fert. 61:179-184.

24. Wilmut, I., C.J. Ashworth, A.J. Springbett and D.I. Sales. 1988. Effect of variation in embryo age on the establishment of pregnancy, and embryo survival and growth in ewes with two embryos. J. Reprod. Fert. 83:233-237.

25. Pope, W.F., R.R. Maurer and F. Stormshak. 1982. Survival of porcine embryos after asynchronous transfer. Proc. Soc. Exptl. Biol. Med. 171:179-183.

26. Lawson, R.A.S and L.P. Cahill. 1983. Modification of the embryo-maternal relationship in ewes by progesterone treatment in the early oestrous cycle. J. Reprod. Fert. 67:473-475.

27. Polge, C. 1982. Embryo transplantation and preservation, pp. 277-291. in Control of Pig Reproduction, D.J.A. Cole and G.R. Foxcroft, eds., Butterworth Scientific, London.

28. Wildt, D.E., A.A. Culver, C.B. Marcomb and W.R. Dukelow. 1976. Effect of progesterone on litter size in pigs. J. Reprod. Fert. 48:209-211.

29. Geisert, R.D., R.H. Renegar, W.W. Thatcher, R.M. Roberts and F.W. Bazer. 1982. Establishment of pregnancy in the pig: I Interrelationships between preimplantation development of the pig blastocyst and uterine endometrial secretions. Biol. Reprod. 27:925-949.

30. Geisert, R.D., W.W. Thatcher, R.M. Roberts and F.W. Bazer. 1982. Establishment of pregnancy in the pig: III Endometrial secretory response to estradiol valerate administered on day 11 of the estrous cycle. Biol. Reprod. 27:957-965.

31. Morgan, G.L., R.D. Geisert, M.T. Zavy and A.T. Fazleabas. 1987. Development and survival of pig blastocysts after estrogen administration on day 9 or days 9 and 10 of pregnancy. J. Reprod. Fert. 80:133-141.

32. Diekman, M.A. and C.G. Long. 1984. Termination of pregnancy in gilts consuming zearalenone on days 7-10 post-mating. J. Anim. Sci. 59 Suppl.1):123.

33. Bell, S.C. 1980. Secretory endometrial and decidual proteins: studies and clinical significance of a maternally-derived group of pregnancy-associated serum proteins. Hum. Reprod. 1:129-143.

34. Huhtala, M.C., N. Kalkkinen, P. Palomaki, M. Julkunen, H. Bohn and M. Seppala. 1986. Amino and sequence similarity between placental protein 14 and β-lactoglobulins from various species. Tumor Biol. 7:273(Abstr).

35. Miele, L., E. Cordella-Miele and A.B. Mukherjee. 1987. Uteroglobin: structure, molecular biology, and new perspectives on its function as a phospholipase A2 inhibitor. Endocrine Reviews 8:474-490.

36. Pollard, J.W., A. Bartocci, R. Arceci, A. Orlofsky, M.B. Ladner and E.R. Stanley. 1987. Apparent role of the macrophage growth factor, CSF-1, in placental development. Nature 330:484-486.

37. Roberts, R.M., T.J. Raub, and F.W. Bazer. 1980. Role of uteroferrin in transplacental iron transport in the pig. Red. Proc. 45:2513-2518.

38. Simmen, R.C.M., Y. Ko, X.H. Liu, M.H. Wilde, W.F. Pope and F.A. Simmen. 1988. A uterine cell mitogen distinct from epidermal growth factor in porcine uterine luminal fluids: characterization and partial purification. Biol. Reprod. 38:551-561.

39. Koistinen, R., W. Kalkkinen, M.-L. Huhtala, M. Seppala, H. Bohn and E.-M. Rutanen. 1986. Placental protein 12 is a decidual protein that binds somatomedin and has an identical N-terminal amino acid sequence with somatomedin-binding protein from human amniotic fluid. Endocrinol. 118:1375-1378.

40. Pentecost, B.J. and C.T. Teng. 1987. Lactotransferrin is the major estrogen inducible protein of mouse uterine secretions. J. Biol. Chem. 262:10134-10139.

41. Murray, M.K., H.G. Verhage and R.C. Jaffe. 1986. Quantification of an estrogen-dependent cat uterine protein (CUPED) in uterine flushings of estrogen- and progesterone-treated ovariectomized cats by radioimmunoassay. Biol. Reprod. 35:531-536.

42. Takeda, A., N. Takahashi and S. Shimizu. 1988. Identification and characterization of an estrogen-inducible glycoprotein (uterine secretory protein-1) synthesized and secreted by rat uterine epithelial cells. Endocrinol. 122:105-113.

43. Adiga, P.R., P.B. Seshagiri, P.V. Malathy and S.V. Sandhga. 1986. Reproduction sepcific vitamin carrier proteins involved in transplacental vitamin transport in mammals, including primates, pp. 317-329. in Pregnancy Proteins in Animals, J. Hau, ed., Walter de Grayter & Co., Berlin.

44. Hearn, J.P. 1986. The embryo-maternal dialogue during early pregnancy in primates. J. Reprod. Fert. 76:806-819.

45. Bazer, F.W., J.L. Vallet, R.M. Roberts, D.C. Sharp and W.W. Thatcher. 1986. Role of conceptus secretory products in establishment of pregnancy. J. Reprod. Fert. 76:841-850.

46. Godkin, J.D., F.W. Bazer, J. Moffatt, F. Sessions and R.M. Roberts. 1982. Purification and properties of a major low molecular weight protein released by the trophoblast of sheep blastocysts on days 13-21. J. Reprod. Fertil. 65:141-150.

47. Imakawa, K., R.V. Anthony, M. Kazemi, K.R. Marotti, H.G. Polites and R.M. Roberts. 1987. Interferon-like sequence of ovine trophoblast protein secreted by embryonic trophectoderm. Nature 330:377-379.

48. Roberts, R.M., K. Imakawa, Y. Niwano, M. Kazemi, P.V. Malathy, T.R. Hansen, A.A. Glass and L.H. Kronenberg. 1988. Interferon production by the preimplantation sheep embryo. J. Interferon. Res. 9:175-187.

49. Imakawa, K., T.R. Hansen, P.V. Malathy, R.V. Anthony, H.G. Polites, K.R. Marotti and R.M. Roberts. 1988. Molecular cloning and characterization of cDNA's corresponding to bovine trophoblast protein-1. A comparison with ovine trophoblast protein-1 and bovine interferon- II. Molec. Endocrinol. 3:127-139.

50. Head, J.R., B.L. Drake and F. Zuckerman. 1987. Major histocompatibility antigens on trophoblast and their regulations: implications in the maternal-fetal relationship. Amer. J. Reprod. Immunol. 15:12-18.

51. Rogers, J.C. and B.L. Drake. 1987. The enigma of the fetal graft. Amer. Scientist 75:51-57.

52. Anderson, G.B. 1988. Interspecific pregnancy: barriers and prospects. Biol. Reprod. 38:1-16.

53. Kronenberg, L.H. 1985. Interferons: Manufacture and application in medicine, pp. 451-462. in Biotechnology: Applications and Research, P. Cheresmisinoff and R. Ouellette, eds., Technomic, Lancaster, Pennsylvania.

54. Friedman, R.M., T. Merigan and T. Sreevalsan. 1985. Interferons as cell growth inhibitors and antitumor factors. UCLA symposia on Molecular and Cellular Biology, New Series, Vol. 50, Alan R. Liss, New York.

55. Friedman, R.M. and S.N. Vogel. 1983. Interferons with special emphasis on the immune system. Advances in Immunol. 34:97-140.

56. Pope, W.F., M.H. Wilde and S. Xie. 1988. Effect of electrocautery of nonovulated day 1 follicles on subsequent morphological variation among day 11 porcine embryos. Biol. Reprod. 39:882-887.

57. Pope, W.F., M.S. Lawyer, B.S. Nara and N.L. First. 1986. Effect of asynchronous superinduction on embryo survival and range of blastocyst development in swine. Biol. Reprod. 35:133-137.

Legend to Figure

Fig. 1 Early development of the embryo in eutherian mammals up to the time of implantation (Adapted from Fig. 1.1 of Ref. 14).

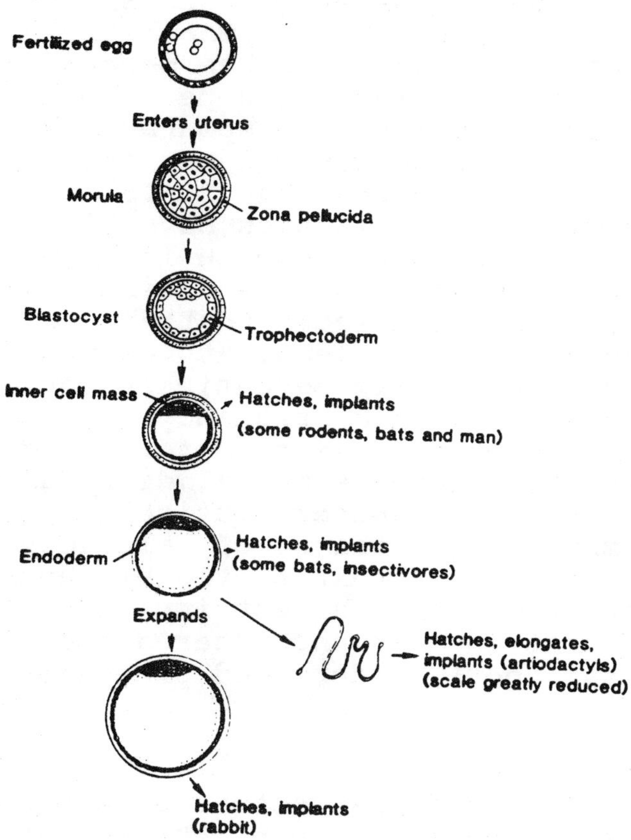

TOWARDS A CELLULAR AND MOLECULAR UNDERSTANDING OF IMPLANTATION IN THE HUMAN: IMPLICATIONS FOR ASSISTED REPRODUCTIVE TECHNOLOGIES

Christos Coutifaris, M.D., Ph.D., Jerome F. Strauss, III, M.D.,Ph.D., Harvey Kliman, M.D., Ph.D.

INTRODUCTION

Successful in vitro fertilization programs report fertilization rates, of 60-75% yet clinical pregnancy rates approximate 20-30% per transfer (Ben Rafael et al., 1986, 1987; Laufer et al, 1984; Lopata 1983). The large gap between fertilization and pregnancy rates has generally been attributed to failure in the peri-implantation period, a time of wastage following natural conception. In spontaneously conceived pregnancies early embryonic wastage may be as high as 31% (Wilcox et al, 1988). It is estimated that 20-25% of human embryos derived from in vitro fertilization have recognizable chromosomal anomalies which are incompatible with life (Plachot et al., 1987). It can be assumed that an additional group of embryos may have other lethal genetic defects that are not detectable by current techniques. Nevertheless, at least 50% of the embryos transferred to the uterus may be entirely normal, and yet their transfer does not result in a pregnancy. There are undoubtedly multiple causes of pregnancy wastage following embryo transfer, some of which may relate to technical difficulties. The method of transcervical embryo transfer routinely employed could contribute to embryo loss because of malplacement or trauma. However, it is also likely that subtle biochemical abnormalities in the embryo and the uterus affect implantation competency. At this point, we cross into an unchartered area since implantation is one of the most poorly understood processes in reproductive biology.

The initial steps of embryo implantation involve fundamental issues of cell-cell and cell-substratum recognition and adhesion prior to and during the invasion of the trophoblast into the endometrium. The composition of the endometrial cell surface and extracellular matrix is rigorously controlled such that implantation can take place only after structural and biochemical changes have occurred under the influence of gonadal steroids and, perhaps, other factors (Aplin et al., 1988; Carpin et al., 1985; Faber et al., 1986; Yamaguchi et al., 1985). Although the hormonal conditions required for uterine receptivity have been extensively studied(Glasser, 1986; Glasser and McCormack, 1980; Glasser and Clark, 1975; Psychoyos, 1973, Weitlauf, 1988), much of the valuable information assembled from the study of laboratory and domestic animals cannot be directly applied to subhuman primates and humans since there are major differences in the morphology, and apparently the mechanisms by which the morphology, and apparently the mechanisms by which implantation occurs among various species. Several different modes of implantation have been identified including 1) fusion implantation where trophoblasts fuse with uterine epithelial cells, as occurs in the rabbit (Schlafke and Enders, 1975); 2) displacement implantation where uterine epithelial cells are dislodged and the trophoblast interacts with the basal lamina, which occurs in rodents (Finn and Bredl, 1973; El-Shershaby and Hinchliffe, 1978; and Pijnenborg et al., 1981; Sherman and Wudl, 1976; and 3) intrusive implantation where trophoblasts insinuate between epithelial cells prior to initiating frank invasion. The latter is the apparent mechanism by which human implantation is initiated based upon the few specimens of early human implantation sites available for study (Enders et al., 1983; Hata et al., 1981 a, b; Pierce et al., 1964).

A major obstacle in the study of implantation in the human has been the lack of adequate in vitro systems, since it is presently not possible to explore the process in a refined manner in vivo. Human endometrial tissue is frequently available, but human zygotes are rarely accessible for experimentation. However, methods to study the function of purified human trophoblasts in culture have been developed in several laboratories (Hall et al., 1977; Kliman et al., 1986; Stromberg et al., 1978;). These techniques now make possible the exploration of the interactions between trophoblast and endometrium in vitro. Here we will review our observations on the characterization of placental trophoblasts, which we propose as a substitute for the human blastocyst in in vitro studies on implantation. We also will present preliminary data from studies in which the interactions of purified cytotrophoblasts with extracellular matrix and endometrial tissue have been examined. These findings are interpreted in light of the existing information on implantation and a working model for nidation in the human is offered.

Preparation and Characterization of Human Cytotrophoblasts

By modifying a previously reported method for the enzymatic dispersion of term placental tissue with trypsin-DNase, by addition of a Percoll gradient centrifugation step, we have been able to prepare highly purified cytotrophoblasts (Figure 1, Kliman et al., 1986; Kliman et al., 1987). This procedure can also be used with first trimester placental tissue, although contamination of the cytotrophoblasts with non-trophoblastic cell types is greater. The isolated cytotrophoblasts undergo striking morphological changes in culture resulting in the formation of functional syncytial trophoblasts.

Trophoblast Aggregation

Using the method described above, it has been possible to establish that mononuclear cytotrophoblasts are precursors of the terminally differentiated syncytial trophoblast. Cytotrophoblasts isolated from term placentae initially aggregate and then fuse to form large multinucleated syncytia in vitro. Time lapse cinematography has unequivocally documented this two step sequence. It can be hypothesized that this homotypic recognition is mediated by unique glycoprotein cell adhesion molecules (CAMs, Edelman, 1988). Using both cytotrophoblasts and JEG-3 choriocarcinoma cells, we have started to characterize the factors responsible for the specificity of trophoblast aggregation. Freshly isolated cytotrophoblasts aggregate in suspension culture over a 24 to 48 h period. Dispersed JEG-3 cells also aggregate in suspension culture, forming tissue-like masses. The extent of aggregation of both cytotrophoblasts and JEG-3 cells in suspension is related to cell density, with greater aggregation occurring with increasing cell concentrations (Kliman et al., 1989). These findings suggest that one determinant of aggregation is the frequency of random cell contact.

The aggregation of trypsinized cells requires the synthesis of proteins since JEG-3 cells fail to coalesce when incubated in the presence of cycloheximide. However, if cycloheximide is removed by washing the cells, aggregation occurs during a subsequent incubation (Kliman et al., 1989). Calcium is required as well since JEG-3 aggregation is impeded when cells are incubated in calcium and magnesium-free medium (Babalola et al., 1989). These results suggest that trypsin-EDTA treatment cleaves calcium-dependent CAMs from the cell surface. We speculate that a trophoblast CAM must be synthesized and inserted into the plasma membrane in order for aggregation of dispersed cytotrophoblasts or JEG-3 cells to occur.

Trophoblast Fusion

The mechanism of fusion of cytotrophoblasts to form syncytial structures has not been elucidated. It is a specific process in that cytotrophoblasts will not fuse with non-trophoblastic cell types (e.g., fibroblasts or fetal liver and kidney cells). The specificity could be imparted by the CAMs mediating cell aggregation.

Several fundamental questions regarding the fusion competency of trophoblastic cells remain unanswered. What prevents cytotrophoblasts, which lie directly under syncytial trophoblasts in the chorionic villi, from fusing? Why do JEG-3 cells aggregate in culture but rarely fuse to form syncytia? One possible explanation may be the existence of factors in cytotrophoblasts and JEG-3 cells which restrain membrane fusion. A 34 kilodalton protein, once thought to be a growth factor but now known to be a member of the lipocortin family, is elaborated by cytotrophoblasts (Choudhury-Roy et al., 1988). Lipocortins bind to phospholipids and are inhibitors of phospholipase A2 (Hayashi et al., 1987; Tait et al., 1988). Since phospholipases have been implicated in the process of membrane fusion, it is possible that the lipocortin-like 34 kd protein plays a role in controlling membrane union.

Trophoblast Interactions With Extracellular Matrix

Cytotrophoblast aggregation and fusion occurs when the cells are cultured under standard conditions in serum-supplemented medium (Kliman et al., 1986). In contrast, if cytotrophoblasts are cultured in serum-free medium without coating of the culture surface with an extracellular matrix protein, the cells remain solitary and do not aggregate or fuse (Feinman et al., 1986; Kao et al., 1988). However, if the culture surface is pre-coated with a matrix protein including types I, IV, and V collagen, fibronectin, or laminin, but not albumin, cytotrophoblast aggregation and fusion take place, even in serum-free medium (Kao et al., 1988). It should also be noted that cytotrophoblasts aggregate in suspension culture in both the absence or presence of serum (Babalola et al., 1989). Thus, the requirement for serum in standard tissue culture appears, at least in part, to be for the attachment factors such as fibronectin. In the absence of these factors, cytotrophoblasts are incapable of firm attachment, spreading and motion. This limits opportunities for cell contact and, thus, aggregation.

It is worth mentioning that cytotrophoblasts synthesize fibronectin (Ulloa-Aguirre et al., 1987) but they clearly cannot utilize the endogenously produced protein for effective interaction with the culture surface under serum-free conditions. Fibronectin produced by placental tissue differs both chemically and functionally from plasma fibronectin: placental fibronectin has a different carbohydrate composition and binds to gelatin with lesser affinity than plasma fibronectin (Zhu et al., 1984; Zhu and Laine, 1985). These differences may account for the failure of cytotrophoblasts to undergo the normal morphologic changes in serum-free culture in the absence of exogenous matrix proteins.

The fact that a variety of extracellular matrix proteins support the morphologic differentiation of human cytotrophoblasts in vitro, as well as the outgrowth of mouse blastocysts (Armant et al., 1986), indicates that the trophoblast expresses a variety of substrate adhesion molecules (SAM's). Since these receptors are frequently coupled to the cellular cytoskeleton (Edelman, 1988), they may have functions other than simply serving as anchors for the cytotrophoblasts (Horwitz et al., 1986). Indeed, these receptors could be part of a transducing system which signals to the cell the nature of its immediate environment. Preliminary immunocytochemical studies suggest that the extracellular matrix protein on which cytotrophoblasts are cultivated affects the immunocytochemical distribution of fibronectin and laminin (Kliman and Strauss, unpublished observations). These findings raise the possibility of the extracellular matrix influencing cellular synthesis and/or secretion of matrix components, presumably through the intermediacy of cell surface substrate adhesion molecules.

Relationship Between Morphologic and Functional Differentiation of The Trophoblast

Cytotrophoblasts do not normally express endocrine activities characteristic of the syncytial trophoblast including chorionic gonadotropin (hCG) and placental lactogen secretion. The morphologic differentiation of cytotrophoblasts to form syncytial structures can be dissociated from functional (i.e., endocrine) differentiation. Agents which stimulate adenylate cyclase or the cAMP analog, 8-bromo-cAMP, stimulate hCG production by cytotrophoblasts by increasing accumulation of the hCG subunit mRNAs (Feinman et al., 1986; Ulloa-Aguirre et al. 1987; Nulsen et al., 1988; Ringler et al., 1989). This stimulation is observed in mononuclear cytotrophoblasts as well as multicellular aggregates. Moreover, it is also seen when cytotrophoblasts are cultured under conditions which prevent aggregation and fusion (e.g., culture in serum-free medium) (Feinman et al., 1986; Kao et al., 1988).

Interactions Between Trophoblast And Endometrium

Several laboratories have proposed <u>in vitro</u> models for implantation in which blastocysts from laboratory animals have been cultured on extracellular matrices, endometrial tissue and lens capsule (Table 1). We are developing systems to explore the interactions of human trophoblasts with human endometrial explants and purified endometrial glandular epithelial cells in culture (Coutifaris et al., 1988, 1989; Kishimoto et al., 1987; Kliman et al., 1988, 1989). The explant system has the advantage of maintaining intact the cellular architecture of the endometrium, but the survival of the endometrial explants is limited to several days. The isolated endometrial cells in standard culture offer the opportunity of examining interactions between specific cell types. However, the absence of stromal-epithelial relationships may affect the way trophoblasts associate with the endometrial glandular epithelium.

Our in vitro models for human implantation assume that isolated cytotrophoblasts, obtained either from first trimester or term placentae, behave like trophectoderm of the blastocyst and that endometrial explants or isolated endometrial cells can retain the characteristics of peri-implantation endometrial tissue in vivo. While it is far too early to know whether these assumptions are legitimate, preliminary data discussed below encourage us to press on with these studies.

Cytotrophoblast Interactions With Endometrial Explants

Cytotrophoblasts isolated from term placentae and first trimester placental tissue bind to endometrial explants in suspension co-culture during a 24 hour incubation. In preliminary observations, cytotrophoblasts attach to the epithelial cells of secretory endometrium, but they do not associate with the epithelial surface of peritoneum or fallopian tube. The cytotrophoblasts do, however, attach to cut surfaces of proliferative <u>and</u> secretory endometrium and fallopian tube, where stroma and extracellular matrix proteins are exposed. After 24-48 hours of co-incubation, a zone of tissue necrosis can be observed at the junction between the attached trophoblastic elements and the endometrium (Fig. 2; Kliman et al., 1988). Histologically, this zone resembles Nitabuch's layer, which is made up of fibrinoid material and separates the cytotrophoblast cell columns seen in normal human implantation sites. Moreover, some cytotrophoblasts penetrate into the endometrial explants and can be clearly identified in tissue sections by the use of immunocytochemistry with antibodies against hCG subunits. Unlike cytotrophoblasts, melanoma cells, endothelial cells and amniocytes do not induce a zone of necrosis.

Cytotrophoblast Interactions With Endometrial Glandular Epithelium And Stroma

Gurpide and colleagues (Schatz et al., 1986; Schatz et al., 1984; Satyaswaroop et al., 1979) developed methods to prepare human endometrial cells and maintain them in culture. These techniques have permitted us and others to examine the interaction of cytotrophoblasts with enriched endometrial cell types. Isolated endometrial glands form nests of epithelial cells after being established in culture. Cytotrophoblasts bind to the collections of glandular epithelial cells and then penetrate into the islands ultimately causing them to detach from the culture surface (Figure 3; Coutifaris et al., 1988; Kishimoto et al, 1987). In contrast, melanoma cells surround the nests of glandular epithelium but do not interact or invade them. Cytotrophoblasts, JEG-3 choriocarcinoma cells and melanoma cells, all adhere to established cultures of endometrial stromal cells and co-mingle with them without causing cellular detachment (Coutifaris et al., 1989).

Recently, Lindenberg and co-workers (1986) incubated hatched human blastocysts onto established monolayer cultures of human endometrial epithelial cells and observed trophoblastic cell attachement and outgrowth. Ultimately, the endometrial cells were displaced and allowed the trophoblasts to come in contact with the culture dish.

Taken together, the findings reviewed above suggest that cytotrophoblasts have a proclivity for endometrial epithelial cells and are capable of penetrating through them in a process which appears to resemble intrusive implantation. Cytotrophoblasts are also capable of associating with stromal cells and a variety of extracellular matrix proteins. As discussed earlier, a number of matrix proteins permit trophoblast attachment, flattening and syncytium formation. We speculate that the specificity (both spatial and temporal) of attachment of trophoblast could be determined by the endometrial epithelium while the underlying extracellular matrix is always permissive. Therefore, under normal circumstances, implantation might occur only in the presence of a receptive endometrial epithelium. However, if the epithelium is eroded, exposing stroma and extracellular matrix, implantation could occur in a variety of locations (e.g., sites where the oviductal epithelium is denuded).

Our speculations raise several key questions. Does endometrial epithelial receptivity result from the expression of unique CAMs recognized by the trophoblast? Is the "window" for human implantation determined by the temporal pattern of expression of these CAM's? The existing literature suggests that the answers to these two questions is affirmative. Studies in laboratory animals are consistent with a role for CAM-like molecules in blastocyst attachment. Cell surface charge is known to be important in blastocyst attachment in rodents (Morris and Potter, 1984; and Nilsson and Hjerten, 1982). In addition, inhibition of glycoprotein synthesis impedes trophoblast binding to the uterus and the lectin, concanavalin A, blocks blastocyst adhesion in the mouse (Surani, 1979; Wu and Chang, 1978). Moreover, a calcium dependent CAM, p-cadherin, has been identified in mouse implantation sites (Nose and Takeichi, 1986). These data are all consistent with a role for CAMs in the implantation process. Recent analyses of endometrial histology and pregnancy rates in women with premature ovarian failure treated with exogenous hormones to prepare the uterus for transfer of a fertilized donor oocyte suggest a discrete period of endometrial receptivity (Rosenwaks, 1987;Navot et al., 1988). In these series, the optimal time of embryo transfer was day 17-19 while no pregnancies were achieved when embryos were transferred on or after day 20. These observations are consistent with a discrete period of endometrial receptivity.

Trophoblast Invasion Into The Endometrium

The mechanism by which trophoblastic elements penetrate into the human endometrium are not known. Studies in laboratory animals suggest that a variety of proteases are involved in implantation, including plasminogen activator (Dabich and Andray, 1974; Denker, 1981; Glass, 1983; Strickland et al., 1976). Axelrod (1985) reported that the t^{w73} mouse produces blastocysts which are less invasive than controls and have an associated diminished production of plasminogen activator activity. Other proteases which degrade extracellular matrix such as stromelysin and collagenases probably participate in trophoblast invasion, since a variety of proteins must be hydrolyzed in concert or sequence during implantation. However, documentation that these enzymes play critical roles in nidation is scant. Since trophoblast invasion must be controlled, mechanisms to localize the site of protease action and to limit the activity of the enzymes are also expected to play a key role in implantation.

Purified human cytotrophoblasts elaborate several proteases capable of digesting gelatin, among them being urokinase (Martin and Arias-Stella, 1982; Queenan et al., 1987). Urokinase may have a direct role in the degradation of fibronectin, as well as activating other enzymes, including plasmin, which hydrolyze matrix proteins (Fisher et al., 1985). The expressed activity of urokinase is determined not only by the amount of enzyme protein but also by levels of plasminogen activator inhibitors which covalently bind to and inhibit the enzyme (Blasi et al., 1987). The trophoblast produces at least two different plasminogen activator inhibitors, plasminogen activator inhibitor types 1 and 2 (PAI-1 and PAI-2). PAI-1 is localized by immunocytochemistry to primarily trophoblasts invading into the endometrium whereas PAI-2 is found predominantly in the syncytial trophoblast of the chorionic villi (Feinberg et al., 1989). The significance of this differential production of PAI-1 and PAI-2 by trophoblastic elements remains to be determined.

The coordinated regulation of urokinase and PAIs provides one means by which trophoblast invasion could be tightly regulated (Feinberg et al., 1988). Furthermore, urokinase is produced as an inactive proenzyme which must be activated by limited proteolysis (Blasi, et al., 1987). Urokinase is also known to bind to cell surface receptors, fixing the site of its action. Thus, there are several levels at which urokinase activity can be modulated including 1) enzyme protein; 2) enzyme activation; 3) cell surface receptors for enzyme and 4) inactivation of enzyme by specific inhibitors. Urokinase is not the only protease with multiple loci for control. Stromelysin and collagenases are also produced as proenzymes and a specific inhibitor of these metalloproteinases (TIMP: Tissue inhibitor of metalloproteinases) is present in the uterine environment (amniotic fluid, Bunning et al., 1984; Chin et al., 1985).

There are many apparent similarities between the process of implantation and the degradation of the extracellular matrix that occurs in association with tumor invasion (Tryggvason et al., 1987). However, in the former case penetration is restricted. Hence, the elucidation of the ways in which trophoblast invasion is controlled will undoubtedly provide insight into the abnormal behavior of choriocarcinoma cells as well as other malignancies.

Paracrine Interactions During Implantation

There is compelling evidence that paracrine dialogue occurs between the blastocyst and endometrium in the peri-implantation period. Physical contact of the blastocyst with the endometrium and carbon dioxide produced by the conceptus have been proposed as "signals" to the endometrium in rodents, but more compelling evidence indicates that histamine, prostanoids and locally generated steroids play the critical roles in the initial events in nidation (Weitlauf, 1988). It remains to be determined whether these substances have similar functions in human pregnancy. We do know from our in vitro studies that progesterone produced by the trophoblast can have a local effect on the adjacent endometrium. Co-incubations of cytotrophoblasts and endometrial explants, there is local induction of morphological changes in the glandular epithelium indicative of secretory activity (e.g., subnuclear vacuolization) by cytotrophoblasts (Kliman and Strauss, unpublished observations).

The conceptus also elaborates polypeptide factors, including monokines and growth factors, which effect endometrial function (Adamson, 1987; Flint et al., 1988; see Roberts et al. in this volume for additional references). Finally, the endometrium, both epithelium and decidua, are capable of producing growth factors, monokines, and growth factor binding proteins, which can either act directly on the trophoblast or modify the action of available growth factors on the trophoblast (Bartocci et al., 1986; Han et al., 1987; Bell and Keyte, 1988; Ringler et al., 1989). These paracrine relationships may modulate the immune system, endometrial activity and trophoblast function and growth (Mogil and Wegman, 1988).

A Working Model For Implantation In The Human And Future Directions For Research

From our studies and the existing literature, we propose a two step model for implantation in the human which entails 1) the binding of trophoblast to a specific endometrial epithelial CAM, the expression of which governs the implantation window; and 2) trophoblast penetration through the epithelial layer with attachment and outgrowth on extra-cellular matrix, the invasion being mediated by proteases checked by the action of specific inhibitors.

It remains to be determined whether the in vitro systems introduced in this paper are adequate to characterize the sequence of events in human nidation. The primary assumptions regarding the equivalency of cytotrophoblasts to the blastocyst trophectoderm and the maintenance of normal endometrial function under in vitro conditions are still open to question. Perhaps hybrid systems in which endometrial explants and cytotrophoblast interactions occur in an animal host (i.e., nude mouse or rat) may also be useful. The evaluation of our working model will most certainly require refinements in the in vitro techniques with which the cell and molecular biology of implantation are studied. None-the-less, should further investigation prove that in vitro systems do indeed reflect the events of implantation, investigators will be in an excellent position to explore the fundamental aspects of nidation. It should be recognized, however, that in vitro systems will not permit experimental approaches to questions relating to the apposition of the blastocyst to the uterine lining (i.e., spatial considerations). This would require the maintenance of the geometry of a normal uterine cavity. Thus, the identification of a suitable sub-human primate model for implantation should be a major goal for future research so that emerging concepts can be evaluated using in vivo approaches.

Our knowledge of the biochemistry of the human endometrium during the peri-implantation period is deficient. We are still in a descriptive phase of research in which the morphology and various biochemical parameters such as steroid receptor levels and distribution are being defined (Jacobs et al., 1987; Garcia et al., 1988; Lessey et al, 1988). A detailed knowledge of the endometrial epithelial cell, the endometrial glandular secretions and the composition of the endometrial stroma are needed as part of the definition of the "receptive" endometrium. Whether stromal-epithelial interactions are critical to the "receptive" state and the extent to which dysynchrony in stromal-epithelial function can perturb the capacity of the endometrium to receive the conceptus remain to be explored.

Several strategies are available to identify putative CAMs involved in blastocyst binding to the endometrium including the generation of monoclonal antibodies to endometrial epithelial cells which could be screened for ability to interfere with cytotrophoblast attachment to endometrial explants or isolated epithelial cells. Full length cDNA clones could be isolated from expression libraries prepared from human endometrial epithelial cell RNA using these immunologic probes. In this way the amino acid sequence of the CAMs could be deduced and additional antibodies prepared from knowledge of the primary amino acid sequences. These antibodies could provide important information in the evaluation of endometrial biopsies, as they could help identify the appropriate hormonal regimens for preparation of the uterus for embryo transfer. A knowledge of the distribution of these CAMs throughout the uterine cavity could provide important clues as to the spatial specificity of implantation, which usually occurs on the posterior fundal endometrium.

The identification of molecules involved in placental morphogenesis and the factors regulating protease and protease inhibitor production by trophoblasts may shed light on pregnancy failure during the peri-implantation period. Moreover, toxemia of pregnancy is associated with a failure of trophoblastic invasion and remodeling of uterine spiral arteries (Rushton, 1984; Khong et al., 1986). Thus, an understanding of the factors controlling trophoblast invasion may provide new insight into the pathophysiology of a major pregnancy-associated disorder. Such knowledge may also shed light on other pathologic conditions including placenta accreta and gestational trophoblastic disease. Immunocytochemistry and _in situ_ hybridization for detection of protease and protease inhibitor protein and mRNAs in normal and abnormal implantation sites will provide useful descriptive information which could yield clues as to the functions of these proteins. With specific enzyme inhibitors in conjunction with the co-culture systems described here, it may be possible to define the actions of each of these proteases in nidation.

The discovery of the various paracrine substances involved in implantation, the elucidation of their sequence of action and feedback relationships represent formidable challenges for future research. Central issues relating to how the early conceptus, particularly the trophoblast, grows and differentiates remain to be addressed. Although it is recognized that the trophoblast is enriched in growth factors and growth factor receptors, the specific functions of many of these proteins are still unknown. _In vitro_ systems appear to be the most amenable experimental approach for clarifying the situation.

Finally, important data relating to the process of human implantation could be obtained from a national or perhaps an international registry of in vitro fertilization/embryo transfer programs. The incidence of abnormal implantation, including early wastage (e.g., "chemical pregnancies"), and its correlation with various regimens for follicular recruitment and post-retrieval endocrine support as well as the mode of embryo transfer needs to be determined. These data coupled with emerging concepts from in vitro experimentation will advance our knowledge of this critical and still mysterious event in human reproduction.

ACKNOWLEDGEMENTS

The authors wish to acknowledge the assistance of their colleagues Drs. Lee-Chuan Kao, Ronald F. Feinberg, G. O. Babalola, John T. Queenan, Jr., Guy E. Ringler, Michael A. Feinman, and Emiliano A. Soto who have contributed to various aspects of the work described in this paper. Special thanks are offered to Julia Haimowitz and Peter Wu for technical assistance and Mrs. Barbara McKenna for their help in the preparation of this manuscript.

REFERENCES

Adamson, E. D. (1987). Expression of proto-oncogenes in the placenta. Placenta 8: 449-466.

Aplin, J. D., A. K. Charlton, S. Ayad (1988). An immunohistochemical study of human endometrial extracellular matrix during the menstrual cycle and first trimester of pregnancy. Cell and Tissue Research 253: 231-240.

Armant, D. R., H. A. Kaplan, W. J. Lennarz (1986). Fibronectin and laminin promote in vitro attachment and outgrowth of mouse blastocysts. Dev. Biol. 116, 510.

Axelrod, H. R. (1985) Altered trophoblast functions in implantation-defective mouse embryos. Dev. Biol. 108: 185-190.

Babalola, G. O., E. A. Soto, C. Coutifaris, H.J. Kliman, J. F. Strauss III (1989). Morphogenesis of the human placenta: characterization of the aggregation of cytotrophoblastic cells. 36th Annual Meeting of the Society for Gynecologic Investigation, San Diego, CA. Abstract 45.

Bartocci, A., J. W. Pollard and E. R. Stanley (1986). Regulation of colony-stimulating factor 1 during pregnancy. J. Exp. Med. 164: 956-961.

Bell, S. C. and J. W. Keyte (1988). N-terminal amino acid sequence of human pregnancy-associated endometrial a_1-globulin, an endometrial insulin-like growth factor (IGF) binding protein-evidence for two small molecular weight IGF binding proteins. Endocrinology 123: 1202-1204.

Ben-Rafael Z., Benadiva CA, M. Ausmanas, B. Barber, L. Blasco, G. L. Flickinger, L. Mastroianni Jr (1987). Dose of human menopausal gonadotropin influences the outcome of an in vitro fertilization program. Fertil. Steril. 48: 964.

Ben-Rafael Z., G. S. Kopf, L. Blasco, R. Tureck, L. Mastroianni, Jr.(1986). Fertilization and Cleavage after reinsemination of human oocytes in vitro. Fertil. Steril 45: 58.

Blasi, F., J-D Vassalli, D. Keld (1987). Urokinase-type plasminogen activator: proenzyme, receptors, and inhibitors. J. Cell. Biol. 104: 801-804.

Bulletti, C., V. M. Jasonni, S. Tabanelli, L. Gianaroli, P. M. Ciotti, A. P. Ferraretti, C. Flamigni (1988). Early human pregnancy in vitro utilizing an artificially perfused uterus. Fertil. Steril. 49: 991-996.

Bunning, R. A., G. Murphy, S. Kumar, P. Phillips, J. Reynolds (1984). Metalloproteinase inhibitors from bovine cartilage and body fluids. E. J. Biochem 139: 75-80.

Cammarata, P.R., L. Oakford, D. Canta-Crouch and R. Wordinger (1987). Attachment of blastocysts to lens capsule: A model system for trophoblast-epithelial cell interactions on a natural basement membrane. Cell Tissue Res. 250: 633-640.

Carpin C., F. Kopp, N. Pourreau-Schneider, J. C. Lissitzky, M.N. Lavaut, P.M. Martin, M. Toga (1985). Laminin distribution in human decidua and immature placenta. Am. J. Obstet. Gynecol 151: 822-826.

Carson, D. D., J. P. Tang, S. Gay (1988). Collagens support embryo attachment and outgrowth in vitro: effects of the Arg-Gly-Asp sequence. Dev. Biol. 127(2): 368-375.

Chin, J. R., G. Murphy, Z. Werb (1985). Stromelysin, a connective tissue-degrading metalloendopeptidase secreted by stimulated rabbit synovial fibroblasts in parallel with collagenase. Biosynthesis, isolation, characterization, and substrates. J. Biol. Chem. 260: 12367-76.

Choudhury-Roy S, A. Sen-Majumdar, V. Murthy, V. S. Mishra, H. J. Kliman, J. E. Nestler, J. F. Strauss III, M. Das (1988). Biosynthesis and turnover of a 34-kDA protein growth factor in human cytotrophoblasts. Eur. J. Biochem. 172: 777-783.

Coutifaris, C., H. J. Kliman and J. F. Strauss III (1988). Development of an in vitro model system for human embryo implantation. Abstracts of the 44th Annual Meeting of the American Fertility Society, Atlanta, GA. Abstract 039.

Coutifaris C., H.J. Kliman, P. Wu, J. F. Strass III (1989). Specificity of trophoblast-endometrial interactions in a human in vitro implantation model system.36th Annual Meeting of the Society for Gynecologic Investigation, San Diego, CA. Abstract 489.

Dabich, D. and T. J. Andray (1974). Prevention of blastocyst implantation in mice with proteinase inhibitors. Fert. Sterility 25: 954-957.

Denker, H-W (1981). Proteinases and implantation. J. Reprod. Fert., Suppl. 29: 183-186.

Edelman, G. M. (1988). Morphoregulatory molecules. Biochemistry 27: 3533-3543.

El-Shershaby, A.M., J. R. Hinchliffe (1975). Epithelial autolysis during implantation of the mouse blastocyst our ultrastructural study. J. Embryol. Exp. Morph. 33: 1067.

Enders, A. C., A. G. Hendricks and S. Schlafke (1983). Implantation in the Rhesus monkey: initial penetration of endometrium. Am. J. Anat. 167: 275-298.

Faber, M., U. M. Wewer, J. G. Berthelsen, L.A. Liotta and R. Albrechtsen (1986). Laminin production by human endometrial stromal cells relates to cyclic and pathologic state of the endometrium. Am. J. Path. 124: 384-398.

Farach, M. C., J. P. Tang, G. L. Decker, D. D. Carson (1987). Heparin/heparan sulfate is involved in attachment and spreading of mouse embryos in vitro. Dev. Biol. 123(2): 401-10.

Feinberg, R. F., L-C Kao, G. Ringler, S. Murray, J. T. Queenan Jr., H. J. Kliman, D. Cines, T-C Wun, J. F. Strauss III (1988). Coordinate regulation of urokinase and plasminogen activator inhibitors in human cytotrophoblast. 35th Annual Meeting of the Society for Gynecologic Investigation, Baltimore, MD. Abstract 344.

Feinberg, R.F., J.F. Strauss III, T-C Wun, H.J. Kliman (1989). Plasminogen activators (PAs) and Plasminogen activator inhibitors (PAIs) in human trophoblasts: markers of trophoblast invasion.36th Annual Meeting of the Society for Gynecologic Investigation, San Diego, CA. Abstract 487.

Feinman, M.A., H. J. Kliman, S. Caltabiano and J. F. Strauss III (1986). 8-Bromo-3'5'-adenosine monophosphate simulates the endocrine activity of human cytotrophoblasts in culture. J. Clin. Endocrinol. Metab. 63, 1211-1217.

Finn, C. A., J. C. S. Bredl (1973). Studies on the development of the implantation reaction in the mouse uterus influence of Actinomycin D. J. Reprod. Fertil. 34: 247.

Fisher, S. J., M. S. Leitch, M.S. Kantor, C. B. Basbnaum and R. H. Kramer (1985). Degradation of extracellular matrix by the trophoblastic cells of first-trimester human placentas. J. Cell. Biochem. 27: 31-41.

Flint, A. P. F., G. E. Lamming and H. J. Stewart (1988). A role for interferons in the maternal recognition of pregnancy. Mol. Cell. Endocrinol. 58: 109-111.

Garcia, E., P. Bouchard, J. De Brux, J. Berdah, R. Frydman, G. Schaison, E. Milgrom, M. Perrot-Applanat (1988). Use of immunocytochemistry of progesterone and estrogen receptors for endometrial dating. J. Clin. Endocrinol. Metab. 67: 80-87.

Glass, R. H. (1983). Degradation of extracellular matrix by mouse trophoblast outgrowth: a model for implantation. J. Cell. Biol. 96: 1108-1116.

Glass, R. H., A. I. Spindle and R. A. Pedersen (1979). Mouse embryo attachment to substratum and interaction of trophoblast with cultured cells. J. Exp. Zool. 208: 327.

Glasser, S. R.(1986). Current concepts of implantation and decidualization. In: The Physiology and Biochemistry of the Uterus in Pregnancy and Labor. G. Hoszar (ed) CRC Press, pp. 127-154.

Glasser, S. R. and J. H. Clark (1985). A determinant role for progesterone in the development of uterine sensitivity to decidualization and ovoimplantation. In: The Developmental Biology of Reproduction, C Market and J. Papaconstantinou (eds). Academic Press, New York, NY, p. 311.

Glasser, S. R., S. A. McCormack (1980). Functional development of rat trophoblast and decidual cells during establishment of the hemochorial placenta. Adv. Biosci. 25: 165.

Glenister, T. W. (1961). Organ culture as a new method for studying the implantation of mammalian blastocysts. Proc. Royal Soc. B 154: 428-431.

Grant, P. S., I. Ljungkvist and O. Nilsson (1975). The hormonal control and morphology of blastocyst invasion in the mouse uterus in vitro. J. Embryol. Exp. Morphol. 34: 310.

Hall, C. St. G., T. E. James, C. Goodyer, C. Branchaud, H. Guyda and C.J.P. Giroud (1977). Short-term culture of human midterm and term placenta: parameters and hormonogenesis. Steroids 30: 569-580.

Han, V. K. M., E. S. Hunter III, R. M. Pratt, J. G. Zendegui, and D.C. Lee (1987). Expression of rat transforming growth factor alpha mRNA during development occurs predominantly in the maternal decidua. Mol. Cell. Biol. 7: 2335-2343.

Hata, T., K. Ohkawa and K. Uchida (1981a). Contact patterns between cytotrophoblast and decidual cells in human implantation site. Acta Obst. Gynaec. Jpn. 35: 529-536.

Hata, T., K. Ohkawa, M. Tomita and M. Kishino (1981b). Phagocytosis of human cytotrophoblast cell invading into decidual tissue in early stage of gestation. Acta. Obst. Gynaec. Jpn. 33: 537-544.

Hayashi, H., M. K. Owada, S. Sonobe, T. Kakunaga, H. Kawakatsu and J. Yano (1987). A 32-kDA protein associated with phospholipase A2-inhibitory activity from human placenta FEBS Lett. 223: 267-272.

Horwitz, A., K. Duggan, C. Buck, M. C. Beckerle, K. Burridge (1986). Interaction of plasma membrane fibronectin receptor with talin - a transmembrane linkage.Nature 320: 531-533.

Jacobs, M. H., J. Balasch, J. M., Gonzalez-Merlo, J. A. Vanrell, C. Wheeler, J. F. Strauss III, L. Blasco, J. E. Wheeler and C. R. Lyttle (1987). Endometrial cytosolic and nuclear progesterone receptors in the luteal phase defect. J. Clin. Endocrinol. Metab. 64: 472-475.

Jenkinson, E. J. (1978). The in vitro blastocyst outgrowth system as a model for the analysis of peri-implantation development, in: Development in Mammals, Vol. 2 (M. Johnson, ed.), p. 157, North-Holland, Amsterdam.

Kao, L-C, S. Caltabiano, S. Wu, J. F. Strauss III and H. J. Kliman (1988). The human villous cytotrophoblast: interactions with extracellular matrix proteins, endocrine function and cytoplasmic differentiation in the absence of syncytium formation. Dev. Biol. 130: 693-702.

Khong, T. Y., F. DeWolf, W. B. Robertson and I. Brosens (1986). Inadequate maternal vascular response to placentation in pregnancies complicated by pre-eclampsia and by small-for-gestational age infants. Br. J. Obstet. Gynaecol. 93: 1049-1059.

Kishimoto Y, T. Tominaga, T. Aso, M. Kinoshita, T. Mori.(1987). Human trophoblast and endometrial interactions in vitro. Acta Obst. Gynaec. Jpn. 39: 463.

Kliman, H. J., C. Coutifaris, G. O. Babalola, E. A. Soto, L-C Kao, J. T. Queenan, Jr., R. F. Feinberg and J. F. Strauss III (1989). The Human Trophoblast: Homotypic and heterotypic cell-cell interactions. In: Development of preimplantation embryos and their environment. Y. Yoshinaga, editor. Alan Liss Inc., New York, NY in press.

Kliman, H. J., J. E. Nestler, E. Sermasi, J. M. Sanger and J. F. Strauss III (1986). Purification, characterization and in vitro differentiation of cytotrophoblasts from human term placentae. Endocrinology 118, 1567-1582.

Kliman, H. J., C. Coutifaris, R. F. Feinberg, and J. F. Strauss III (1988). Interactions between human term trophoblasts and endometrium in an in vitro suspension culture system. Presented at the Internal Workshop on Maternal Recognition of Pregnancy, Jerusalem, Israel, March 20-24, 1988.

Kliman, H. J., M.A. Feinman, and J. F. Strauss III (1987). Differentiation of human cytotrophoblasts into syncytiotrophoblasts in culture. In "Trophoblast Research" (R. Miller and H. Thied, eds.)., Vol. 2, pp. 407-421. Plenum Medical, New York, NY.

Laufer, N., Tarlatzis, B. L., Naftolin F(1984). In vitro fertilization state of the art. Semin. Reprod. Endocrinol. 2:2.

Lessey, B. A., A. P. Killam, D. A. Metzger, A. F. Haney, G. L. Greene, K. S. McCarty Jr. (1988). Immunohistochemical analysis of human uterine estrogen and progesterone receptors throughout the menstrual cycle. J. Clin. Endocrinol. Metab. 67: 334-340.

Lindenberg S., P. Hyttel, S. Lenz, P. V. Holmes (1986). Ultrastructure of the early human implantation in vitro. Hum. Reprod. 1: 533-538.

Lopata, A. Concepts in human in vitro fertilization and embryo transfer(1983). Fertil. Steril 40: 289.

Martin, O. and F. Arias-Stella (1982). Plasminogen activator production by trophoblast cells in vitro. Effect of steroid hormones an protein synthesis inhibitors. Am. J. Obstet. Gynecol. 142: 402-409.

Mogil, R. J. and T. G. Wegman (1988). Maternal immune reactivity as a determinant of placental function and fetal survival. Seminars in Reproductive Endocrinology 6: 145.

Morris, J. E., S.W. Potter (1984). A comparison of developmental changes in surface charge in mouse blastocyts and uterine epithelium using DEAE beads and dextran sulfate in vitro. Develop. Biol. 103: 190.

Navot D., K. Droesch, H. C. Liu, D. Keriner, L. Veeck, K. Steingold, S. Muasher, Z. Rosenwaks.(1988) Efficiency of human conception in vitro related to the window of implantation. Abst. 419, 35th Annual Meeting of the Society for Gynecologic investigation, Baltimore, Maryland.

Nilsson, B. O. and S. H. Jerten (1982). Electrophoretic quantification of the changes in the average net negative surface charge density of mouse blastocyts implanting in vivo and in vitro. Biol. Reprod. 27: 485.

Nose, A. and M. Takeichi (1986). A novel cadherin cell adhesion molecule: Its expression patterns associated with implantation and organogenesis of mouse embryos. J. Cell. Biol. 103: 2649-2658.

Nulsen, J. C., M. J. Woolkalis, G. S. Kopf, and J. F. Strauss III (1988). Adenylate cyclase in human cytotrophoblasts: characterization and its role in modulating human chorionic gonadotropin secretion. J. Clin. Endocrinol. Metab. 66: 258-265.

Pierce, G. B., Jr., A.R. Midgley, Jr., T. F. Beals (1964). An ultrastructural study of differentiation and maturation of trophoblasts of the monkey. Lab Invest. 13: 451.

Pijnenborg, R., W. B. Robertson, I. Brosens, G. Dixon (1981). Trophoblast invasion and the establishment of hemochorial placentation in man and laboratory animals. Placenta 2: 71.

Plachot, M., A.M. Junca, J. Mandelbam, J. de Grouchy, J. Salat-Baroux, J. Cohen (1987). Chromosome investigations in early life. II. Human Preimplantation Embryos. Human Reprod. 2: 29.

Psychoyos, A. Endocrine control of egg implantation (1973). In: Handbook of Physiology, Volume 2, Section 7, R. O. Greep and E. B. Astwood (eds). American Physiological Society, Washington, DC, p. 187.

Queenan, J. T., Jr., L-C Kao, C. E. Arboleda, A. Ulloa-Aguirre, T. G. Golos, D. B. Cines, J. F. Strauss (1987) Regulation of urokinase type plasminogen activator production by culture human cytotrophoblasts. J. Biol. Chem. 262: 10903-10906.

Ringler, G. E., C. Coutifaris, J. F. Strauss III, and M. Geier (1989). Accumulation of colony stimulating factor-1 in amniotic fluid during human pregnancy. Am. J. Obstet. Gynecol. (in press).

Ringler, G. E., L-C Kao, W. L. Miller and J. F. Strauss III (1988a). Effects of 8-bromo-cAMP on expression of endocrine function by cultured human trophoblast cells. Regulation of specific mRNAs. Mol. Cell. Endocrinol. in press.

Rosenwaks, Z (1987). The donor oocyte program in Norfolk Fifth World Congress on in vitro fertilization and embryo transfer. PS-072. Norfolk, Virginia.

Rushton, D. I. (1984). Placenta as a reflection of maternal disease. In: Pathology of the Placenta (E. Perrin, ed.) Churchill Livingston, New York, pp 57-87.

Satyaswaroop P. G., R. S. Bressler, M. M. de la Pena and E. Gurpide (1979). Isolation and culture of human endometrial glands. J. Clin. Endocrinol. Metab. 48: 639-641.

Satyaswaroop, P. G., H. Fleming, R. S. Bressler, E. Gurpide (1978). Human endometrial cancer call cultures for hormonal studies. Cancer Research 38: 4367.

Schatz, F., L. Markiewicz, E. Gurpide (1984). Effects of estriol on PGF2s output by cultures of human endometrium and endometrial cells. J. Steroid Biochem 20: 999-1003.

Schatz, F., L. Markiewicz, E. Gurpide (1986). Hormonal effects of PGF2s output by cultures of epithelial and stromal cells in human endometrium. J. Steroid Biochem 24: 297.

Schlafke, S., A. C. Enders (1975). Cellular basis of interaction between trophoblast and uterus at implantation. Biol. Reprod. 12: 41.

Sengupta, J., R. L. Given, J. B. Carey, and H. M. Weitlauf (1986). Primary culture of mouse endometrium on floating collagen gels: a potential in vitro model for implantation. Ann NY Acad. Sci. 976: 75-94.

Sherman, M. I. , L. W. Wudl (1976). The implanting mouse blastocyst in: The cell surface in animal embryogenesis and development. Poste, G. and Nicolson, G. L. (Eds). Elsevier/North-Holland, Amsterdam, p. 81.

Strickland, S., E. Reich, M. I. Sherman (1976). Plasminogen activator in early embryogenesis: enzyme production by trophoblast and parietal endoderm cell. Cell 9: 231-240.

Stromberg, K., J. C. Azizkhan, and K. V. Speeg (1978). Isolation of functional human trophoblast cells and their partial characterization in primary culture. In Vitro 14: 631-638.

Surani, M. (1979). Glycoprotein synthesis and inhibition of glycosylation by tunicamycin and preimplantation mouse embryos and adhesion. Cell 18: 217-222.

Tait, J. F., M. Sakata, B. A. McMullen, C. H. Miao, T. Funakoshi, L.E. Hendrickson, K. Fujikawa (1988). Placental anticoagulant proteins: Isolation and comparative characterization of four members of the lipocortin family. Biochem 27: 6268-6276.

Tryggrason, K., M. Hoyhtya and T. Salo (1987). Proteolytic degradation of extracellular matrix in tumor invasion. Biochim. Biophys. Acta. 907: 191-217.

Van Blerkom, J. and D. J. Chavez (1981). Morphodynamics of outgrowths of mouse trophoblast in the presence and absence of a monolayer of uterine epithelium. Am. J. Anat. 167: 143-155.

Weitlauf, H. M. (1988) Biology of Implantation in the Physiology of Reproduction (E. Knobil and J. Neill, eds.) pp. 231-262 Raven press, New York.

Wilcox, A. J., C. R. Weinberg, J. F. O'Connor, D. Baird, T. P. Schlatterer, R. E. Canfield, I. G. Armstrong, B.C. Nisula(1988). Incidence of early loss of pregnancy. New England J. Med. 319: 189.

Wu, J. T., M. C. Chang (1978). Increase in concanavalin-A binding sites in mouse blastocysts during implantation. J. Exp. Zool. 205: 447.

Yamaguchi, Y., IM. Isemura, Z. Yosizawa, K. Kurosawa, K. Yoshinaga, A. Sato, M. Suzuki (1985). Changes in the distribution of fibronectin in the placenta during normal human pregnancy. Am. J. Obstet. Gynecol. 152: 715-718.

Ulloa-Aguirre, A.M. August, T. G. Golos, L-C Kao, N. Sakuragi, H. J. Kliman and J. F. Strauss (1987). 8-Bromo-3'5' adenosine monophosphate regulates expression of chorionic gonadotropin and fibronectin in human cytotrophoblasts. J. Clin. Endocrinol. Metab. 64, 1002-1009.

Zhu, B. C., S. R. Fisher, H. Pande, J. Calaycay, J. E. Shively and R. A. Laine (1984). Human placental (fetal) fibronectin, increased glycosylation and higher protease resistance than plasma fibronectin. J. Biol. Chem. 259, 3962-3970.

Zhu, B. C., R. A. Laine (1985). Polylactosamine glycosylation on human fetal placental fibronectin weakens the binding affinity of fibronectin to gelatin. J. Biol. Chem. 260, 4041-4044.

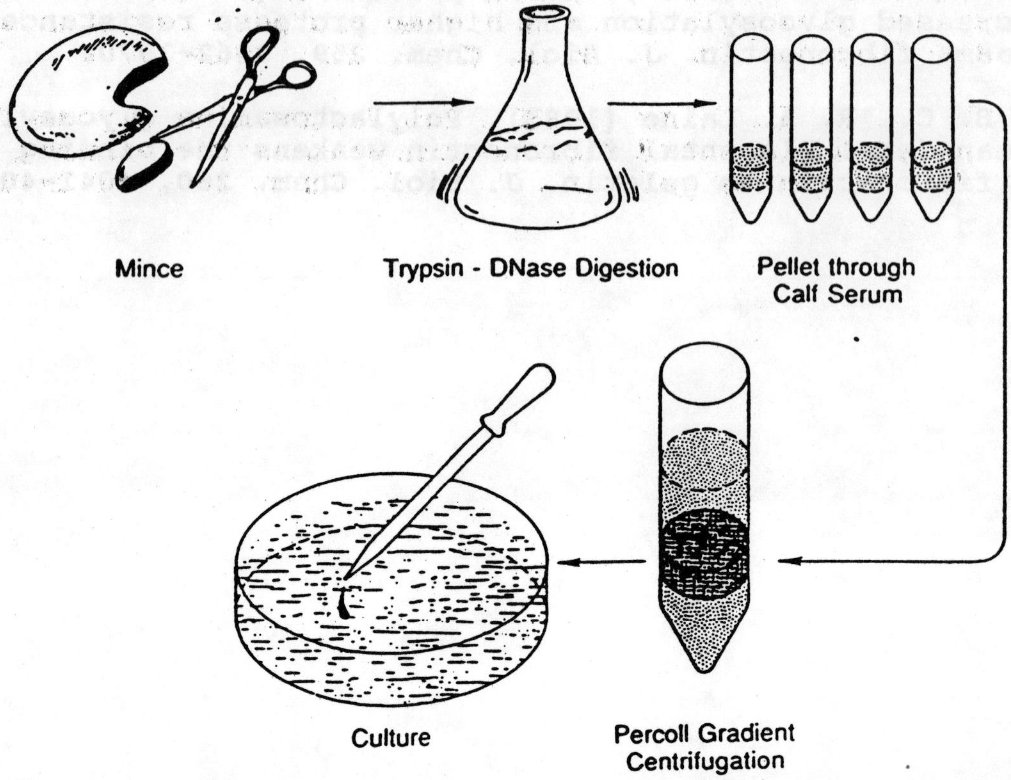

Figure 1 Isolation of human cytotrophoblasts.

Figure 2 Contact necrosis. Cytotrophoblasts from a term placenta were co-cultured with mid-proliferative phase endometrium (E) for 24 h, the tissue was fixed in Bouin's solution, processed for light microscopy and stained with hematoxylin and eosin. The trophoblasts (T) have attached to the surface of the endometrium and have induced a zone of necrosis (arrow heads). This zone was only present at points of trophoblast attachment. Note that some of the cytotrophoblasts have formed syncytial structures (arrows) during the 24 hour culture period. The bar represents 20 um.

Figure 3 Interactions of trophoblasts with human endometrial glands in vitro. Purified human endometrial glands were co-cultured with human cytotrophoblasts for 24 h, fixed with Bouin's solution and immunocytochemically stained with antibodies against alpha- hCG using DAB as chromagen. Several darkly stained trophoblast groups can be seen around the endometrial gland (E). Note a trophoblast group making initial contact with the gland (arrow head). Two large syncytial trophoblast groups can be seen adherent to the gland's edge (large arrows). In addition, several trophoblasts have penetrated the glandular group (small arrows). The bar represent 50 um.

Table 1: In vitro models for the study of implantation.

MODELS	DESCRIPTION	REFERENCES	
Blastocyst Outgrowth	Animal blastocysts cultured on ECM coated surfaces	Jenkinson Armant, et al. Farach, et al. Carson, et al.	1978 1986 1987 1988
	Animal blastocysts cultured on cell monolayers	Glass, et al.	1979
Endometrial Floating Collagen Gels	Endometrial epithelium cultured on floating collagen gels	Sengupta, et al.	1986
Attachment Model	Animal blastocyts cultured on lens capsule	Cammarata, et al.	1987
Endometrial Explant Co-culture	Animal blastocysts cultured with animal endometrial strips	Glenister, et al.	1961
Endometrial Organ Co-culture	Animal blastocysts cultured with whole animal uteri	Grant, et al.	1975
	Human blastocysts cultured with whole perfused human uteri	Bulletti, et al.	1988
Endometrial Monolayer Co-Culture	Human chorionic villi cultured on endometrial gland monolayers	Kishimoto, et al.	1987
	Human blastocysts cultured on endometrial gland monolayers	Lindenberg, et al.	1986
	Human trophoblasts cultured on endometrial gland monolayers	Coutifaris, et al.	1988
Endometrial-trophoblast Suspension Co-culture	Human trophoblasts co-cultured in suspension with human endometrial explants	Kliman, et al.	1988